I HAVE LANDED

Stephen Jay Gould is Alexander Agassiz Professor of Zoology and Professor of Geology at Harvard University. His books include *Ever Since Darwin*, *Wonderful Life*, *Questioning the Millennium*, *Leonardo's Mountain of Clams and the Diet of Worms*, *The Lying Stones of Marrakech* and, most recently, *Rocks of Ages*.

ALSO BY STEPHEN JAY GOULD

Ontogeny and Phylogeny
Ever Since Darwin
The Panda's Thumb
The Mismeasure of Man
Hen's Teeth and Horse's Toes
The Flamingo's Smile
An Urchin in the Storm
Time's Arrow, Time's Cycle
Illuminations (with R. W. Purcell)
Wonderful Life
Bully for Brontosaurus
Finders Keepers (with R. W. Purcell)
Eight Little Piggies
Dinosaur in a Haystack
Full House
Questioning the Millennium
Leonardo's Mountain of Clams and the Diet of Worms
Rocks of Ages
The Lying Stones of Marrakech
Crossing Over (with R. W. Purcell)
The Structure of Evolutionary Theory

Stephen Jay Gould

I HAVE LANDED

Splashes and Reflections in Natural History

VINTAGE

Published by Vintage 2003

2 4 6 8 10 9 7 5 3 1

First published in Great Britain in 2002 by Jonathan Cape

The essays in this book were previously published in magazines and newspapers.

Vintage
Random House, 20 Vauxhall Bridge Road,
London SW1V 2SA

Random House Australia (Pty) Limited
20 Alfred Street, Milsons Point, Sydney
New South Wales 2061, Australia

Random House New Zealand Limited
18 Poland Road, Glenfield,
Auckland 10, New Zealand

Random House (Pty) Limited
Endulini, 5A Jubilee Road, Parktown 2193,
South Africa

The Random House Group Limited Reg. No. 954009
www.randomhouse.co.uk

A CIP catalogue record for this book
is available from the British Library

ISBN 0 09 974971 8

Papers used by Random House are natural, recyclable products made from wood grown in sustainable forests. The manufacturing processes conform to the environmental regulations of the country of origin

Printed and bound in Great Britain by
Cox & Wyman Limited, Reading, Berkshire

TO MY READERS

Fellow members of the ancient and universal
(and vibrantly continuing) Republic of Letters

Contents

I HAVE LANDED

Preface

A Suffix to Begin a Preface

THE HEADING for this opening paragraph sounds contradictory, but stands as a true description of sad necessity and proper placement. The material is suffixial, both in actual chronology and in obviously unintended, but eerily seamless, fit as an unavoidable ending, knitting this book together by recursion to the beginning essay and title piece. I wrote the following preface in the summer of 2001, beginning with some musings about numerical coincidences of my own career, including the completion of this series of essays at an even 300, fortuitously falling in the millennial month of January 2001, also the centennial of the year that my family began its American journey with my grandfather's arrival at Ellis Island—as he wrote in the English grammar book that he purchased, at age thirteen and just off the boat: "I have landed. September 11, 1901." I need say no more for now, as no one, living and sentient on the day, will ever forget the pain and transformation of September 11, 2001. I have added—primarily because duty compelled in the most general and moral sense, but also because the particular joy and hope of Papa Joe's words in 1901 must not be extinguished by the opposite event of spectacular evil on the exact day of his centennial in 2001—a closing section of four short pieces tracing my own emotional odyssey, and the message of tragic hope that an evolutionary biologist might legitimately locate amidst the rubble and tears of our present moment.

The Preface Itself

In 1977, and quite by accident, my first volume of essays in *Natural History* for general readers *(Ever Since Darwin)* appeared at the same time as my first technical book for professional colleagues in evolutionary theory *(Ontogeny and Phylogeny)*. The *New York Times,* viewing this conjunction as highly unusual, if not downright anomalous, featured me in their *Book Review* section as a "freak" of literary nature on this account—and I cannot deny that this article helped to propel a career then mired in infancy. I suppose that I also viewed this conjunction as both strange and fortuitous. (The technical book, for reasons beyond my control, had been delayed by more than a year, causing me only frustration, untempered by any inkling of potential advantage in simultaneity along these different pathways.)

Now, exactly twenty-five years later, and again with frustration rather than intention (this time entirely of my own making, for I failed to finish the technical book in time for an intended and truly millennial appearance in 2000 or 2001, and had to settle instead for the merely palindromic 2002*), this tenth and last volume of essays for general readers from my completed series in *Natural History* magazine also appears at the same time as the technical "life work" of my mature years, twenty years in the making and 1,500 pages in the printing (*The Structure of Evolutionary Theory,* Harvard University Press). But I have undergone a significant change in attitude during this quarter-century (as well, I trust, as an equally significant improvement in writing and thinking). I no longer view this conjunction of technical and "popular" as anomalous, or even as interesting or unusual (at least in principle if not in frequency of realization among my colleagues). For, beyond some obvious requirements of stylistic tuning to expected audiences—avoidance of technical jargon in popular essays as the most obvious example—I have come to believe, as the primary definition of these "popular" essays, that the conceptual depth of technical and general writing should not differ, lest we disrespect the interest and intelligence of millions of potential readers who lack advanced technical training in science, but who remain just as fascinated as any professional, and just as well aware of the importance of science to our human and earthly existence.

Coincidence and numerology exert an eerie fascination upon us, in large part because so many people so thoroughly misunderstand probability, and therefore believe that some deep, hidden, and truly cosmic significance must

*Although Jesse, my autistic son and fabulous day-date calculator, has since pointed out to me the fascinating patterns of palindromic years. They "bunch" only at the ends of millennia—so we have enjoyed two in 1991 and 2002. Our ancestors did even better in 999 and 1001. But our descendants must now wait more than a century for 2112, as calendars revert to the usual pattern of more than a century between palindromic years. So these backward-forward years are indeed rarer and more special than I realized.

attend such "unexpected" confluences as the death of both John Adams and Thomas Jefferson (no great pals during most of their lives) on the same day, July 4, 1826, coincidentally the fiftieth anniversary of the United States as well; or the birth of Charles Darwin and Abraham Lincoln on the same day of February 12, 1809. Scholars can also put such coincidences to good use—as Jacques Barzun did in a famous book *(Darwin, Marx, Wagner)* by centering a contrast among these three key figures upon a focal work completed by each in the same year of 1859, a technique that I have borrowed in a smaller way herein (see essay 5) by joining Darwin with a great painter and a great naturalist through an even more tightly coordinated set of events in the same year of 1859. And yet, I would argue that these numerological coincidences remain fascinating precisely because they can boast *no* general or cosmic meaning whatsoever (being entirely unremarkable at their observed frequency under ordinary expectations of probability), and therefore can only embody the quirky and personal meaning that we choose to grant them.

Thus, when I realized that my three-hundredth monthly essay for *Natural History* (written since January 1974, without a single interruption for cancer, hell, high water, or the World Series) would fall fortuitously into the millennial issue of January 2001, the inception of a year that also marks the centenary of my family's arrival in the United States, I did choose to read this coincidence of numerological "evenness" as a sign that this particular forum should now close at the equally portentous number of ten volumes (made worthy of mention only by the contingency of our decimal mathematics. Were I a Mayan prince, counting by twenties, I would not have been so impressed, but then I wouldn't have been writing scientific essays either). When I then felt the double whammy of an equally "exact" and notable twenty-five years (a quarter of the square of our decimal base) between two odd and fortuitous conjunctions in life's passage—the yoking of my first essay and first technical book in 1977, followed by a similar duality in 2002 of this tenth and last essay book from *Natural History* and my life's major technical "monstergraph" (as we tend to call overly long monographs in the trade), well then, despite my full trust and knowledge of probability, how could I deny that something must be beaming me a marching order to move on to other scholarly and literary matters (but never to slow the pace or lose an iota of interest—for no such option exists within my temperament).

In trying to epitomize what I have learned during these twenty-five years and ten volumes, I can only make a taxonomic analogy based upon locating a voice by finer subdivision toward my own individuality. That is, I grew and differentiated from a nondescript location on the bough of a biggish category on the tree of writers into unique occupancy of a little twiglet of my own true self. At the very start, for reasons both ethical and practical (for otherwise I would have experienced neither pleasure nor learning), I opted for the *family* of "no conceptual simplification," as stated above—in other words, for the great humanistic tradition of treating readers as equals and not as consumers of "easy listening" at drive time under cruise control. I then, if you will, entered a distinctive *genus* of such general writers, a taxon that I have long called Galilean—the intellectual puzzle solvers as opposed to the lyrical exalters of nature, or Franciscans. I then cast my allegiance to a distinctive *species* within the Galilean genus—writers who try to integrate their scientific themes into humanistic contexts and concerns, rather than specializing in logical clarity for explaining particular scientific puzzles. (Incidentally, when I claim that I no longer like my regrettably still popular first book, *Ever Since Darwin,* I say so not primarily because much of its content has been invalidated (a necessary consequence of scientific health and progress for any book written twenty-five years ago), or because its stylistic juvenilities now embarrass me, but rather because I now find these essays too generic in lacking the more personal style that I hope I developed later.)

If I have succeeded in finding a distinctive voice for a *subspecies* of humanistic natural history, my interest in how people actually do science guided my long and tortuous path. How do scientists and other researchers blast and bumble toward their complex mixture of conclusions (great factual discoveries of enduring worth mixed with unconscious social prejudices of astonishing transparency to later generations)? When my method works, I fancy that I can explain complex interfaces between human foibles and natural realities through the agency of what might be designated as "mini intellectual biography"—the distilled essences of the central motivations and concepts of interesting and committed scholars and seekers from all our centuries and statuses—from the greatest physician of his age (essay 11) who could only name, but could not cure or characterize, the new scourge of syphilis (Fracastoro in the sixteenth century), to an unknown woman with a wondrous idea for reconciling scripture and paleontology with all the fervor of Victorian evangelicalism (Isabelle Duncan of essay 7), to solving the mystery of why the greatest stuffed shirt of Edwardian biology had, as a young man, attended Karl Marx's funeral as the only English scientist present (essay 6), to a biological view, then legitimate but now disproved, that led Sigmund Freud to some truly weird speculations about the course of human phylogeny (essay 8). Each mini intellectual biography tells an interesting story of a person and (if successful) elucidates an important scientific concept as well.

The eight categories for parsing the 31 essays of this final volume follow the general concerns of the entire series, but with some distinctive twists. (Perhaps the author protesteth too much, but I am always happily surprised to find, when the time comes to collate these essays into a volume, that they fall into a tolerably coherent order of well-balanced sets of categories, even though I write each piece for itself, with no thought of a developing superstructure made of empty rooms crying out for verbal furniture.) The first essay, and title-bearer of the book, stands alone, as an end focused upon a beginning, all to exalt the continuity of personal life through family lineages and of earthly life through evolution.

The second group expresses my explicit commitment to meaningful joinings between the facts, methods, and concerns of science and the humanistic disciplines, one per essay in this case: to literature in essay 2, history in essay 3, music and theater in essay 4, and art in essay 5. The third group includes three of my mini-intellectual biographies, each in this case devoted to a person and a controlling idea that Darwin's revolution made compelling and relevant. In the fourth group, I try to apply the same basically biographical strategy to the much "stranger" and (for us) difficult intellectual approach to the natural world followed by thinkers of the sixteenth and seventeenth centuries, before "the scientific revolution" (the term generally used by professional historians of science) of Newton's generation fully established the notions of empiricism and experimentation that continue to feel basically familiar to us today. By grappling with this "intellectual paleontology" of fascinating, potent, but largely extinct worldviews held by folks with exactly the same basic mental equipment that we possess today, we can learn more about the flexibility and limitation of the mind than any study of any modern consensus can provide.

Part V explores the different genre of the op-ed format, limited to 1,000 words or fewer. Essays 12 and 13 provide two different takes—one for the fully vernacular audience of *Time,* the other for the professional readers of *Science*—on creationist attacks upon the study of evolution. The remaining four short pieces, from the *New York Times* op-ed page and from *Time* magazine, show how strongly evolution intrudes into our public lives, perhaps more so (in a philosophical and intellectual rather than a purely practical or technological sense) than any other set of scientific concepts.

Each essay in part VI then discusses a truly basic or definitional concept in evolutionary theory (the meaning of the word itself, the nature and limitations of creation stories in general, the meaning of diversity and classification, the direction—or nondirection—of life's history). I use a variety of tactics as organizing devices, ranging from my biographical interests (21 on Linnaeus, 22 on Agassiz, Von Baer, and Haeckel), to a more conventional account of organisms (23 on feathered dinosaurs or early bipedal ground birds), to a personal tale about why this evolutionary biologist felt so comfortable spending the millennial day of January 1, 2000, singing in a performance of Haydn's *The Creation.*

Part VII treats the social implications, utilities, and misutilities of evolution, as seen through the ever-troubling lens of claims for false and invidious innate distinctions of worth among organisms, ranging from native versus introduced plants (essay 24) to supposedly inferior and superior races of humans, with three optimistic final essays on three worthy scientists, from the seventeenth, eighteenth, and nineteenth centuries respectively, who stood among the infrequent defenders of natural equality.

The short pieces of parts V and VIII first appeared as editorials or op-ed commentaries. The full-length essays of all other sections represent the final entries in a series of three hundred written for *Natural History* magazine from January 1974 to January 2001—with five exceptions from other fora: essay 2, on Nabokov, from an exhibition catalog by antiquarian bookseller Paul Horowitz; essay 4, on Gilbert and Sullivan, from *The American Scholar;* essay 5 from the exhibition catalog for a retrospective of Frederick Church's great landscape paintings, displayed at the National Gallery of Art in Washington, D.C.; essay 24, on native plants, from the published proceedings of a conference on the architecture of landscape gardening held at Dunbarton Oaks; and essay 26 from *Discover* magazine.

In closing (but bear with me for an extended final riff), I cannot begin to express the constant joy that writing these essays has brought me since I began late in 1973. Each has taught me something new and important, and each has given me human contact with readers who expressed a complete range of opinion from calumny to adulation, but always with feeling and without neutrality—so God bless them, every one. In return for this great gift that I could not repay in a thousand lifetimes, at least I can promise that, although I have frequently advanced wrong, or even stupid, arguments (in the light of later discoveries), at least I have never been lazy, and have never betrayed your trust by cutting corners or relying on superficial secondary sources. I have always based these essays upon original works in their original languages (with only two exceptions, when Fracastoro's elegant Latin verse and Beringer's foppish Latin pseudocomplexities eluded my imperfect knowledge of this previously universal scientific tongue).

Moreover, because I refuse to treat these essays as lesser, derivative, or dumbed-down versions of technical or scholarly writing for professional audiences, but insist upon viewing them as no different in conceptual depth (however distinct in language) from other genres of original research, I have not hesitated to present, in this format, genuine discoveries, or at least distinctive interpretations, that would conventionally make their first appearance in a technical journal for professionals. I confess that I have often been frustrated by the disinclinations, and sometimes the downright refusals, of some (in my judgment) overly parochial scholars who will not cite my essays (while they happily quote my technical articles) because the content did not see its first published light of day in a traditional, peer-reviewed publication for credentialed scholars. Yet I have frequently placed into these essays original findings that I regard as more important, or even more complex, than several items that I initially published in conventional scholarly journals. For example, I believe that I made a significant discovery of a previously unknown but pivotal annotation that Lamarck wrote into his personal copy of his first published work on evolution. But I presented this discovery in an essay within this series (essay 6 in my previous book *The Lying Stones of Marrakech*), and some scholars will not cite this source in their technical writing.

By following these beliefs and procedures, I can at least designate these essays as distinctive or original, rather than derivative or summarizing—however execrable or wrongheaded (or merely eminently forgettable) any individual entry may eventually rank in posterity's judgment. In scholars' jargon, I hope and trust that my colleagues will regard these essays as *primary rather than secondary sources.* I would defend this conceit by claiming originality on four criteria of descending confidence, from a first category of objective novelty to a fourth that detractors may view as little more than a confusion of dotty idiosyncrasy with meaningful or potentially enlightening distinctiveness.

By my first criterion, some essays present original discoveries about important documents in the history of science—either in the location of uniquely annotated copies (as in Agassiz's stunning marginalia densely penciled into his personal copy of the major book of his archrival Haeckel, essay 22), or in novel analysis of published data (as in my calculation of small differences in mean brain sizes among races, explicitly denied by the author of the same data, essay 27).

In this category of pristine discovery, I can claim no major intellectual or theoretical significance for my own favorite among the true novelties of this book. But when I found the inscription of a great woman, begun as a dedication in a gift from a beautiful young fiancée to her future husband, Thomas Henry Huxley, in 1849, and then completed more than sixty years later, as a grandmother and elderly family matriarch, to Julian Huxley—the sheer human beauty in this statement of love across generations, this wonderfully evocative symbol of continuity (in dignity and decency) within a world of surrounding woe, struck me as so exquisitely beautiful, and so ethically and aesthetically "right," that I still cannot gaze upon Henrietta Huxley's humble page of handwriting without tears welling in my eyes (as, I confess, they flow even now just by writing the thought!). I am proud that I could find, and make public, this little precious gem, this pearl beyond price, of our human best.

By the second criterion, I reach new interpretations, often for material previously unanalyzed at all (bypassed in total silence or relegated to an embarrassingly perfunctory footnote)—as in my all-time favorite evolutionary historical puzzle of why E. Ray Lankester appeared as the only native Englishman at Karl Marx's funeral (essay 6); my first-ever modern exegesis of the bizarre, but internally compelling, reconciliation of Genesis and geology presented by the unknown Isabelle Duncan, even though the chart accompanying her book has become quite famous as an early "scene from deep time" (essay 7); the first analysis based on proper biological understanding of the Lamarckian and recapitulatory theories needed to justify the particular claims of a newly found essay by Sigmund Freud (essay 8); the recognition that the apparently trivial addition of a fifth race (Malay) permitted Blumenbach, in devising a system that became almost universal in application, to make a fundamental alteration in the geometry of racial classification from unranked geographic location to two symmetrical departures from maximal Caucasian beauty (essay 26); and the first published analysis of the first extensive set of fossils ever drawn and printed from a single locality, the 1598 treatise of Bauhin, invoked as a model for inevitable errors in depicting empirical objects when no well-formulated theory of their origin and meaning guides the enterprise (essay 10).

A third category indulges my personal conviction that the joining of two overtly disparate people (in time, temperament, or belief), or two apparently different kinds of events, in a legitimate union based on some deeper commonality, often provides our best insight into a generality transcending the odd yoking. Thus, Church, Darwin, and Humboldt do shout a last joint hurrah in 1859 (essay 5); the stunning anti-Semitism stated so casually and *en passant* in the preface of a famous seventeenth-century pharmacopoeia does link to old ways of thinking, the little-known classification of fossils, and the famous case of the weapon salve (essay 9); the Latin verses used by Fracastoro to name and characterize syphilis in 1530 do contrast in far more than obvious ways to the recent decipherment of the bacterial genome causing the disease (essay 11); Bill Buckner's legs in the 1986 World Series do deeply link to Jim Bowie's Alamo letter, hidden in plain sight, as both become foci for a universal tendency to tell our historical tales in predictably distorted ways (essay 3); and the totally different usage and meaning of the same word *evolution* by astronomers discussing the history of stars and biologists narrating the history of lineages does starkly identify two fundamentally different styles of explanation in science (essay 18).

In a fourth, last, and more self-indulgent category, I can only claim that a purely (and often deeply) personal engagement supplies a different, if quirky, theme for treating an otherwise common subject, or gaining an orthogonal insight into an old problem. Thus, a different love for Gilbert and Sullivan at age ten (when the entire corpus fell into my permanent memory by pure imbibition) and age fifty leads me to explore a different argument about the general nature of excellence (essay 4); I could develop some arguments that far more learned literary critics had missed about Nabokov and his butterflies because they did not know the rules and culture of professional taxonomy, the great novelist's other (and original) profession (essay 2); the unlikely conjunction of a late-twentieth-century ballplayer with a dying hero at the Alamo does reinforce an important principle about the abstraction behind the stark differences, but how could the linkage even suggest itself, absent a strong amateur's (in the best and literal sense of a lover's) affection for both baseball and history (essay 3); and, finally, T. H. Huxley's dashing wife, passing a grandmother's torch to grandson Julian two generations later does mirror the first words written in terror and exhilaration by my grandfather at age thirteen, fresh off the boat and through Ellis Island, but unknowingly "banked" for a potential realization that required two generations of spadework, and then happened to fall upon me as the firstborn of the relevant cohort—a tale that could not be more personal on the one hand, but that also, at the opposite end of a spectrum toward full generality, evokes the most important evolutionary and historical principle of all the awe and necessity of unbroken continuity (essay 1, my *ave atque vale*).

Finally—and how else could I close—if I found a voice and learned so much in three hundred essays (literally "tries" or "attempts"), I owe a debt that cannot be overstated to the corps of readers who supplied will and synergy in three indispensable ways, making this loneliest of all intellectual activities (writing by oneself) a truly collective enterprise. First, for showing me that, contrary to current cynicism and mythology about past golden ages, the abstraction known as "the intelligent layperson" does exist—in the form of millions of folks with a passionate commitment to continuous learning (indeed to a virtual definition of life as the never-ending capacity so to do); we may be a small minority of Americans, but we still form multitudes in a nation 300 million strong.

Second, for the simple pleasure of fellowship in the knowledge that a finished product, however satisfactory to its author, will not slide into the slough of immediate erasure and despond, but will circulate through dentists' offices, grace the free-magazine shelf of the BosWash shuttle flights, and assume an honored place on the reading shelf (often just the toilet top) of numerous American bathrooms.

Third, and most gratifying, for the practical virtues of interaction: As stated explicitly in two of these essays (1 and 7), I depend upon readers to solve puzzles that my research failed to illuminate. Time and time again, and unabashedly, I simply ask consumers for help—and my reward has always arrived, literally posthaste (quite good enough, for the time scale of this enterprise does not demand e-mail haste). As the first and title essay proves—for the piece itself could not have been written otherwise—I have also received unsolicited information of such personal or intellectual meaning to me that tears became the only appropriate response.

In previous centuries of a Balkanized Western world, with any single nation sworn to enmity toward most others, and with allegiances shifting as quickly as the tides and as surprisingly as the tornado, scholars imagined (and, for the most part, practiced in their "universal" Latin) the existence of a "Republic of Letters" freely conveying the fruits of scholarship in full generosity across any political, military, or ethnic divide. I have found that such a Republic of Letters continues, strong and unabated, allowing me to participate in something truly ecumenical and noble. And, for this above all, I love and admire you all, individually and collectively. I therefore dedicate this last volume "to my readers."

I

Pausing in Continuity

I

I Have Landed

As a young child, thinking as big as big can be and getting absolutely nowhere for the effort, I would often lie awake at night, pondering the mysteries of infinity and eternity—and feeling pure awe (in an inchoate, but intense, boyish way) at my utter inability to comprehend. How could time begin? For even if a God created matter at a definite moment, then who made God? An eternity of spirit seemed just as incomprehensible as a temporal sequence of matter with no beginning. And how could space end? For even if a group of intrepid astronauts encountered a brick wall at the end of the universe, what lay beyond the wall? An infinity of wall seemed just as inconceivable as a never-ending extension of stars and galaxies.

I will not defend these naïve formulations today, but I doubt that I have come one iota closer to a personal solution since those boyhood ruminations so long ago. In my philosophical moments—and not only as an excuse for personal failure, for I see no sign that others have succeeded—I rather suspect that the evolved powers of the human mind may not include the wherewithal for posing such questions in answerable ways (not that we ever would, should, or could halt our inquiries into these ultimates).

However, I confess that in my mature years I have embraced the Dorothean dictum: yea, though I might roam through the pleasures of eternity and the palaces of infinity (not to mention the valley of the shadow of death), when a body craves contact with the brass tacks of a potentially comprehensible reality, I guess there's no place like home. And within the smaller, but still tolerably ample, compass of our planetary home, I would nominate as most worthy of pure awe—a metaphorical miracle, if you will—an aspect of life that most people have never considered, but that strikes me as equal in majesty to our most spiritual projections of infinity and eternity, while falling entirely within the domain of our conceptual understanding and empirical grasp: the continuity of *etz chayim,* the tree of earthly life, for at least 3.5 billion years, without a single microsecond of disruption.

Consider the improbability of such continuity in conventional terms of ordinary probability: Take any phenomenon that begins with a positive value at its inception 3.5 billion years ago, and let the process regulating its existence proceed through time. A line marked zero runs along below the current value. The probability of the phenomenon's descent to zero may be almost incalculably low, but throw the dice of the relevant process billions of times, and the phenomenon just has to hit the zero line eventually.

For most processes, the prospect of such an improbable crossing bodes no permanent ill, because an unlikely crash (a year, for example, when a healthy Mark McGwire hits no home runs at all) will quickly be reversed, and ordinary residence well above the zero line reestablished. But life represents a different kind of ultimately fragile system, utterly dependent upon unbroken continuity. For life, the zero line designates a permanent end, not a temporary embarrassment. If life ever touched that line, for one fleeting moment at any time during 3.5 billion years of sustained history, neither we nor a million species of beetles would grace this planet today. The merest momentary brush with voracious zero dooms all that might have been, forever after.

When we consider the magnitude and complexity of the circumstances required to sustain this continuity for so long, and without exception or forgiveness in each of so many components—well, I may be a rationalist at heart, but if anything in the natural world merits a designation as "awesome," I nominate the continuity of the tree of life for 3.5 billion years. The earth experienced severe ice ages, but never froze completely, not for a single day. Life fluctuated through episodes of global extinction, but never crossed the zero line, not for one millisecond. DNA has been working all this time, without an hour of vacation or even a moment of pause to remember the extinct brethren of a billion dead branches shed from an evergrowing tree of life.

When Protagoras, speaking inclusively despite the standard translation, defined "man" as "the measure of all things," he captured the ambiguity of our feelings and intellect in his implied contrast of diametrically opposite interpretations: the expansion of humanism versus the parochiality of limitation. Eternity and infinity lie too far from the unavoidable standard of our own bodies to secure our comprehension; but life's continuity stands right at the outer border of ultimate fascination: just close enough for intelligibility by the measure of our bodily size and earthly time, but sufficiently far away to inspire maximal awe.

Moreover, we can bring this largest knowable scale further into the circle of our comprehension by comparing the macrocosm of life's tree to the microcosm of our family's genealogy. Our affinity for evolution must originate from the same internal chords of emotion and fascination that drive so many people to trace their bloodlines with such diligence and detail. I do not pretend to know why the documentation of unbroken heredity through generations of forebears brings us so swiftly to tears, and to such a secure sense of rightness, definition, membership, and meaning. I simply accept the primal emotional power we feel when we manage to embed ourselves into something so much larger.

Thus, we may grasp one major reason for evolution's enduring popularity among scientific subjects: our minds must combine the subject's sheer intellectual fascination with an even stronger emotional affinity rooted in a legitimate comparison between the sense of belonging gained from contemplating family genealogies, and the feeling of understanding achieved by locating our tiny little twig on the great tree of life. Evolution, in this sense, is "roots" writ large.

To close this series of three hundred essays in *Natural History,* I therefore offer two microcosmal stories of continuity—two analogs or metaphors for this grandest evolutionary theme of absolutely unbroken continuity, the intellectual and emotional center of "this view of life."* My stories descend in range and importance from a tale about a leader in the founding generation of Darwinism to a story about my grandfather, a Hungarian immigrant who rose from poverty to solvency as a garment worker on the streets of New York City.

---

*This essay, number 300 of a monthly entry written without a break from January 1974 to January 2001 and appearing in *Natural History* magazine, terminates a series titled "This View of Life." The title comes from Darwin's poetic statement about evolution in the last paragraph of the *Origin of Species:* "There is grandeur in this view of life . . ."

Our military services now use the blandishments of commercial jingles to secure a "few good men" (and women), or to entice an unfulfilled soul to "be all that you can be in the army." In a slight variation, another branch emphasizes external breadth over internal growth: join the navy and see the world.

In days of yore, when reality trumped advertisement, this motto often did propel young men to growth and excitement. In particular, budding naturalists without means could attach themselves to scientific naval surveys by signing on as surgeons, or just as general gofers and bottle washers. Darwin himself had fledged on the *Beagle,* largely in South America, between 1831 and 1836, though he sailed (at least initially) as the captain's gentleman companion rather than as the ship's official naturalist. Thomas Henry Huxley, a man of similar passions but lesser means, decided to emulate his slightly older mentor (Darwin was born in 1809, Huxley in 1825) by signing up as assistant surgeon aboard HMS *Rattlesnake* for a similar circumnavigation, centered mostly on Australian waters, and lasting from 1846 to 1850.

Huxley filled these scientific *Wanderjahre* with the usual minutiae of technical studies on jellyfishes and grand adventures with the aboriginal peoples of Australia and several Pacific islands. But he also trumped Darwin in one aspect of discovery with extremely happy and lifelong consequences: he met his future wife in Australia, a brewer's daughter (a lucrative profession in this wild and distant outpost) named Henrietta Anne Heathorn, or Nettie to the young Hal. They met at a dance. He loved her silky hair, and she reveled in his dark eyes that "had an extraordinary way of flashing when they seemed to be burning—his manner was most fascinating" (as she wrote in her diary).

Huxley wrote to his sister in February 1849, "I never met with so sweet a temper, so self-sacrificing and affectionate a disposition." As Nettie's only dubious trait, Hal mentioned her potential naïveté in leaving "her happiness in the hands of a man like myself, struggling upwards and certain of nothing." Nettie waited five years after Hal left in 1850. Then she sailed to London, wed her dashing surgeon and vigorously budding scientist, and enjoyed, by Victorian standards, an especially happy and successful marriage with an unusually decent and extraordinarily talented man. (Six of their seven children lived into reasonably prosperous maturity, a rarity in those times, even among the elite.) Hal and Nettie, looking back in their old age (Hal died in 1895, Nettie in 1914), might well have epitomized their life together in the words of a later song: "We had a lot of kids, a lot of trouble and pain, but then, oh Lord, we'd do it again."

The young and intellectually restless Huxley, having mastered German, decided to learn Italian during his long hours of boredom at sea. (He read Dante's *Inferno* in the original *terza rima* during a year's jaunt, centered upon

New Guinea.) Thus, as Huxley prepared to leave his fiancée in April 1849 (he would return for a spell in 1850, before the long five-year drought that preceded Nettie's antipodal journey to their wedding), Nettie decided to give him a parting gift of remembrance and utility: a five-volume edition, in the original Italian of course, of *Gerusalemme liberata* by the great Renaissance poet Torquato Tasso. (This epic, largely describing the conquest of Jerusalem by the First Crusade in 1099, might not be deemed politically correct today, but the power of Tasso's verse and narrative remains undiminished.)

Nettie presented her gift to Hal as a joint offering from herself, her half-sister Oriana, and Oriana's husband, her brother-in-law William Fanning. She inscribed the first volume in a young person's hand: "T. H. Huxley. A birthday and parting gift in remembrance of three dear friends. May 4th 1849." And now we come to the point of this tale. For some reason that I cannot fathom but will not question, this set of books sold (to lucky me) for an affordable pittance at a recent auction. (Tasso isn't big these days, and folks may have missed the catalog entry describing the provenance and context.)

So Nettie Heathorn came to England, married her Hal, raised a large family, and lived out her long and fulfilling life well into the twentieth century. As

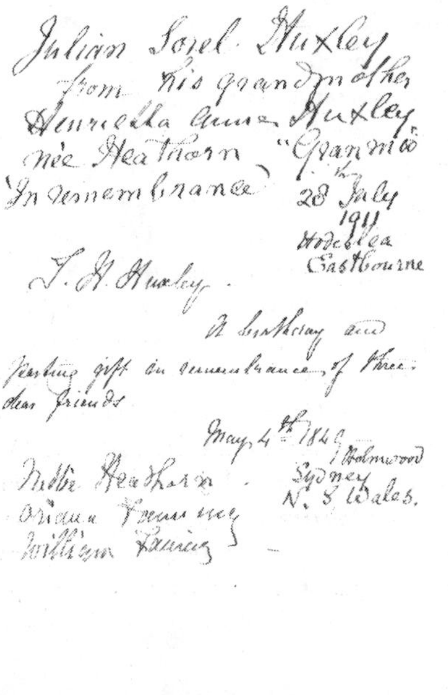

Henrietta Heathorn's inscriptions in a volume of poetry given first to her fiancé, Thomas Huxley, and sixty years later to her grandson Julian Huxley.

she had been blessed with accomplished children, she also enjoyed, in later life, the promise of two even more brilliant grandchildren: the writer Aldous Huxley and the biologist Julian Huxley. In 1911, more than sixty years after she had presented the five volumes of Tasso to Hal, Nettie Heathorn, then Henrietta Anne Huxley, and now Granmoo to her grandson Julian, removed the books from such long residence on her shelf, and passed them on to a young man who would later carry the family's intellectual torch with such distinction. She wrote below her original inscription, now in the clear but shaky hand of an old woman, the missing who and where of the original gift: "Holmwood. Sydney, N.S. Wales. Nettie Heathorn, Oriana Fanning, William Fanning."

Above her original words, penned sixty years before in youth's flower, she then wrote, in a simple statement that needs no explication in its eloquent invocation of life's persistence: "Julian Sorel Huxley from his grandmother Henrietta Anne Huxley née Heathorn 'Granmoo.'" She then emphasized the sacred theme of continuity by closing her rededication with the same words she had written to Hal so many years before: "'In remembrance' 28 July 1911. Hodeslea, Eastbourne."

If this tale of three generations, watched over by a great woman as she follows life's passages from dashing bride to doting grandmother, doesn't epitomize the best of humanity, as symbolized by our continuity, then what greater love or beauty can sustain our lives in this vale of tears and fascination? Bless all the women of this world who nurture our heritage while too many men rush off to kill for ideals that might now be deeply and personally held, but will often be viewed as repugnant by later generations.

My maternal grandparents—Irene and Joseph Rosenberg, or Grammy and Papa Joe to me—loved to read in their adopted language of English. My grandfather even bought a set of Harvard Classics (the famous "five-foot shelf" of Western wisdom) to facilitate his assimilation to American life. I inherited only two of Papa Joe's books, and nothing of a material nature could be more precious to me. The first bears a stamp of sale: "Carroll's book store. Old, rare and curious books. Fulton and Pearl Sts. Brooklyn, N.Y." Perhaps my grandfather obtained this volume from a *landsman,* for I can discern, through erasures on three pages of the book, the common Hungarian name "Imre." On the front page of this 1892 edition of J. M. Greenwood's *Studies in English Grammar,* my grandfather wrote in ink, in an obviously European hand, "Prop. of Joseph A. Rosenberg, New York." To the side, in pencil, he added the presumed date of his acquisition: "1901. Oct. 25th." Just below, also in pencil, he appended the most eloquent of all conceivable words for this context—even though one might argue that he used the wrong tense, confusing the compound past of con-

1901. Oct 25th, Prof. J.
I have landed
Sept. 11th 1901.
Joseph A. Rosenberg
New York.

My grandfather's inscriptions on the title page of the English grammar book that he bought soon after arriving in America as a thirteen-year-old immigrant.

tinuous action with an intended simple past to designate a definite and completed event (not bad for a barely fourteen-year-old boy just a month or two off the boat): "I have landed. Sept. 11, 1901."

Of all that I shall miss in ending this series of essays, I shall feel most keenly the loss of fellowship and interaction with readers. Early in the series, I began—more as a rhetorical device to highlight a spirit of interaction than as a practical tactic for gaining information—to pose questions to readers when my research failed to resolve a textual byway. (As a longtime worshiper at the altar of detail, nothing niggles me more than a dangling little fact—partly, I confess, from a sense of order, but mostly because big oaks do grow from tiny acorns, and one can never know in advance which acorn will reach heaven.)

As the series proceeded, I developed complete faith—not from hope, but from the solid pleasure of invariant success—that any posted question would elicit a host of interesting responses, including the desired factual resolution. How did the Italian word *segue* pass from a technical term in the rarefied world

Grammy and Papa Joe as I knew them in their later life, shown here in the early 1950s.

of classical music into common speech as a synonym for "transition" (resolved by personal testimony of several early radio men who informed me that in the 1920s they had transferred this term from their musical training to their new gigs as disc jockeys and producers of radio plays). Why did seventeenth-century engravers of scientific illustrations usually fail to draw snail shells in reverse on their plates (so that the final product, when printed on paper, would depict the snail's actual direction of coiling), when they obviously understood the principle of inversion and always etched their verbal texts "backwards" to ensure printed readability? Who were Mary Roberts, Isabelle Duncan, and several other "invisible" women of Victorian science writing who didn't even win a line in such standard sources as the *Encyclopaedia Britannica* or the *Dictionary of National Biography*? (See essay 7 for a reader's resolution to this little mystery.)

Thus, when I cited my grandfather's text *en passant* in an earlier essay, I may have wept for joy, but could not feign complete surprise, when I received the most wonderful of all letters from a reader:

> For years now I have been reading your books, and I think I should really thank you for the pleasure and intellectual stimulation I have received from you. But how to make even a small return for your essays? The answer came to me this week. I am a genealogist who specializes in passenger list work. Last Sunday I was rereading that touching essay that features your grandfather, Joseph A. Rosenberg who wrote "I have landed. Sept. 11, 1901." It occurred to me that you might like to see his name on the passenger list of the ship on which he came.

I think I always knew that I might be able to find the manifest of Papa Joe's arrival at Ellis Island. I even half intended to make the effort "some day." But, in honest moments of obeisance to the Socratic dictum "know thyself," I'm pretty sure that, absent this greatest of all gifts from a reader, I never would have found the time or made the move. (Moreover, I certainly hadn't cited Papa Joe's inscription in a lazy and intentional "fishing expedition" for concrete information. I therefore received the letter of resolution with pure exhilaration—as a precious item beyond price, freely given in fellowship, and so gratefully received without any conscious anticipation on my part.)

My grandfather traveled with his mother and two younger sisters on the SS *Kensington,* an American Line ship, launched in 1894 and scrapped in 1910, that could carry sixty passengers first class, and one thousand more in steerage—a good indication of the economics of travel and transport in those days of easy immigration for European workers, then so badly needed for the factories and sweatshops of a booming American economy based on manual labor. The *Kensington* had sailed from Antwerp on August 31, 1901, and arrived in New York, as Papa Joe accurately recorded, on September 11. My page of the "list or manifest of alien immigrants" includes thirty names, Jewish or Catholic by inference, and hailing from Hungary, Russia, Rumania, and Croatia. Papa Joe's mother, Leni, listed as illiterate and thirty-five years of age, appears on line 22 with her three children just below: my grandfather, recorded as Josef and literate at age fourteen, and my dear aunts Regina and Gus, cited as Regine and Gisella (I never knew her real name) at five years and nine months old, respectively. Leni carried $6.50 to start her new life.

I had not previously known that my great-grandfather Farkas Rosenberg (accented on the first syllable, pronounced *farkash,* and meaning "wolf" in Hungarian) had preceded the rest of his family, and now appeared on the manifest as their sponsor, "Wolf Rosenberg of 644 East 6th Street." I do not remember Farkas, who died during my third year of life—but I greatly value the touching tidbit of information that, for whatever reason in his initial flurry of assimilation, Farkas had learned, and begun to use, the English translation of a name that strikes many Americans as curious, or even amusing, in sound—for he later reverted to Farkas, and no one in my family knew him by any other name.

My kind and diligent reader then bestowed an additional gift upon me by locating Farkas's manifest as well. He had arrived, along with eight hundred passengers in steerage, aboard the sister ship SS *Southwark* on June 13, 1900, listed as Farkas Rosenberg, illiterate at age thirty-four (although I am fairly sure that he could at least read and probably write Hebrew) and sponsored by a cousin named Jos. Weiss (but unknown to my family, and perhaps an enabling fiction). Farkas, a carpenter by trade, arrived alone with one dollar in his pocket.

Papa Joe's later story mirrors the tale of several million poor immigrants to a great land that did not welcome them with open arms (despite Lady Liberty's famous words), but also did not foreclose the possibility of success if they could

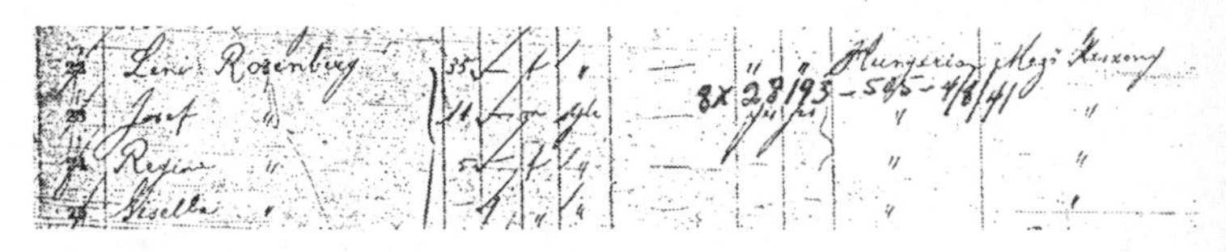
Leni Rosenberg
Regine
Gisella

Partial page of the passenger manifest for my grandfather's arrival in America on September 11, 1901. He is listed (as Josef Rosenberg) along with his mother, Leni, and his two younger sisters (my aunts Gus and Regina).

prevail by their own wits and unrelenting hard work. And who could, or should, have asked for more in those times? Papa Joe received no further schooling in America, save what experience provided and internal drive secured. As a young man, he went west for a time, working in the steel mills of Pittsburgh and on a ranch somewhere in the Midwest (not, as I later found out, as the cowboy of my dreams, but as an accountant in the front office). His mother, Leni, died young (my mother, Eleanor, bears her name in remembrance), as my second book of his legacy testifies. Papa Joe ended up, along with so many Jewish immigrants, in the garment district of New York City, where, after severing his middle finger in an accident as a cloth cutter, he eventually figured out how to parlay his remarkable, albeit entirely untrained, artistic talents into a better job that provided eventual access to middle-class life (and afforded much titillation to his grandchildren)—as a designer of brassieres and corsets.

He met Irene, also a garment worker, when he lived as a boarder at the home of Irene's aunt—for she had emigrated alone, at age fourteen in 1910, after a falling-out with her father, and under her aunt's sponsorship. What else can one say for the objective record (and what richness and depth could one *not* expose, at least in principle, and for all people, at the subjective level of human life, passion, and pure perseverance)? Grammy and Papa Joe married young, and their only early portrait together radiates both hope and uncertainty. They raised three sons and a daughter; my mother alone survives. Two of their children finished college.

Somehow I always knew, although no one ever pressured me directly, that the third generation, with me as the first member thereof, would fulfill the

ה׳ ממית ומחיה מוריד שאול ויעל

(Sam. I. Chap. 2 R. 6.)

Page of memory

for beloved and worthy deceased.

My beloved Mother died Apr. 7-1911

Papa Joe's prayer book recording the death of his mother in 1911.

deferred dream of a century by obtaining an advanced education and entering professional life. (My grandmother spoke Hungarian, Yiddish, German, and English, but could only write her adopted language phonetically. I will never forget her embarrassment when I inadvertently read a shopping list that she had written, and realized that she could not spell. I also remember her joy when, invoking her infallible memory and recalling some old information acquired in her study for citizenship, she won ten dollars on a Yiddish radio quiz for correctly identifying William Howard Taft as our fattest president.)

I loved Grammy and Papa Joe separately. Divorce, however sanctioned by their broader culture, did not represent an option in their particular world. Unlike Hal and Nettie Huxley, I'm not at all sure that they would have done it again. But they stuck together and prevailed, at least in peace, respect, and toleration, perhaps even in fondness. Had they not done so, I would not be here—and for this particular twig of evolutionary continuity I could not be more profoundly grateful in the most ultimate of all conceivable ways. I also loved them fiercely, and I reveled in the absolute certainty of their unquestioned blessing and unvarying support (not always deserved of course, for I really did throw that rock at Harvey, even though Grammy slammed our front door on Harvey's father, after delivering a volley of Yiddish curses amid proclamations that her Stevele would never do such a thing—while knowing perfectly well that I surely could).

The tree of all life and the genealogy of each family share the same topology and the same secret of success in blending two apparently contradictory themes of continuity without a single hair's breadth of breakage, and change without a moment's loss of a potential that need not be exploited in all episodes but must remain forever at the ready. These properties may seem minimal, even derisory, in a universe of such stunning complexity (whatever its inexplicable eternity or infinity). But this very complexity exalts pure staying power (and the lability that facilitates such continuity). Showy statues of Ozymandias quickly become lifeless legs in the desert; bacteria have scuttled around all the slings and arrows of outrageous planetary fortune for 3.5 billion years and running.

I believe in the grandeur of this view of life, the continuity of family lines, and the poignancy of our stories: Nettie Heathorn passing Tasso's torch as Granmoo two generations later; Papa Joe's ungrammatical landing as a stranger in a strange land and my prayer that, in some sense, he might see my work as a worthy continuation, also two generations later, of a hope that he fulfilled in a different way during his own lifetime. I suspect that we feel the poignancy in such continuity because we know that our little realization of an unstated family promise somehow mirrors the larger way of all life, and there-

Grammy and Papa Joe, soon after their wedding, circa 1915.

fore becomes "right" in some sense too deep for words or tears. I can therefore write the final essay in this particular series because I know that I will never run out of unkept promises, or miles to walk; and that I may even continue to sprinkle the journey remaining before sleep with a new idea or two. This view of life continues, flowing ever forward, while the current patriarch of one tiny and insignificant twig pauses to honor the inception of the twig's centennial in a new land, by commemorating the first recorded words of a fourteen-year-old forebear.

Dear Papa Joe, I have been faithful to your dream of persistence, and attentive to a hope that the increments of each worthy generation may buttress the continuity of evolution. You could speak those wondrous words right at the joy and terror of inception. I dared not repeat them until I could fulfill my own childhood dream—something that once seemed so mysteriously beyond any hope of realization to an insecure little boy in a garden apartment in Queens—to become a scientist and to make, by my own effort, even the tiniest addition to human knowledge of evolution and the history of life. But now, with my step 300, so fortuitously coincident with the world's new 1,000 and your own 100,* perhaps I have finally won the right to restate your noble words, and to tell you that their inspiration still lights my journey: I have landed. But I also can't help wondering what comes next!

---

*This essay, the three-hundredth and last of my series, appeared in January 2001—the inception of the millennium by a less popular, but more mathematically sanctioned, mode of reckoning. My grandfather also began the odyssey of my family in America when he arrived from Europe in 1901.

II

Disciplinary Connections: Scientific Slouching Across a Misconceived Divide

2

No Science Without Fancy, No Art Without Facts: The Lepidoptery of Vladimir Nabokov

The Paradox of Intellectual Promiscuity

No one ever accused Francis Bacon of modesty, but when England's lord chancellor proclaimed his "great instauration" of human understanding and vowed to take all knowledge as his province, the stated goal did not seem ludicrously beyond the time and competence of a great thinker in Shakespeare's age. But as knowledge exploded, and then fragmented into disciplines with increasingly rigid and self-policed boundaries, the restless scholar who tried to operate in more than one domain became an object of suspicion—either a boastful pretender across the board ("jack of all and master of none," in the old cliché), or a troublesome dilettante in an alien domain, attempting to impose the methods of his genuine expertise upon inappropriate subjects in a different world.

We tend toward benign toleration when great thinkers and artists pursue disparate activities as a harmless hobby, robbing little time from their fundamental achievements. Goethe (and Churchill, and many others) may have been lousy Sunday painters, but Faust and Werther suffered no neglect thereby. Einstein (or so I have heard from people with direct experience) was an indifferent violinist, but his avocation fiddled little time away from physics.

However, we grieve when we sense that a subsidiary interest stole precious items from a primary enterprise of great value. Dorothy Sayers's later theological writings may please aficionados of religion, but most of her devout fans would have preferred a few more detective novels featuring the truly inimitable Lord Peter Wimsey. Charles Ives helped many folks by selling insurance, and Isaac Newton must have figured out a thing or two by analyzing the prophetic texts of Daniel, Ezekiel, and Revelation—but, all in all, a little more music or mathematics might have conferred greater benefit upon humanity.

Therefore, when we recognize that a secondary passion took substantial time from a primary source of fame, we try to assuage our grief over lost novels, symphonies, or discoveries by convincing ourselves that a hero's subsidiary love must have informed or enriched his primary activity—in other words, that the loss in quantity might be recompensed by a gain in quality. But such arguments may be very difficult to formulate or sustain. In what sense did Paderewski become a better pianist by serving as prime minister of Poland (or a better politician by playing his countryman Chopin)? How did a former career in major-league baseball improve (if we give a damn, in this case) Billy Sunday's evangelical style as a stump preacher? (He sometimes began sermons—I am not making this up—by sliding into the podium as an entering gesture.)

No modern genius has inspired more commentary in this mode than Vladimir Nabokov, whose "other" career as a taxonomist of butterflies has inspired as much prose in secondary criticism as Nabokov ever lavished upon Ada, Lolita, and all his other characters combined. In this case in particular—because Nabokov was no dilettante spending a few harmless Sunday hours in the woods with his butterfly net, but a serious scientist with a long list of publications and a substantial career in entomology—we crave some linkage between his two lives, some way to say to ourselves, "We may have lost several novels, but Nabokov spent his entomological time well, developing a vision and approach that illuminated, or even transformed, his literary work." (Of course, speaking parochially, professional taxonomists, including the author of this essay, might regret even more the loss of several monographs implied by Nabokov's novels!)

To allay any remaining suspicions among the literati, let me assure all readers about a consensus in my professional community: Nabokov was no amateur (in the pejorative sense of the term), but a fully qualified, clearly talented, duly employed professional taxonomist, with recognized "world class" expertise in the biology and classification of a major group, the Latin American Polyommatini, popularly known to butterfly aficionados as "blues."

No passion burned longer, or more deeply, in Nabokov's life than his love for the natural history and taxonomy of butterflies. He began in early childhood, encouraged by a traditional interest in natural history among the upper-class intelligentsia of Russia (not to mention the attendant economic advantages of time, resources, and opportunity). Nabokov stated in a 1962 interview (Zimmer, page 216): "One of the first things I ever wrote in English was a paper on Lepidoptera I prepared at age twelve. It wasn't published because a butterfly I described had been described by someone else." Invoking a lovely entomological metaphor in a 1966 interview, Nabokov spoke of childhood fascination, continuous enthusiasm throughout life, and regret that political realities had precluded even more work on butterflies (Zimmer, page 216):

> But I also intend to collect butterflies in Peru or Iran before I pupate. . . . Had the Revolution not happened the way it happened, I would have enjoyed a landed gentleman's leisure, no doubt, but I also think that my entomological occupations would have been more engrossing and energetic and that I would have gone on long collecting trips to Asia. I would have had a private museum.

Nabokov published more than a dozen technical papers on the taxonomy and natural history of butterflies, mostly during his six years of full employment as Research Fellow (and unofficial curator) in Lepidoptery at the Museum of Comparative Zoology at Harvard University, where he occupied an office three floors above the laboratory that has been my principal scientific home for thirty years. (I arrived twenty years after Nabokov's departure and never had the pleasure of meeting him, although my knowledge of his former presence has always made this venerable institution, built by Louis Agassiz in 1859 and later tenanted by several of the foremost natural historians in America, seem even more special.)

Nabokov worked for Harvard, at a modest yearly salary of about one thousand dollars, between 1942 and 1948, when he accepted a teaching post in literature at Cornell University. He was a respected and recognized professional in his chosen field of entomological systematics. The reasons often given for attributing to Nabokov either an amateur, or even only a dilettante's, status arise from simple ignorance of accepted definitions for professionalism in this field.

First, many leading experts in various groups of organisms have always been "amateurs" in the admirable and literal (as opposed to the opposite and pejorative) sense that their love for the subject has inspired their unparalleled knowledge, and that they do not receive adequate (or any) pay for their work. (Taxonomy is not as expensive, or as laboratory-driven, as many scientific fields. Careful and dedicated local observation from childhood, combined with diligence in reading and study, can supply all the needed tools for full expertise.)

Second, poorly remunerated and inadequately titled (but full-time) employment has, unfortunately, always been *de rigueur* in this field. The fact that Nabokov worked for little pay, and with the vague title Research Fellow, rather than a professorial (or even a curatorial) appointment, does not imply nonprofessional status. When I took my position at the same museum in 1968, several heads of collections, recognized as world's experts with copious publications, worked as "volunteers" for the symbolic "dollar a year" that gave them official status on the Harvard payroll.

Third, and most important, I do not argue that all duly employed taxonomists can claim enduring expertise and righteous status. Every field includes some clunkers and nitwits, even in high positions! I am not, myself, a professional entomologist (I work on snails among the Mollusca), and therefore cannot judge Nabokov's credentials on this crucial and final point. But leading taxonomic experts in the large and complex group of "blues" among the butterflies testify to the excellence of his work, and grant him the ultimate accolade of honor within the profession by praising his "good eye" for recognizing the (often subtle) distinctions that mark species and other natural groups of organisms (see the bibliography to this essay for two articles by leading butterfly taxonomists: Remington; and Johnson, Whitaker, and Balint). In fact, as many scholars have stated, before Nabokov achieved a conventional form of literary success with the publication of *Lolita,* he could have been identified (by conventional criteria of money earned and time spent) as a professional lepidopterist and amateur author!

In conjunction with this collegial testimony, we must also note Nabokov's own continual (and beautifully stated) affirmation of his love and devotion to all aspects of a professional lepidopterist's life. On the joys of fieldwork and collecting, he effuses in a letter to Edmund Wilson in 1942 (quoted in Zimmer, page 30): "Try, Bunny, it is the noblest sport in the world." Of the tasks traditionally deemed more dull and trying—the daily grind of the laboratory and microscope—he waxed with equal ardor in a letter to his sister in 1945, in the midst of his Harvard employment (in Zimmer, page 29):

> My laboratory occupies half of the fourth floor. Most of it is taken up by rows of cabinets, containing sliding cases of butterflies. I am custodian of these absolutely fabulous collections. We have butterflies from all over the world. . . . Along the windows extend tables holding my microscopes, test tubes, acids, papers, pins, etc. I have an assistant, whose main task is spreading specimens sent by collectors. I work on my personal research . . . a study of the classification of American "blues" based on the structure of their genitalia (minuscule sculpturesque hooks, teeth, spurs, etc., visible only under the microscope), which I sketch in with the aid of various marvelous devices, variants of the magic lantern. . . . My work enraptures but utterly exhausts me. . . . To know that no one before you has seen an organ you are examining, to trace relationships that have occurred to no one before, to immerse yourself in the wondrous crystalline world of the microscope, where silence reigns, circumscribed by its own horizon, a blindingly white arena—all this is so enticing that I cannot describe it.

Nabokov worked so long and so intensely in grueling and detailed observation of tiny bits of insect anatomy that his eyesight became permanently compromised—thus placing him in the company of several of history's most famous entomologists, especially Charles Bonnet in the eighteenth century and August Weismann in the nineteenth, who sacrificed their sight to years of eye-straining work. In a television interview in 1971, Nabokov stated (Zimmer, page 29):

> Most of my work was devoted to the classification of certain small blue butterflies on the basis of their male genitalic structure. These studies required the constant use of a microscope, and since I devoted up to six hours daily to this kind of research my eyesight was impaired forever; but on the other hand, the years at the Harvard Museum remain the most delightful and thrilling in all my adult life.

Nonetheless, and as a touching, final testimony to his love and dedication to entomology, Nabokov stated in a 1975 interview (Zimmer, page 218) that his enthusiasm would still pull him inexorably in ("like a moth to light" one is tempted to intone) if he ever allowed impulse to vanquish bodily reality:

> Since my years at the Museum of Comparative Zoology in Harvard, I have not touched a microscope, knowing that if I did, I would drown again in its bright well. Thus I have not, and probably never shall, accomplish the greater part of the entrancing research work I had imagined in my young mirages.

Thus, in conclusion to this section, we cannot adopt the first solution to "the paradox of intellectual promiscuity" by arguing that Nabokov's lepidoptery represents only the harmless diversion of an amateur hobbyist, ultimately stealing no time that he might realistically have spent writing more novels. Nabokov loved his butterflies as much as his literature. He worked for years as a fully professional taxonomist, publishing more than a dozen papers that have stood the test of substantial time.

Can we therefore invoke the second solution by arguing that time lost to literature for the sake of lepidoptery nonetheless enhanced his novels, or at least distinguished his writing with a brand of uniqueness? I will eventually suggest a positive answer, but by an unconventional argument that exposes the entire inquiry as falsely parsed. I must first, however, show that the two most popular versions of this "second solution" cannot be defended, and that the paradox of intellectual promiscuity must itself be rejected and identified as an impediment to proper understanding of the relationships between art and science.

Two False Solutions to a Nonproblem

In surveying commentaries written by literary scholars and critics about Nabokov's work on butterflies, I have been struck by their nearly universal adherence to either of two solutions for the following supposed conundrum: Why did one of the greatest writers of our century spend so much time working and publishing in a markedly different domain of such limited interest to most of the literate public?

The Argument for Equal Impact

In this first solution, Nabokov's literary fans may bemoan their losses (just as any lover of music must lament the early deaths of Mozart and Schubert). Still, in seeking some explanation for legitimate grief, we may find solace in claiming that Nabokov's transcendent genius permitted him to make as uniquely innovative and distinctive a contribution to lepidoptery as to literature. However much we may wish that he had chosen a different distribution for his time, we can at least, with appropriate generosity, grant his equal impact and benefit upon natural history. Adherents to this solution have therefore tried to develop arguments for regarding Nabokov's lepidoptery as specially informed by his general genius, and as possessing great transforming power for natural history.

But none of these claims can be granted even a whisper of plausibility by biologists who know the history of taxonomic practice and evolutionary theory. Nabokov, as documented above, was a fully professional and highly competent taxonomic specialist on an important group of butterflies—and for this fine work he gains nothing but honor in my world. However, no natural historian has ever viewed Nabokov as an innovator, or as an inhabitant of what humanists call the "vanguard" (not to mention the avant-garde) and scientists the "cutting edge." Nabokov may have been a major general of literature, but he can only be ranked as a trustworthy, highly trained career infantryman in natural history.

Vladimir Nabokov practiced his science as a conservative specialist on a particular group of organisms, not in any way as a theorist or a purveyor of novel ideas or methods. He divided and meticulously described; he did not unify or generalize. (I will explain in the next section why a natural historian can make such a judgment without intending any condescension or lack of respect.) Nonetheless, four arguments have been advanced again and again by literary commentators who seem driven by a desire to depict Nabokov as a revolutionary spirit in natural history as well.

1. *The myth of innovation.* Many critics have tried, almost with an air of desperation, to identify some aspect of Nabokov's methodology that might be labeled as innovative. But taxonomic professionals will easily recognize these claims as fallacious—for the putative novelty represents either a fairly common (if admirable) practice, or else an idiosyncrasy (a "bee in the bonnet") that Nabokov surely embraced with great ardor, but that cannot be regarded as a major issue of scientific importance.

As a primary example, many critics have stressed Nabokov's frequent complaints about scientists who fail to identify the original describers when citing the formal Latin name of a butterfly—either in listing species in popular field guides, or in identifying subspecies in technical publications. Zimmer (page 10), for example, writes: "A growing number of non- and semi-scientific publications nowadays omit the author. Nabokov called it 'a deplorable practice of commercial origin which impairs a number of recent zoological and botanical manuals in America.'"

By the rules of nomenclature, each organism must have a binomial designation consisting of a capitalized genus name *(Homo)* and a lowercase "trivial" name *(sapiens),* with the two together forming the species name *(Homo sapiens).* (Linnaean taxonomy is called "binomial" in reference to these two parts of a species's name.) It is also customary, but not required, to add (not in italics) the name of the first describer of the species after the binomial designation—as in *Homo sapiens* Linnaeus. This custom certainly helps specialists by permitting easier tracing of the history of a species's name. But this practice is also extremely time-consuming (locating the original describer is often tedious and difficult; I don't know the first authors for several of the snail species most central to my own research). Moreover, when hundreds of names are to be listed (as in popular field guides), rigid adherence to this custom requires a great deal of space for rather limited benefit.

Therefore, popular publications (especially the manuals of Nabokov's ire above) generally omit the names of describers. In addition, and for the same reason, technical publications often compromise by including describers' names for species, but omitting them for subspecies (trinomial names for geographically defined subgroups within a species). Honorable people can argue either side of this issue; I tend to agree with Nabokov's critics in this case—but I cannot generate much personal passion over this relatively minor issue.

In another example, Boyd (*The American Years,* page 128) praises Nabokov's methods: "Nabokov's mode of presentation was ahead of his time. Instead of showing a photograph of a single specimen of a butterfly species or a diagram of the genitalia of a single specimen, he presented when necessary a range of specimens of certain subspecies in nine pages of crowded plates." Here I side entirely with Nabokov and his proper recognition of natural history's primary subject matter: variation and diversity at all levels. But Nabokov did not proceed in either a unique or an unusually progressive manner in illustrating multiple specimens (I rather suspect that his decision reflected his fussy and meticulous thoroughness more than any innovative theoretical vision about the nature of variation.) This issue has provoked a long history of discussion and varying practice in taxonomy—and many other specialists have stood with Nabokov on the right side (as I would say) of this question.

2. *The myth of courage.* As an adjunct (or intensification) to claims for innovation, many literary critics have identified Nabokov as theoretically coura-

geous (and forward-looking) in his expressed doubts about Darwinian orthodoxies, particularly on the subject of adaptive value for patterns of mimicry in butterfly wings.

In this context, a remarkable passage from *Speak, Memory* has often been cited. Nabokov apparently wrote, but never published, an extensive scientific article (see Remington, page 282) in an attempt to refute natural selection as the cause of mimicry by denying the purely adaptive value of each component of resemblance. (Darwinians have assumed that mimicry—the evolution, in one butterfly species, of striking resemblance, generally in color patterns of the wings, to another unrelated form—arises for adaptive benefit, usually for permitting a "tasty" species to gain protection by simulating a noxious species that predators have learned to avoid). This paper has been lost, except for the following fragment that Nabokov included in *Speak, Memory:*

> "Natural selection," in the Darwinian sense, could not explain the miraculous coincidence of imitative aspect and imitative behavior, nor could one appeal to the theory of "the struggle for life" when a protective device was carried to a point of mimetic subtlety, exuberance, and luxury far in excess of a predator's power of appreciation. I discovered in nature the nonutilitarian delights that I sought in art. Both were a form of magic, both were a game of intricate enchantment and deception.

An understandable prejudice of intellectual life leads us to view tilters at orthodoxy as courageous front-line innovators. Nonetheless, one may also attack a common view for opposite reasons of conservative allegiance to formerly favored ideas. On Nabokov's forcefully expressed doubts about Darwinian interpretations of mimicry, two observations identify his stance as more traditionally conservative than personally innovative or particularly courageous. First, when Nabokov wrote his technical papers in the 1940s, the modern Darwinian orthodoxy had not yet congealed, and a Nabokovian style of doubt remained quite common among evolutionary biologists, particularly among taxonomists immersed in the study of anatomical detail and geographic variation (see Robson and Richards, 1936, for the classic statement; see Gould, 1983, and Provine, 1986, for documentation that a hard-line Darwinian orthodoxy only coalesced later in the 1950s and 1960s). Thus Nabokov's views on mimicry represent a common attitude among biologists in his time, a perspective linked more to earlier consensuses about non-Darwinian evolution than to legitimate modern challenges. (I am, by the way and for my sins, well recognized, and often reviled, for my own doubts about Darwinian orthodoxies, so I do not make this judgment of Nabokov while acting as *defensor fidei.*)

Second, although we must always struggle to avoid the primary error of historiography—the anachronistic use of later conclusions to judge the cogency of an earlier claim—in assessing Nabokov's views on mimicry, we may still fairly note that Nabokov's convictions on this subject have not withstood the standard scientific test of time (*veritas filia temporis,* to cite Bacon once again). The closing words of a world's expert on the evolutionary biology of butterflies, and a firm admirer of Nabokov's science, may be cited here. My colleague Charles Lee Remington writes (page 282):

> Impressive though the intellectual arguments are . . . it would be unreasonable to take them very seriously in science today. Mimicry and other aspects of adaptive coloration and shape involve such superb and elaborate resemblances that various biologists had questioned the Darwinian explanations during the early decades of this century. Subsequent publication of so many elegant experimental tests of mimicry and predator learning . . . and color-pattern genetics . . . has caused the collapse of the basic challenges, in my view as a specialist in the field. However, I do guess that Nabokov had such a strong metaphysical investment in his challenge to natural selection that he might have rejected the evolutionary conclusions for his own satisfaction. He was an excellent naturalist and could cite for himself very many examples of perfect resemblances, but he may have been too untrained in the complexities of modern population genetics.

Finally, I must also note that several other prime components of Nabokov's biological work would now be viewed as superseded rather than prescient, and would also be judged as a bit antiquated in their own time, rather than innovative or even idiosyncratic. In particular, as a practical taxonomist, Nabokov advocated a definition of species based only on characters preserved in specimens of museum collections. Today (and, for the most part, in Nabokov's time as well), most evolutionary biologists would strongly insist that species be recognized as "real" and discrete populations in nature, not as units defined by identifiable traits in artificially limited data of human collections. Many species owe their distinction to genetic and behavioral features that maintain the cohesion of a population in nature, but may not be preserved in museum specimens. Nonetheless, Nabokov explicitly denied that such populations should be recognized as species—a view that almost all naturalists would now reject. Nabokov wrote in one of his technical papers (cited in Zimmer, page 15): "For better or worse our present notions of species in Lepidoptera is based solely on the checkable structure of dead specimens, and if Forster's Furry cannot be distinguished from the Furry Blue except by its chromosome number, Forster's Furry must be scrapped."

3. *The myth of artistry.* Nabokov made many drawings of butterflies, both published, and as charming, often fanciful illustrations in copies of his books presented to friends and relatives, especially to his wife, Vera. These drawings are lovely, and often quite moving in their sharp outlines and naïve brightnesses—but, putting the matter diplomatically, the claim (sometimes made) that these drawings should be judged either as unusual in their accuracy or as special in their beauty can only be labeled as kindly hagiographical, especially in the light of a truly great tradition for wonderful and sensitive art among the best natural history illustrators, from Maria Merian to Edward Lear (who wrote limericks as a hobby, but worked as a skilled illustrator for a profession).

4. *The myth of literary quality.* Some critics, recognizing the merely conventional nature of Nabokov's excellence in taxonomy, have stated that, at least, he wrote his non-innovative descriptions in the most beautifully literate prose ever composed within the profession. Zaleski (page 36), for example, extolls Nabokov for writing, in technical papers, "what is surely the most polished prose even applied to butterfly studies." Again, such judgments can only be subjective—but I have spent a career reading technical papers in this mode, while applying at least a serious amateur's eye to literary style and quality. Nabokov's descriptive prose flows well enough, but I find nothing distinctive in his contributions to this highly restricted genre, where rules and conventions of spare and "objective" writing offer so little opportunity to spread one's literary wings.

The Argument for Literary Illumination

Once we debunk, for Nabokov's case, two false solutions to the paradox of intellectual promiscuity—the argument, refuted above, that his lepidoptery represented a harmless private passion, robbing no substantial time from his literary output; and the claim, rejected in the first part of this section, that his general genius at least made his lepidoptery as distinctive and as worthy as his literature—only one potential source for conventional solace remains: the proposition that although time spent on lepidoptery almost surely decreased his literary output, the specific knowledge and the philosophical view of life that Nabokov gained from his scientific career directly forged (or at least strongly contributed to) his unique literary style and excellence.

We can cite several important precedents for such a claim. Jan Swammerdam, the greatest entomologist of the seventeenth century, devoted the last part of his life to evangelical Christianity, claiming that a fundamental entomological metaphor had directed his developing religious views: the life cycle of a butterfly as an emblem for the odyssey of a Christian soul, with the caterpillar (larva) representing our bodily life on earth, the pupa denoting the period of the soul's waiting after bodily death, and the butterfly marking a glorious resurrection.

In another example, one that would be viewed as more fruitful by most contemporary readers, Alfred Kinsey spent twenty years working as an entomologist on the taxonomy of the gall-wasp *Cynips* before turning to the surveys of human sexual behavior that would mark his notoriety as a pivotal figure in the social history of the twentieth century. In a detailed preface to his first great treatise, *Sexual Behavior in the Human Male* (1948), Kinsey explained how a perspective gained from insect taxonomy upon the nature of populations—particularly the copious variation among individuals, and the impossibility of marking one form as normal and the others as deviant—had directly informed and inspired his research on sexual behavior. Kinsey wrote:

> The techniques of this research have been taxonomic, in the sense in which modern biologists employ the term. It was born out of the senior author's longtime experience with a problem in insect taxonomy. The transfer from insect to human material is not illogical, for it has been a transfer of a method that may be applied to the study of any variable population.

We know that Nabokov made continual and copious reference to entomological subjects, particularly to butterflies, in all his literary productions—in passages ranging from the minutely explicit to the vaguely cryptical, to the broadly general. Several scholars have tabulated and annotated this rich bounty. Nabokov's critics could therefore scarcely avoid the potential hypothesis, especially given the precedents of Swammerdam and Kinsey, that Nabokov's lepidoptery shaped his literature in direct and crucial ways.

Literary scholars have often ventured such a claim, particularly by asserting that Nabokov used his knowledge of insects as a rich source for metaphors and symbols. In the strongest version, most, if not nearly all, citations of butterflies convey a level of deep symbolic meaning in Nabokov's prose. For example, Joann Karges wrote in her book on Nabokov's Lepidoptera (cited in Zimmer, page 8): "Many of Nabokov's butterflies, particularly pale and white ones, carry the traditional ageless symbol of the anima, psyche, or soul . . . and suggest the evanescence of a spirit departed or departing from the body."

Two arguments, one a specific denial of this search for symbolism, and the other a more general statement about art and science, strongly refute this last hope for the usual form of literary solace in Nabokov's dedication to science—a claim that the extensive time thus spent strongly improved his novels. For the first (quite conclusive and specific) argument, Nabokov himself vehemently insisted that he not only maintained no interest in butterflies as literary symbols, but that he would also regard such usage as a perversion and desecration of his true concerns. (Artists, and all of us, of course, have been known to dissemble, but I see no reason to gainsay Nabokov's explicit and heartfelt comments on this subject.) For example, he stated in an interview (quoted in Zimmer, page 8): "That in some cases the butterfly symbolizes something (e.g., *Psyche*) lies utterly outside my area of interest."

Over and over again, Nabokov debunks symbolic readings in the name of respect for factual accuracy as a primary criterion. For example, he criticizes Poe's symbolic invocation of the death's-head moth because Poe didn't describe the animal and, even worse, because he placed the species outside its true geographic range: "Not only did he [Poe] not visualize the death's-head moth, but he was also under the completely erroneous impression that it occurs in America" (in Zimmer, page 186). Most tellingly, in a typical Nabokovian passage in *Ada,* he playfully excoriates Hieronymus Bosch for including a butterfly as a symbol in his *Garden of Earthly Delights,* but then depicting the wings in reverse by painting the gaudy top surface on an insect whose folded wings should be displaying the underside!

> A tortoiseshell in the middle panel, placed there as if settled on a flower—mark the "as if," for here we have an example of exact knowledge of the two admirable girls, because they say that actually the *wrong* side of the bug is shown, it should have been the underside, if seen, as it is, in profile, but Bosch evidently found a wing or two in the corner cobweb of his casement and showed the prettier upper surface in depicting his incorrectly folded insect. I mean I don't give a hoot for the esoteric meaning, for the myth behind the moth, for the masterpiece-baiter who makes Bosch express some bosh of his time, I'm allergic to allegory.

Finally, when Nabokov does cite a butterfly in the midst of a metaphor, he attributes no symbolic meaning to the insect, but only describes an accurate fact to carry his more general image. For example, he writes in *Mary* (cited in Zimmer, page 161): "Their letters managed to pass across the terrible Russia of that time—like a cabbage white butterfly flying over the trenches."

Second, and more generally, if we wish to argue that Nabokov's lepidoptery gave direct substance, or set the style, of his literature, then we must face a counterclaim—for the best case of explicit linkage led Nabokov into serious error. (And I surely will not propagate the smug scientist's philistine canard that literary folks should stick to their lasts and leave us alone because they always screw up our world with their airy-fairy pretensions and insouciance about accuracy.) If I wanted to advance a case for direct linkage, I would have to emphasize a transfer from Nabokov's artistic vision to his science, not vice versa—unfortunately, in this instance, to the detriment of natural history. Nabokov frequently stated that his non-Darwinian interpretation of mimicry flowed directly from his literary attitude—as he tried to find in nature "the nonutilitarian delights that I sought in art" (see page 37 for a full citation of this passage). And, as argued previously, this claim represents the most serious general error in Nabokov's scientific writing.

The Solution of Accuracy

In standard scientific practice, when tests of a favored hypothesis have failed, and one is beating one's head against a proverbial wall, the best strategy for reclaiming a fruitful path must lie in the empirical record, particularly in scrutinizing basic data for hints of a pattern that might lead to a different hypothesis. In Nabokov's case, both his explicit statements and his striking consistency of literary usage build such a record and point clearly to an alternative solution. The theme has not been missed by previous critics, for one can hardly fail to acknowledge something that Nabokov emphasized so forcefully. But I feel that most published commentary on Nabokov's lepidoptery has failed to grasp the centrality of this argument as a primary theme for understanding his own concept of the relationship between his literary and scientific work—primarily, I suppose, because we have been befogged by a set of stereotypes about conflict and difference between these two great domains of human understanding.

Conventional solutions fail because they have focused on too specific a level—that is, to the search for how one domain, usually science in this case, impacted the other. But the basic source of relationship may be hiding at a deeper level (deeper, that is, in a geometric sense, not in any claim about morality or greater importance). Perhaps the major linkage of science and literature lies in some distinctive, *underlying* approach that Nabokov applied *equally* to both domains—a procedure that conferred the same special features upon *all* his efforts. In this case we should not posit a primary and directional impact of one domain upon the other. Rather, we should investigate the hypothesis that Nabokov's art and science both benefited, in like measure, from his application of a method, or a mode of mental functioning, that exemplifies the basic character of his particular genius.

All natural historians know that "replication with difference" builds the best test case for a generality—for how can we prove a coordinating hypothesis unless we can apply it to multiple cases, and how can we be confident in our conclusion unless these cases be sufficiently different in their immediate context to demonstrate that any underlying commonality must lie in a single mental approach applied to disparate material? Among great twentieth-century thinkers, I know no better case than Nabokov's for testing the hypothesis that an underlying unity of mental style (at a level clearly meriting the accolade of genius) can explain one man's success in extensive and fully professional work in two disciplines conventionally viewed as maximally different, if not truly opposed. If we can validate this model for attributing interdisciplinary success to a coordinating and underlying mental uniqueness, rather than invoking the conventional argument about overt influence of one field upon another, then Nabokov's story may teach us something important about the unity of creativity, and the falsity (or at least the contingency) of our traditional separation, usually in mutual recrimination, of art from science.

Above all else—and why should we not take him at his word?—Nabokov vociferously insisted that he cherished meticulous accuracy in detail as the defining feature of all his productions (as illustrated in the passage quoted on page 41 from *Ada*). All commentators have noted these Nabokovian claims (for one could hardly fail to mention something stated so frequently and forcefully by one's principal subject). Previous critics have also recognized that a commitment to detailed accuracy not only defines Nabokov's maximally rich and meticulously careful prose, but might also be greatly valued for professional work in the description of butterfly species. Unfortunately, however, most commentary then follows a lamentable stereotype about science (particularly for such "low status" fields as descriptive natural history), and assumes that Nabokov's commitment to accuracy must have imposed opposite qualities upon his work in these two professions—thus, and again lamentably, reinforcing the conventional distinction of art and science as utterly different and generally opposed. Such detail, we are told, enriches Nabokov's literature, but also brands his science as pedestrian, unimaginative, and "merely" descriptive (as in the cliché about folks who never see forests because they only focus on distinctive features of individual trees). The stereotype of the taxonomist as a narrow-minded, bench-bound pedant then reconfirms this judgment. Zaleski (page 38), for example, sums up his article on Nabokov's lepidoptery by writing:

> In both books and butterflies, Nabokov sought ecstasy, and something beyond. He found it in the worship of detail, in the loving articulation of organic flesh and organized metaphor. . . . He was perfectly suited as a master novelist and a laboratory drudge.

Zaleski goes on to report that Nabokov importuned his Cornell students with a primary motto: "Caress the details, the divine details." "In high art and pure science," he stated, "detail is everything." Indeed, Nabokov often praised the gorgeous detail of meticulous taxonomic language as inherently literary in itself, speaking of "the precision of poetry in taxonomic description" (in Zimmer, page 176). He also, of course, extolled precision in anatomical description for its scientific virtue. He wrote a letter to Pyke Johnson in 1959, commenting upon a proposed jacket design for his *Collected Poems* (cited in Remington, page 275):

> I like the two colored butterflies on the jacket but they have the bodies of ants, and no stylization can excuse a simple mistake. To stylize adequately one must have complete knowledge of the thing. I would be the laughing stock of my entomological colleagues if they happened to see these impossible hybrids.

In reading through all Nabokov's butterfly references (in his literary works) as preparation for writing this essay, I was struck most of all by his passion for accuracy in every detail of anatomy, behavior, or location. Even his poetical or metaphorical descriptions capture a common visual impression—as when he writes in "The Aurelian," a story from 1930, about "an oleander hawk [moth] . . . its wings vibrating so rapidly that nothing but a ghostly nimbus was visible about its streamlined body." Even his occasional fantasies and in-jokes, accessible only to a few initiates (or readers of such study guides as Zimmer's) build upon a strictly factual substrate. For example, Nabokov thought he had discovered a new species of butterfly during his Russian boyhood. He wrote a description in English and sent the note to a British entomologist for publication. But the English scientist discovered that Nabokov's species had already been named in 1862 by a German amateur collector named Kretschmar, in an obscure publication. So Nabokov bided his time and finally chose a humorous form of revenge in his novel *Laughter in the Dark* (quoted in Zimmer, page 141): "Many years later, by a pretty fluke (I know I should not point out these plums to people), I got even with the first discoverer of my moth by giving his own name to a blind man in a novel."

Literary critics sometimes chided Nabokov for his obsessive attention to detail. Nabokov, in true form, described these attacks with a witty (and somewhat cryptic) taxonomic reference—speaking in *Strong Opinions* (quoted in Zimmer, page 175) of detractors "accusing me of being more interested in the subspecies and the subgenus than in the genus and the family." (Subspecies and subgenera represent categories for fine subdivision of species and genera. The rules of nomenclature recognize these categories as available for convenience, but not required in practice. That is, species need not be divided into subspecies, nor genera into subgenera. But genera and families represent basic and more inclusive divisions that must be assigned to all creatures. That is, each species must belong to a genus, and each genus to a family.)

Nabokov generalized his defense of meticulous detail beyond natural history and literature to all intellectual concerns. In a 1969 interview, he scornfully dismissed critics who branded such insistence upon detail as a form of pedantry (my translation from Nabokov's French, as cited in Zimmer, page 7): "I do not understand how one can label the knowledge of natural objects or the vocabulary of nature as pedantry." In annotating his personal copy of the French translation of *Ada,* Nabokov listed the three unbreakable rules for a good translator: intimate knowledge of the language from which one translates; experience as a writer of the language into which one translates; and (the third great dictate of detail) "that one knows, in both languages, the words designating concrete objects (natural and cultural, the flower and the clothing)" (my translation from Nabokov's French original, cited in Zimmer, page 5).

Zimmer (page 8) epitomizes the central feature of Nabokov's butterfly citations: "They are all real butterflies, including the invented ones which are mimics of real ones. And they usually are not just butterflies in general, but precisely the ones that would occur at that particular spot, behaving exactly the way they really would. Thus they underscore, or rather help constitute, the veracity of a descriptive passage." In an insightful statement, Zimmer (page 7) then generalizes this biological usage to an overarching Nabokovian principle with both aesthetic and moral components:

> Both the writer of fiction and the naturalist drew on a profound delight in precise comparative observation. For Nabokov, a work of nature was like a work of art. Or rather it *was* a profound work of art, by the greatest of all living artists, evolution, and as much a joy to the mind and a challenge to the intellect as a Shakespeare sonnet. Hence it deserved to be studied like it, with never ending attention to detail and patience.

But perhaps the best summary of Nabokov's convictions about the ultimate value of accurate detail can be found in "A Discovery," a short poem written in 1943:

> Dark pictures, thrones, the stones that pilgrims kiss
> Poems that take a thousand years to die
> But ape the immortality of this
> Red label on a little butterfly.

(Again, some taxonomic exegesis must be provided to wrest general understanding from the somewhat elitist—scarcely surprising given his social background—and not always user-friendly Nabokov. Museum curators traditionally affix red labels only to "holotype" specimens—that is, to individuals chosen as official recipients of the name given to a new species. The necessity for such a rule arises from a common situation in taxonomic research. A later scientist may discover that the original namer of a species defined the group too broadly by including specimens from more than one genuine species. Which specimens shall then keep the original name, and which shall be separated out to receive a different designation for the newly recognized species? By official rules, the species of the designated holotype specimen keeps the original name, and members of the newly recognized species must receive a new name. Thus, Nabokov tells us that no product of human cultural construction can match the immortality of the permanent name-bearer for a genuine species in nature. The species may become extinct, of course, but the name continues forever to designate a genuine natural population that once inhabited the earth. The holotype specimen therefore becomes our best example of an immortal physical object. And the holotype specimen bears a red label in standard museum practice.)

Nabokov's two apparently disparate careers therefore find their common ground on the most distinctive feature of his unusual intellect and uncanny skill—the almost obsessive attention to meticulous and accurate detail that served both his literary productions and his taxonomic descriptions so well, and that defined his uncompromising commitment to factuality as both a principle of morality and a guarantor and primary guide to aesthetic quality. Science and literature therefore gain their union on the most palpable territory of concrete things, and on the value we attribute to accuracy, even in smallest details, as a guide and an anchor for our lives, our loves, and our senses of worth.

This attitude expresses a general belief and practice in science (at least as an ideal, admittedly not always achieved due to human frailty). Of all scientific subfields, none raises the importance of intricate detail to such a plateau of importance as Nabokov's chosen profession of taxonomic description for small and complex organisms. To function as a competent professional in the systematics of Lepidoptera, Nabokov really had no choice but to embrace such attention to detail, and to develop such respect for nature's endless variety.

But this attitude to detail and accuracy carries no ineluctable status in literature—so Nabokov's unaltered skills and temperament, now applied to his second profession, conferred distinction, if not uniqueness, upon him. The universal and defining excellence of a professional taxonomist built a substrate for the uncommon, and (in Nabokov's case) transcendent, excellence of a writer. After all, the sheer glory of voluminous detail does not ignite everyone's muse in literature. Some folks can't stand to read every meandering and choppy mental detail of one day in the life of Leopold Bloom, but others consider *Ulysses* the greatest novel of the twentieth century. I ally myself with the second group. I also love *Parsifal*—and the writing of Vladimir Nabokov. I have always been a taxonomist at heart. Nothing matches the holiness and fascination of accurate and intricate detail. How can you appreciate a castle if you don't cherish all the building blocks, and don't understand the blood, toil, sweat, and tears underlying its construction?*

I could not agree more with Nabokov's emphasis upon the aesthetic and moral—not only the practical and factual—value of accuracy and authenticity in intricate detail. This sensation, this love, may not stir all people so ardently (for *Homo sapiens,* as all taxonomists understand so well, includes an especially wide range of variation among individuals of the species). But such a basic aesthetic, if not universal, surely animates a high percentage of humanity, and must evoke something very deep in our social and evolutionary heritage. May

*Incidentally, Nabokov represented an intractable mystery to me until I learned that he grew up trilingual in Russian, English, and French—a common situation among the Russian upper classes in his day. Even as a teenager reading *Lolita,* I couldn't understand how anyone who learned English as a second tongue could become such a master of linguistic detail. Indeed, one cannot. Conrad narrated wonderful stories, but could never play with his adopted language as Nabokov did with one of his native tongues.

I mention just one true anecdote to represent this general argument? The head of the National Air and Space Museum in Washington, D.C., once hosted a group of blind visitors to discuss how exhibits might be made more accessible to their community. In this museum the greatest airplanes of our history—including the Wright Brothers' biplane from Kitty Hawk and Lindbergh's *Spirit of St. Louis*—hang from the ceiling, entirely outside the perception of blind visitors. The director apologized, and explained that no other space could be found for such large objects, but then asked his visitors whether a scale model of the Spirit of St. Louis, made available for touch, would be helpful. The blind visitors caucused and returned with their wonderful answer: Yes, they responded, we would appreciate such a model, but it must be placed directly under the unperceptible original. If the aesthetic and moral value of genuine objects can stir us so profoundly that we insist upon their presence even when we can have no palpable evidence thereof, but only the assurance that we stand in the aura of reality, then factual authenticity cannot be gainsaid as a fundamental desideratum of the human soul.

This difficult and tough-minded theme must be emphasized in literature (as the elitist and uncompromising Nabokov understood so well), particularly to younger students of the present generation, because an ancient, and basically anti-intellectual, current in the creative arts has now begun to flow more strongly than ever before in recent memory—the tempting Siren song of a claim that the spirit of human creativity stands in direct opposition to the rigor in education and observation that breeds both our love for factual detail and our gain of sufficient knowledge and understanding to utilize this record of human achievement and natural wonder.

No more harmful nonsense exists than this common supposition that deepest insight into great questions about the meaning of life or the structure of reality emerges most readily when a free, undisciplined, and uncluttered (read, rather, ignorant and uneducated) mind soars above mere earthly knowledge and concern. The primary reason for emphasizing the supreme aesthetic and moral value of detailed factual accuracy, as Nabokov understood so well, lies in our need to combat this alluring brand of philistinism if we wish to maintain artistic excellence as both a craft and an inspiration. (Anyone who thinks that success in revolutionary innovation can arise *sui generis,* without apprenticeship for basic skills and education for understanding, should visit the first [chronological] room of the Turner annex at the Tate Gallery in London—to see the early products of Turner's extensive education in tools of classical perspective and representation, the necessary skills that he had to master before moving far beyond into a world of personal innovation.)

This Nabokovian argument for a strictly *positive* correlation (as opposed to the usual philistine claim for negative opposition) between extensive training and potential for creative innovation may be more familiar to scientists than to creative artists. But this crucial key to professional achievement must be actively promoted within science as well. Among less thoughtful scientists, we often encounter a different version of the phony argument for disassociation of attention to detail and capacity for creativity—the fallacy embedded in Zaleski's statement (cited on page 44) that Nabokov's obsessive love of detail made him a "laboratory drudge," even while opening prospects of greatness in literature.

The false (and unstated) view of mind that must lie behind this assertion—and that most supporters of the argument would reject if their unconscious allegiance were made explicit—assumes a fixed and limited amount of mental "stuff" for each intellect. Thus, if we assign too much of our total allotment to the mastery of detail, we will have nothing left for general theory and integrative wonder. But such a silly model of mental functioning can only arise from a false metaphorical comparison of human creativity with irrelevant systems based on fixed and filled containers—pennies in a piggy bank or cookies in a jar.

Many of the most brilliant and revolutionary theoreticians in the history of science have also been meticulous compilers of detailed evidence. Darwin developed his theory of natural selection in 1838, but prevailed because, when he finally published in 1859, he had also amassed the first credible factual compendium (overwhelming in thoroughness and diversity) for the evolutionary basis of life's history. (All previous evolutionary systems, including Lamarck's, had been based on speculation, however cogent and complex the theoretical basis.) Many key discoveries emerged and prevailed because great theoreticians respected empirical details ignored by others. In the most familiar example, Kepler established the ellipticity of planetary orbits when he realized that Tycho Brahe's data yielded tiny discrepancies from circularity that most astronomers would have disregarded as "close enough"—whereas Kepler knew that he could trust the accuracy of Tycho's observations.

I do not deny that some scientists see trees but not forests, thereby functioning as trustworthy experts of meticulous detail, but showing little interest or skill in handling more general, theoretical questions. I also do not deny that Nabokov's work on butterfly systematics falls under this rubric. But I strenuously reject the argument that Nabokov's attention to descriptive particulars, or his cherishing of intricate factuality, precluded strength in theory on principle. I do not understand Nabokov's psyche or his ontogeny well enough to speculate about his conservative approach to theoretical questions, or his disinclination to grapple with general issues in evolutionary biology. We can only, I suspect, intone some clichés about the world's breadth (including the domain of science), and about the legitimate places contained therein for people with widely divergent sets of skills.

I therefore strongly reject any attempt to characterize Nabokov as a laboratory drudge for his love of detail and his lack of attention to theoretical issues. The science of taxonomy has always honored, without condescension, professionals who develop Nabokov's dedication to the details of a particular group, and who establish the skills and "good eye" to forge order from nature's mire of confusing particulars. Yes, to be frank, if Nabokov had pursued only butterfly taxonomy as a complete career, he would now be highly respected in very limited professional circles, but not at all renowned in the world at large. But do we not honor the dedicated professional who achieves maximal excellence in an admittedly restricted domain of notoriety or power? After all, if Macbeth had been content to remain Thane of Cawdor—a perfectly respectable job—think of the lives and grief that would thus have been spared. But, of course, we would then have to lament a lost play. So let us celebrate Nabokov's excellence in natural history, and let us also rejoice that he could use the same mental skills and inclinations to follow another form of bliss.

An Epilogue on Science and Literature

Most intellectuals favor a dialogue between professionals in science and the arts. But we also assume that these two subjects stand as polar opposites in the domain of learning, and that diplomatic contact for understanding between adversaries sets the basic context for such a dialogue. At best, we hope to dissipate stereotypes and to become friends (or at least neutrals), able to put aside our genuine differences for temporary bonding in the practical service of a few broader issues demanding joint action by all educated folk.

A set of stereotypes still rules perceptions of "otherness" in these two domains—images based on little more than ignorance and parochial fear, but powerful nonetheless. Scientists are soulless dial-twirlers; artists are arrogant, illogical, self-absorbed blowhards. Dialogue remains a good idea, but the two fields, and the personalities attracted to them, remain truly and deeply different.

I do not wish to forge a false union in an artificial love feast. The two domains differ, truly and distinctly, in their chosen subject matter and established modes of validation. The magisterium (teaching authority) of science extends over the factual status of the natural world, and to the development of theories proposed to explain why these facts, and not others, characterize our universe. The magisteria of the arts and humanities treat ethical and aesthetic questions about morality, style, and beauty. Since the facts of nature cannot, in logic or principle, yield ethical or aesthetic conclusions, the domains must remain formally distinct on these criteria.

But many of us who labor in both domains (if only as an amateur in one) strongly feel that an overarching mental unity builds a deeper similarity than disparate subject matter can divide. Human creativity seems to work much as a coordinated and complex piece, whatever the different emphases demanded by disparate subjects—and we will miss the underlying commonality if we only stress the distinctions of external subjects and ignore the unities of internal procedure. If we do not recognize the common concerns and characteristics of all creative human activity, we will fail to grasp several important aspects of intellectual excellence—including the necessary interplay of imagination and observation (theory and empirics) as an intellectual theme, and the confluence of beauty and factuality as a psychological theme—because one field or the other traditionally downplays one side of a requisite duality.

Moreover, we must use the method of "replication with difference" if we wish to study and understand the human quintessence behind our varying activities. I cannot imagine a better test case for extracting the universals of human creativity than the study of deep similarities in intellectual procedure between the arts and sciences.

No one grasped the extent of this underlying unity better than Vladimir Nabokov, who worked with different excellences as a complete professional in both domains. Nabokov often insisted that his literary and entomological pursuits shared a common mental and psychological ground. In *Ada,* while invoking a common anagram for "insect," one of Nabokov's characters beautifully expresses the oneness of creative impulse and the pervasive beauty of chosen subject matter: "'If I could write,' mused Demon, 'I would describe, in too many words no doubt, how passionately, how incandescently, how incestuously—*c'est le mot*—art and science meet in an insect.'"

Returning to his central theme of aesthetic beauty in both the external existence and our internal knowledge of scientific detail, Nabokov wrote in 1959 (quoted in Zimmer, page 33): "I cannot separate the aesthetic pleasure of seeing a butterfly and the scientific pleasure of knowing what it is." When Nabokov spoke of "the precision of poetry in taxonomic description"—no doubt with conscious intent to dissipate a paradox that leads most people to regard art and science as inexorably distinct and opposed—he used his literary skills in the service of generosity (a high, if underappreciated, virtue underlying all attempts to unify warring camps). He thus sought to explicate the common ground of his two professional worlds, and to illustrate the inevitably paired components of any integrated view that could merit the label of our oldest and fondest dream of fulfillment—the biblical ideal of "wisdom." Thus, in a 1966 interview, Nabokov broke the boundaries of art and science by stating that the most precious desideratum of each domain must also characterize any excellence in the other—for, after all, truth is beauty, and beauty truth. I could not devise a more fitting title for this essay, and I can imagine no better ending for this text:

> The tactile delights of precise delineation, the silent paradise of the camera lucida, and the precision of poetry in taxonomic description represent the artistic side of the thrill which accumulation of new knowledge, absolutely useless to the layman, gives its first begetter. . . . There is no science without fancy, and no art without facts.

Bibliography

Boyd, B. 1990. *Valdimir Nabokov: The American Years.* Princeton, N.J.: Princeton University Press.

Gould, S. J. 1983. The Hardening of the Modern Synthesis. In Marjorie Greene, ed., *Dimensions of Darwinism.* Cambridge, England: Cambridge University Press.

Johnson, K., G. W. Whitaker, and Z. Balint. 1996. Nabokov as lepidopterist: An informed appraisal. *Nabokov Studies.* Volume 3, 123–44.

Karges, J. 1985. *Nabokov's Lepidoptera: Genres and Genera.* Ann Arbor, Mich.: Ardis.

Kinsey, A. C., W. B. Pomeroy, and C. E. Martin. 1948. *Sexual Behavior in the Human Male.* Philadelphia: W. B. Saunders.

Provine, W. 1986. *Sewall Wright and Evolutionary Biology.* Chicago: University of Chicago Press.

Remington, C. R. 1990. Lepidoptera studies. In the *Garland Companion to Vladimir Nabokov,* 274–82.

Robson, G. C., and O. W. Richards. 1936. *The Variation of Animals in Nature.* London: Longmans, Green & Co.

Zaleski, P., 1986, Nabokov's blue period. *Harvard Magazine,* July–August, 34–38.

Zimmer, D. E. 1998. *A Guide to Nabokov's Butterflies and Moths.* Hamburg.

3

Jim Bowie's Letter and Bill Buckner's Legs

Charlie Croker, former football hero of Georgia Tech and recently bankrupted builder of the new Atlanta—a world of schlock and soulless office towers, now largely unoccupied and hemorrhaging money—seeks inspiration, as his world disintegrates, from the one item of culture that stirs his limited inner self: a painting, originally done to illustrate a children's book ("the only book Charlie could remember his father and mother ever possessing"), by N. C. Wyeth of "Jim Bowie rising up from his deathbed to fight the Mexicans at the Alamo." On "one of the happiest days of his entire life," Charlie spent $190,000 at a Sotheby's auction to buy this archetypal scene for a man of action. He then mounted his treasure in the ultimate shrine for successful men of our age—above the ornate desk on his private jet.

Tom Wolfe describes how his prototype for redneck moguls (in his novel *A Man in Full*) draws strength from his inspirational painting:

> And so now, as the aircraft roared and strained to gain altitude, Charlie concentrated on the painting of Jim Bowie . . . as he had so many times before. . . . Bowie, who was already dying, lay on a bed. . . . He had propped himself up on one elbow. With his other hand he was brandishing his famous Bowie knife at a bunch of Mexican soldiers. . . . It was the way Bowie's big neck and his jaws jutted out towards the Mexicans and the way his eyes blazed defiant to the end, that made it a great painting. Never say die, even when you're dying, was what that painting said. . . . He stared at the indomitable Bowie and waited for an infusion of courage.

Nations need heroes, and Jim Bowie did die in action at the Alamo, along with Davy Crockett and about 180 fighters for Texian independence (using the *i* then included in the name), under the command of William B. Travis, an articulate twenty-six-year-old lawyer with a lust for martyrdom combined with a fearlessness that should not be disparaged, whatever one may think of his judgment. In fact, I have no desire to question Bowie's legitimate status as a hero at the Alamo at all. But I do wish to explicate his virtues by debunking the legend portrayed in Charlie Croker's painting, and by suggesting that our admiration should flow for quite different reasons that have never been hidden, but that the legend leads us to disregard.

The debunking of canonical legends ranks as a favorite intellectual sport for all the usual and ever-so-human reasons of one-upmanship, aggressivity within a community that denies itself the old-fashioned expression of genuine fisticuffs, and the simple pleasure of getting details right. But such debunking also serves a vital scholarly purpose at the highest level of identifying and correcting some of the most serious pitfalls in human reasoning. I make this somewhat grandiose claim for the following reason:

The vertebrate brain seems to operate as a device tuned to the recognition of patterns. When evolution grafted consciousness in human form upon this organ in a single species, the old inherent search for patterns developed into a propensity for organizing these patterns as stories, and then for explaining the surrounding world in terms of the narratives expressed in such tales. For universal reasons that probably transcend the cultural particulars of individual groups, humans tend to construct their stories along a limited number of themes and pathways, favored because they grant both useful sense and satisfying meaning to the confusion (and often to the tragedy) of life in our complex surrounding world.

Stories, in other words, only "go" in a limited number of strongly preferred ways, with the two deepest requirements invoking, first, a theme of directionality (linked events proceeding in an ordered sequence for definable reasons, and not as an aimless wandering—back, forth, and sideways—to nowhere); and, second, a sense of motivation, or definite reasons propelling the sequence (whether we judge the outcomes good or bad). These motivations will be rooted directly in human purposes for stories involving our own species. But tales about nonconscious creatures or inanimate objects must also provide a surrogate for valor (or dishonorable intent for dystopian tales)—as in the virtue of evolutionary principles that dictate the increasing general complexity of life, or the lamentable inexorability of thermodynamics in guaranteeing the eventual burnout of the sun. In summary, and at the risk of oversimplification, we like to explain pattern in terms of directionality, and causation in terms of valor. The two central and essential components of any narrative—pattern and cause—therefore fall under the biasing rubric of our mental preferences.

I will refer to the small set of primal tales based upon these deep requirements as "canonical stories." Our strong propensity for expressing all histories, be they human, organic, or cosmic, in terms of canonical stories would not entail such enormous problems for science—but might be viewed, instead, as simply humorous in exposing the foibles of *Homo sapiens*—if two properties of mind and matter didn't promote a potentially harmless idiosyncrasy into a pervasive bias actively derailing our hopes for understanding events that unfold in time. (The explanation of temporal sequences defines the primary task of a large subset among our scientific disciplines—the so-called "historical sciences" of geology, anthropology, evolutionary biology, cosmology, and many others. Thus, if the lure of "canonical stories" blights our general understanding of historical sequences, much of what we call "science" labors under a mighty impediment.)

As for matter, many patterns and sequences in our complex world owe their apparent order to the luck of the draw within random systems. We flip five heads in a row once every thirty-two sequences on average. Stars clump into patterns in the sky because they are distributed effectively at random (within constraints imposed by the general shape of our Milky Way galaxy) with respect to the earth's position in space. An absolutely even spacing of stars, yielding no perceivable clumps at all, would require some fairly fancy, and obviously nonexistent, rules of deterministic order. Thus, if our minds obey an almost irresistible urge to detect patterns, and then to explain these patterns in the causal terms of a few canonical stories, our quest to understand the sources (often random) of order will be stymied.

As for mind, even when we can attribute a pattern to conventional nonrandom reasons, we often fail to apprehend both the richness and the nature of those causes because the lure of canonical stories leads us to entertain only a small subset among legitimate hypotheses for explaining the recorded events. Even worse, since we cannot observe everything in the blooming and buzzing confusion of the world's surrounding richness, the organizing power of canonical stories leads us to ignore important facts readily within our potential sight, and to twist or misread the information that we do manage to record. In other words, and to summarize my principal theme in a phrase, canonical stories predictably "drive" facts into definite and distorted pathways that validate the outlines and necessary components of these archetypal tales. We therefore fail to note important items in plain sight, while we misread other facts by forcing them into preset mental channels, even when we retain a buried memory of actual events.

This essay illustrates how canonical stories have predictably relegated crucial information to misconstruction or invisibility in two great folk tales of American history: Bowie's letter and Buckner's legs, as oddly (if euphoniously) combined in my title. I will then extend the general message to argue that the allure of canonical stories acts as the greatest impediment to better understanding throughout the realm of historical science—one of the largest and most important domains of human intellectual activity.

Jim Bowie's Letter

How the canonical story of "all the brothers were valiant, and all the sisters virtuous" has hidden a vital document in plain sight. (This familiar quotation first appears on the tomb of the Duchess of Newcastle, who died in 1673 and now lies in Westminster Abbey.)

The Alamo of San Antonio, Texas, was not designed as a fortress, but as a mission church built by eighteenth-century Spaniards. Today the Alamo houses exhibits and artifacts, most recalling the death of all Texian defenders in General Santa Anna's assault, with a tenfold advantage in troops and after nearly two weeks of siege, on March 6, 1836. This defeat and martyrdom electrified the Texian cause, which triumphed less than two months later when Sam Houston's men captured Santa Anna at the Battle of San Jacinto on April 21, and then forced the Mexican general to barter Texas for his life, his liberty, and the return of his opium bottle.

The Alamo's exhibits, established and maintained by the Daughters of the Republic of Texas, and therefore no doubt more partisan than the usual (and, to my mind, generally admirable) fare that the National Park Service provides in such venues, tells the traditional tale, as I shall do here. (Mexican sources, no doubt, purvey a different but equally traditional account from another perspective.) I shall focus on the relationship of Bowie and Travis, for my skepticism about the canonical story focuses on a fascinating letter, written by Bowie and prominently displayed in the Alamo, but strangely disregarded to the point of invisibility in the official presentation.

In December 1835, San Antonio had been captured by Texian forces in fierce fighting with Mexican troops under General Cos. On January 17, 1836, Sam Houston ordered Jim Bowie and some thirty men to enter San Antonio, destroy the Alamo, and withdraw the Texian forces to more defendable ground. But Bowie, after surveying the situation, disagreed for both strategic and symbolic reasons, and decided to fortify the Alamo instead. The arrival, on February 3, of thirty additional men under the command of William B. Travis strengthened Bowie's decision.

An illustration of the conventional myth about Jim Bowie's death at the Alamo. Although bedridden and dying, Bowie still manages to kill several Mexican soldiers before his inevitable defeat.

But tension inevitably developed between two such different leaders, the forty-year-old, hard-drinking, fearlessly independent, but eminently practical and experienced Bowie, and the twenty-six-year-old troubled and vainglorious Travis, who had left wife and fortune in Alabama to seek fame and adventure on the Texian frontier. (Mexico had encouraged settlement of the Texian wilderness by all who would work the land and swear allegiance to the liberal constitution of 1824, but the growing Anglo majority had risen in revolt, spurred by the usual contradictory motives of lust for control and love of freedom, as expressed in anger at Santa Anna's gradual abrogation of constitutional guarantees.)

Bowie commanded the volunteers, while Travis led the "official" army troops. A vote among the volunteers overwhelmingly favored Bowie's continued leadership, so the two men agreed upon an uneasy sharing of authority, with all orders to be signed by both. This arrangement became irrelevant, and Travis assumed full command, when Bowie fell ill with clearly terminal pneumonia and a slew of other ailments just after the siege began on February 23. In fact, Charlie Croker's painting notwithstanding, Bowie may have been comatose, or even already dead, when Mexican forces broke through on March 6. He may have made his legendary last "stand" (in supine position), propped up in his bed with pistols in hand, but he could not have mounted more than a symbolic final defense, and his legendary knife could not have reached past the Mexican bayonets in any case.

The canonical story of valor at the Alamo features two incidents, both centered upon Travis, with one admitted as legendary by all serious historians, and the other based upon a stirring letter, committed to memory by nearly all Texas schoolchildren ever since. As for the legend, when Travis realized that no reinforcements would arrive, and that all his men would surely die if they defended the Alamo by force of arms (for Santa Anna had clearly stated his terms of no mercy or sparing of life without unconditional surrender), he called a meeting, drew a line in the sand, and then invited all willing defenders of the Alamo to cross the line to his side, while permitting cowards and doubters to scale the wall and make their inglorious exit (as one man did). In this stirring legend, Jim Bowie, now too weak to stand, asks his men to carry his bed across the line.

Well, Travis may have made a speech at the relevant time, but no witness and survivor (several women and one slave) ever reported the story. (The tale apparently originated about forty years later, supposedly told by the single man who had accepted Travis's option to escape.)

As for the familiar letter, few can read Travis's missive with a dry eye, while even the most skeptical of Alamo historians heaps honor upon this document of February 24, carried by a courier (who broke through the Mexican lines) to potential reinforcements, but addressed to "The People of Texas and All Americans in the World." (For example, Ben H. Proctor describes Travis as "egotistical, proud, vain, with strong feelings about his own destiny, about glory and personal mission . . . trouble in every sense of the word," but judges this missive as "one of the truly remarkable letters of history, treasured by lovers of liberty everywhere." (See Proctor's pamphlet, *The Battle of the Alamo* [Texas State Historical Association, 1986].)

> I am besieged, by a thousand or more of the Mexicans under Santa Anna—I have sustained a continual bombardment and cannonade for 24 hours and have not lost a man—The enemy has demanded a surrender at discretion, otherwise, the garrison are to be put to the sword, if the fort is taken—I have answered the demand with a cannon shot, and our flag still waves proudly from the walls—*I shall never surrender or retreat.* Then, I call on you in the name of Liberty, of patriotism and everything dear to the American character, to come to our aid, with all dispatch—The enemy is receiving reinforcements daily and will no doubt increase to three or four thousand in four or five days. If this call is neglected, I am determined to sustain myself as long as possible and die like a soldier who never forgets what is due to his own honor and that of his country—VICTORY OR DEATH.

Although a small group of thirty men did arrive to reinforce the Alamo, their heroic presence as cannon and bayonet fodder could not alter the course of events, while a genuine force that could have made a difference, several hundred men stationed at nearby Goliad, never came to Travis's aid, for complex reasons still under intense historical debate. Every Texian fighter died in Santa Anna's attack on March 6. According to the usual legend, all the men fell in action. But substantial, if inconclusive, evidence indicates that six men may have surrendered at the hopeless end, only to be summarily executed by Santa Anna's direct order. The probable presence of Davy Crockett among this group accounts for the disturbing effect and emotional weight of this persistent tale.

As something of an Alamo buff, and a frequent visitor to the site in San Antonio, I have long been bothered and intrigued by a crucial document, a letter by the Alamo's other leader, Jim Bowie, that seems to provide quite a different perspective upon the siege, but doesn't fit within the canonical legend and hardly receives a mention in any official account at the shrine itself. Bowie's letter thus remains "hidden in plain sight"—sitting in its own prominent glass case, right in the main hall of the on-site exhibition. This curious feature of "prominently displayed but utterly passed over" has fascinated me for twenty years. I have, in three visits to the Alamo, bought every popular account of the battle for sale at the extensive gift shop. I have read these obsessively and can assert that Bowie's letter, while usually acknowledged, receives short shrift in most conventional descriptions.

Let us return to a phrase in Travis's celebrated letter and fill in some surrounding events: "the enemy has demanded a surrender . . . I have answered the demand with a cannon shot." The basic outline has not been disputed: When Santa Anna entered town with his army and began his siege on February 23, he unfurled a blood-red flag—the traditional demand for immediate surrender, with extermination as the consequence of refusal—from the tower of the Church of San Fernando. Travis, without consulting his co-commander, fired the Alamo's largest cannon, an eighteen-pounder, in defiant response—just as he boasted in his famous letter, written the next day.

The complexities that threaten the canonical story now intrude. Although Santa Anna had issued his uncompromising and blustering demand in a public display, many accounts, filled with different details but all pointing in the same credible direction, indicate that he also proposed a parley for negotiation with the Alamo defenders. (Even if Santa Anna didn't issue this call, the canonical story takes its strong hit just from the undisputed fact that Bowie, for whatever reason, thought the Mexicans had suggested a parley. Among the various versions, Santa Anna's forces also raised a white flag—the equally traditional signal for a parley—either accidentally or purposefully, and either before or after Travis's cannon shot; or else that a Mexican soldier sounded the standard bugle call for an official invitation to negotiations.)

In any case, Bowie, who by most accounts was furious at Travis for the impetuous bravado and clearly counterproductive nature of his purely symbolic cannon shot, grabbed a piece of paper and wrote, in Spanish signed with a faltering hand (for Bowie was already ill, but not yet prostrate and still capable of leadership), the "invisible" letter that just won't mesh with the canonical story, and therefore remains hidden on prominent display at the Alamo (I cite the full text of Bowie's letter, in the translation given in C. Hopewell's biography, *James Bowie* [Eakin Press, 1994]):

> Because a shot was fired from a cannon of this fort at the time a red flag was raised over the tower, and soon afterward having been informed that your forces would parley, the same not having

> been understood before the mentioned discharge of cannon, I wish to know if, in effect, you have called for a parley, and with this object dispatch my second aide, Benito James, under the protection of a white flag, which I trust will be respected by you and your forces. God and Texas.

I don't want to exaggerate the meaning of this letter. I cannot assert a high probability for a different outcome if Bowie had remained strong enough to lead, and if Santa Anna had agreed to negotiations. Some facts dim the force of any speculation about a happier outcome that would have avoided a strategically senseless slaughter with an inevitable military result, and would thus have spared the lives of 180 Texians (and probably twice as many Mexicans). For example, Bowie did not display optimal diplomacy in his note, if only because he had originally written "God and the Mexican Federation" in his signatory phrase (indicating his support for the constitution of 1824, and his continued loyalty to this earlier Mexican government), but, in a gesture that can only be termed defiant, crossed out "The Mexican Federation" and wrote "Texas" above.

More important, Santa Anna officially refused the offer of Bowie's courier, and sent back a formal response promising extermination without mercy unless the Texians surrendered unconditionally. Moreover, we cannot be confident that Texian lives would have been spared even if the Alamo's defenders had surrendered without a fight. After all, less than a month after the fall of the Alamo, Santa Anna executed several hundred prisoners—the very men who might have come to Travis's aid—after their surrender at Goliad.

In the confusion and recrimination between the two commands, Travis then sent out his own courier and received the same response, but, according to some sources, with the crucial addition of an "informal" statement that if the Texians laid down their arms within an hour, their lives and property would be spared, even though the surrender must be technically and officially "unconditional." Such, after all, has always been the way of war, as good officers balance the need for inspirational manifestos with their even more important moral and strategic responsibility to avoid a "glory trap" of certain death. Competent leaders have always understood the crucial difference between public proclamations and private bargains.

Thus, I strongly suspect that if Bowie had not become too ill to lead, some honorable solution would eventually have emerged through private negotiations, if only because Santa Anna and Bowie, as seasoned veterans, maintained high mutual regard beneath their strong personal dislike—whereas I can only imagine what Santa Anna thought of the upstart and self-aggrandizing Travis. In this alternate and unrealized scenario, most of the brothers would have remained both valiant and alive. What resolution fits best with our common sensibilities of morality and human decency: more than four hundred men slaughtered in a battle with an inevitable result, thus providing an American prototype for a claptrap canonical story about empty valor over honorable living; or an utterly nonheroic, tough-minded, and practical solution that would have erased a great story from our books, but restored hundreds of young men to the possibilities of a full life, complete with war stories told directly to grandchildren?

Finally, one prominent Alamo fact, though rarely mentioned in this context, provides strong support for the supposition that wise military leaders usually reach private agreements to avoid senseless slaughter. Just three months earlier, in December 1835, General Cos had made his last stand against Texian forces at exactly the same site—within the Alamo! But Cos, as a professional soldier, raised a white flag and agreed to terms with the Texian conquerors: he would surrender, disarm, withdraw his men, retreat southwestward over the Rio Grande, and not fight again. Cos obeyed the terms of his bargain, but when he had crossed the Rio Grande to safety, Santa Anna demanded his return to active duty. Thus, the same General Cos—alive, kicking, and fighting—led one of the companies that recaptured the Alamo on March 6. Travis would have cut such a dashing figure at San Jacinto!

Bill Buckner's Legs

How the canonical story of "but for this" has driven facts that we can all easily recall into a false version dictated by the needs of narrative.

Any fan of the Boston Red Sox can recite chapter and verse of a woeful tale, a canonical story in the land of the bean and the cod, called "the curse of the Bambino." The Sox established one of Major League Baseball's most successful franchises of the early twentieth century. But the Sox won their last World Series way back in 1918. A particular feature of all subsequent losses has convinced Boston fans that their team labors under an infamous curse, initiated in January 1920, when Boston owner Harry Frazee simply and cynically sold the team's greatest player—the best left-handed pitcher in baseball, but soon to make his truly indelible mark on the opposite path of power hitting—for straight cash needed to finance a flutter on a Broadway show, and not for any advantages or compensation in traded players. Moreover, Frazee sold Boston's hero to the hated enemy, the New York Yankees. This man, of course, soon acquired the title of Sultan of Swat, the Bambino, George Herman ("Babe") Ruth.

The Red Sox have played in four World Series (1946, 1967, 1975, and 1986) and several playoff series since then, and they have always lost in the most heartbreaking manner—by coming within an inch of the finish line and then self-destructing. Enos Slaughter of the rival St. Louis Cardinals scored from first on a single in the decisive game of the 1946 World Series. In 1975, the Sox lost Game Seven after a miraculous victory in Game Six, capped by Bernie Carbo's three-run homer to tie the score and won, in extra innings, by Carlton Fisk, when he managed to overcome the laws of physics by body English, and cause a ball that he had clearly hit out of bounds to curve into the left-field foul pole for a home run (as the Fenway Park organist blasted out the "Hallelujah Chorus," well after midnight).

And so the litany goes. But all fans will tell you that the worst moment of utter incredibility—the defeat that defies any tale of natural causality, and must therefore record the operation of a true curse—terminated Game Six in the 1986 World Series. (Look, I'm not even a Sox fan, but I still don't allow anyone to mention this event in my presence; the pain remains too great!) The Sox, leading the Series three games to two and requiring only this victory for their first Ring since 1918, entered the last inning with a comfortable two-run lead. Their pitcher quickly got the first two outs. The Sox staff had peeled the foil off the champagne bottles (but, remembering the curse, had not yet popped the corks). The Mets management had already, and graciously, flashed "congratulations Red Sox" in neon on their scoreboard. But the faithful multitude of fans, known as "Red Sox Nation," remained glued to their television sets in exquisite fear and trembling.

And the curse unfolded, with an intensity and cruelty heretofore not even imagined. In a series of scratch hits, bad pitches, and terrible judgments, the Mets managed to score a run. (I mean, even a batting-practice pitcher, even you or I, could have gotten someone out for the final victory!) Reliever Bob Stanley, a good man dogged by bad luck, came in and threw a wild pitch to bring the tying run home. (Some, including yours truly, would have scored the pitch as a passed ball, but let's leave such contentious irrelevancies aside for the moment.) And now, with two outs, a man on second and the score tied, Mookie Wilson steps to the plate.

Bill Buckner, the Sox's gallant first baseman, and a veteran with a long and truly distinguished career, should not even have been playing in the field. For weeks, manager John McNamara had been benching Buckner for defensive purposes during the last few innings of games with substantial Red Sox leads—for, after a long and hard season, Buckner's legs were shot, and his stride gimpy. In fact, he could hardly bend down. But the sentimental McNamara wanted his regular players on the field when the great, and seemingly inevitable, moment arrived—so Buckner stood at first base.

I shudder as I describe the outcome that every baseball fan knows so well. Stanley, a great sinker-ball pitcher, did exactly what he had been brought in to accomplish. He threw a wicked sinker that Wilson could only tap on the ground toward first base for an easy out to cap the damage and end the inning with the score still tied, thus granting the Sox hitters an opportunity to achieve a comeback and victory. But the ball bounced right through Buckner's legs into the outfield as Ray Knight hurried home with the winning run. Not to the side of his legs, and not under his lunging glove as he dived to the right or left for a difficult chance—but right through his legs! The seventh and concluding game hardly mattered. Despite brave rhetoric, no fan expected the Sox to win (hopes against hope to be sure, but no real thoughts of victory). They lost.

This narration may drip with my feelings, but I have presented the straight facts. The narrative may be good and poignant enough in this accurate version, but such a factual tale cannot satisfy the lust of the relevant canonical story for an evident reason. The canonical story of Buckner's travail must follow a scenario that might be called "but for this." In numerous versions of "but for this," a large and hugely desired result fails to materialize—and the absolutely oppo-

I tremble as I write the caption for this most painful moment in the history of baseball—as Mookie Wilson's easy grounder bounces between Bill Buckner's legs.

site resolution, both factually and morally, unfolds instead—because one tiny and apparently inconsequential piece of the story fails to fall into place, usually by human error or malfeasance. "But for this" can brook no nuancing, no complexity, no departure from the central meaning and poignant tragedy that an entire baleful outcome flows absolutely and entirely from one tiny accident of history.

"But for this" must therefore drive the tale of Bill Buckner's legs into the only version that can validate the canonical story. In short, poor Bill must become the one and only cause and focus of ultimate defeat or victory. That is, if Buckner fields the ball properly, the Sox win their first World Series since 1918 and eradicate the Curse of the Bambino. But if Buckner bobbles the ball, the Mets win the Series instead, and the curse continues in an even more intense and painful way. For Buckner's miscue marks the unkindest bounce of all, the most improbable, trivial little error sustained by a good and admired man. What hath God wrought?

Except that Buckner's error did *not* determine the outcome of the World Series for one little reason, detailed above but all too easily forgotten. When Wilson's grounder bounced between Buckner's legs, the score was already tied! (Not to mention that this game was the sixth and, at worst for the Sox, the penultimate game of the Series, not the seventh and necessarily final contest. The Sox could always have won Game Seven and the entire Series, no matter how the negotiations of God and Satan had proceeded over Bill Buckner as the modern incarnation of Job in Game Six.) If Buckner had fielded the ball cleanly, the Sox would not have won the Series at that moment. They would only have secured the opportunity to do so, if their hitters came through in extra innings.

We can easily excuse any patriotic American who is not a professional historian, or any casual visitor for that matter, for buying into the canonical story of the Alamo—all the brothers were valiant—and not learning that a healthy and practical Bowie might have negotiated an honorable surrender at no great cost to the Texian cause. After all, the last potential eyewitness has been underground for well over a century. We have no records beyond the written reports, and historians cannot trust the account of any eyewitness, for the supposed observations fall into a mire of contradiction, recrimination, self-interest, aggrandizement, and that quintessentially human propensity for spinning a tall tale.

But any baseball fan with the legal right to sit in a bar and argue the issues over a mug of the house product should be able to recall the uncomplicated and truly indisputable facts of Bill Buckner's case with no trouble at all, and often with the force of eyewitness memory, either exulting in impossibly fortuitous joy, or groaning in the agony of despair and utter disbelief, before a television set. (To fess up, I should have been at a fancy dinner in Washington, but I "got sick" instead and stayed in my hotel room. In retrospect, I should not have stood in bed.)

The subject attracted my strong interest because, within a year after the actual event, I began to note a pattern in the endless commentaries that have hardly abated, even fifteen years later—for Buckner's tale can be made relevant by analogy to almost any misfortune under a writer's current examination, and Lord only knows we experience no shortage of available sources for pain. Many stories reported, and continue to report, the events accurately—and why not, for the actual tale packs sufficient punch, and any fan should be able to extract the correct account directly from living and active memory. But I began to note that a substantial percentage of reports had subtly, and quite unconsciously I'm sure, driven the actual events into a particular false version—the pure "end member" of ultimate tragedy demanded by the canonical story "but for this."

I keep a growing file of false reports, all driven by requirements of the canonical story—the claim that, but for Buckner's legs, the Sox would have won the Series, forgetting the inconvenient complexity of a tied score at Buckner's ignominious moment, and sometimes even forgetting that the Series still had another game to run. This misconstruction appears promiscuously, both in hurried daily journalism and in rarefied books by the motley crew of poets and other assorted intellectuals who love to treat baseball as a metaphor for anything else of importance in human life or the history of the universe. (I have written to several folks who made this error, and they have all responded honorably with a statement like: "Omigod, what a jerk I am! Of course the score was tied. Jeez [sometimes bolstered by an invocation of Mary and Joseph as well], I just forgot!")

For example, a front-page story in *USA Today* for October 25, 1993, discussed Mitch Williams's antics in the 1993 Series in largely unfair comparison with the hapless and blameless Bill Buckner:

> Williams may bump Bill Buckner from atop the goat list, at least for now. Buckner endured his nightmare Oct. 25, 1986. His Boston Red Sox were one out away from their first World Series title since 1918 when he let Mookie Wilson's grounder slip through his legs.

Or this from a list of Sox misfortunes, published in the *New York Post* on October 13, 1999, just before the Sox met the Yanks (and lost, of course) in their first full series of postseason play:

> Mookie Wilson's grounder that rolled through the legs of Bill Buckner in Game 6 of the 1986 World Series. That happened after the Red Sox were just one out away from winning the World Series.

For a more poetic view between hard covers, consider the very last line of a lovely essay written by a true poet and devoted fan to introduce a beautifully illustrated new edition of the classic poem about failure in baseball, *Casey at the Bat:*

> Triumph's pleasures are intense but brief; failure remains with us forever, a mothering nurturing common humanity. With Casey we all strike out. Although Bill Buckner won a thousand games with his line drives and brilliant fielding, he will endure in our memories in the ninth inning of the sixth game of a World Series, one out to go, as the ball inexplicably, ineluctably, and eternally rolls between his legs.

But the nasty little destroyer of lovely canonical stories then pipes up in his less mellifluous tones: "But I don't know how many outs would have followed, or who would have won. The Sox had already lost the lead; the score was tied." Factuality embodies its own form of eloquence; and gritty complexity often presents an even more interesting narrative than the pure and archetypal "end member" version of our canonical stories. But something deep within us drives accurate messiness into the channels of canonical stories, the primary impositions of our minds upon the world.

To any reader who now raises the legitimate issue of why I have embellished a book about natural history with two stories about American history that bear no evident relevance to any overtly scientific question, I simply restate my opening and general argument: human beings are pattern-seeking, story-telling creatures. These mental propensities generally serve us well enough, but they also, and often, derail our thinking about all kinds of temporal sequences—in the natural world of geological change and the evolution of organisms, as well as in human history—by leading us to cram the real and messy complexity of life into simplistic channels of the few preferred ways that human stories "go." I call these biased pathways "canonical stories"—and I argue that our preferences for tales about directionality (to explain patterns), generated by motivations of valor (to explain the causal basis of these patterns) have distorted our understanding of a complex reality where different kinds of patterns and different sources of order often predominate.

I chose my two stories on purpose—Bowie's letter and Buckner's legs—to illustrate two distinct ways that canonical stories distort our reading of actual patterns: first, in the tale of Jim Bowie's letter, by relegating important facts to virtual invisibility when they cannot be made to fit the canonical story, even though we do not hide the inconvenient facts themselves, and may even place them on open display (as in Bowie's letter at the Alamo); and, second, in the tale of Bill Buckner's legs, where we misstate easily remembered and ascertainable facts in predictable ways because these facts did not unfold as the relevant canonical stories dictate.

These common styles of error—hidden in plain sight, and misstated to fit our canonical stories—arise as frequently in scientific study as in historical inquiry. To cite, in closing, the obvious examples from our canonical misreadings of the history of life, we hide most of nature's diversity in plain sight when we spin our usual tales about increasing complexity as the central theme and organizing principle of both evolutionary theory and the actual history of life. In so doing, we unfairly privilege the one recent and transient species that has evolved the admittedly remarkable invention of mental power sufficient to ruminate upon such questions.

This silly and parochial bias leaves the dominant and most successful products of evolution hidden in plain sight—the indestructible bacteria that have represented life's mode (most common design) for all 3.5 billion years of the fossil record (while *Homo sapiens* hasn't yet endured for even half a million years—and remember that it takes a thousand million to make a single billion). Not to mention that if we confine our attention to multicellular animal life, insects represent about 80 percent of all species, while only a fool would put money on us, rather than them, as probable survivors a billion years hence.

For the second imposition of canonical stories upon different and more complex patterns in the history of life—predictable distortion to validate preferred tales about valor—need I proceed any further than the conventional tales of vertebrate evolution that we all have read since childhood, and that follow our Arthurian mythology about knights of old and men so bold? I almost wince when I find the first appearance of vertebrates on land, or of insects in the air, described as a "conquest," although this adjective retains pride of place in our popular literature.

And we still seem unable to shuck the image of dinosaurs as born losers vanquished by superior mammals, even though we know that dinosaurs prevailed over mammals for more than 130 million years, starting from day one of mammalian origins. Mammals gained their massively delayed opportunity only when a major extinction, triggered by extraterrestrial impact, removed the dinosaurs—for reasons that we do not fully understand, but that probably bear no sensible relation to any human concept of valor or lack thereof. This cosmic fortuity gave mammals their chance, not because any intrinsic superiority (the natural analog of valor) helped them to weather this cosmic storm, but largely, perhaps, because their small size, a side-consequence of failure to compete with dinosaurs in environments suited for large creatures, gave mammals a lucky break in the form of ecological hiding room to hunker down.

Until we abandon the silly notion that the first amphibians, as conquerors of the land, somehow held more valor, and therefore embody more progress, than the vast majority of fishes that remained successfully in the sea, we will never understand the modalities and complexities of vertebrate evolution. Fish, in any case, encompass more than half of all vertebrate species today, and might well be considered the most persistently successful class of vertebrates. So should we substitute a different canonical story called "there's no place like home" for the usual tale of conquest on imperialistic models of commercial expansion?

If we must explain the surrounding world by telling stories—and I suspect that our brains do stick us in this particular rut—let us at least expand the range of our tales beyond the canonical to the quirky, for then we might learn to appreciate more of the richness out there beyond our pale and usual ken, while still honoring our need to understand in human terms. Robert Frost caught the role and necessity of stories—and the freedom offered by unconventional tales—when he penned a premature epitaph, in 1942, as one of his brilliant epitomes of deep wisdom:

> And were an epitaph to be my story
> I'd have a short one ready for my own.
> I would have written of me on my stone:
> I had a lover's quarrel with the world.

4

The True Embodiment of Everything That's Excellent

On December 8, 1889, the day after the triumphant premiere of *The Gondoliers,* their last successful collaboration, W. S. Gilbert wrote to Sir Arthur Sullivan, in the generous tone so often expressed in his letters, despite the constant tension of their personal relationship: "I must again thank you for the magnificent work you have put into the piece. It gives one a chance of shining right through the twentieth century with a reflected light." John Wellington Wells, the eponym of their first successful full-length comic opera *The Sorcerer* (1877), employed a "resident djinn" who could "prophesy with a wink of his eye, peep with security into futurity." But we usually don't ascribe similar skills to his literary progenitor.

Yet, with the opening of Mike Leigh's *Topsy Turvy**—a wondrous evocation of this complex partnership, the Victorian theatrical world, and the nature of creativity in general, all centered on the composition and first production of *The Mikado,* Gilbert and Sullivan's greatest and most enduring hit, in 1885—Gilbert's short note to Sullivan can only recall his initial characterization of Mr. Wells's djinn as "a very small prophet who brings us unbounded returns." For this film, opening in New York just a few days before our millennial transition, surely marks the fulfillment of Gilbert's predictions for endurance to this very moment.

I must confess to a personal reason for pleasure in this renewed currency and attention for the works of Gilbert and Sullivan—a baker's dozen of comic operas (the music to a fourteenth has been lost), now often dismissed as the silliest and fustiest of (barely) surviving Victorian oddments, and a genuine embarrassment to anyone with modern intellectual pretensions. But I may now emerge from decades of (relative) silence to shout my confession that I love these pieces with all my heart, and that I even regard them as epitomes of absolute excellence for definable reasons that may help us to understand this most rare and elusive aspect of human potential.

In my latency at ages ten to twelve, Gilbert and Sullivan became the passion of my life. I would save my nickels and dimes for several months until I accumulated the $6.66 needed to buy the old London recordings of each opera at Sam Goody's. I listened so often, with a preadolescent capacity for rote retention, that I learned every word and note of the corpus by pure imbibition, without a moment of conscious effort. (I could not expunge this information now, no matter how hard I might try, even though I can't remember anything learned last week—another anomaly and paradox of our mental lives.)

This phase ended, of course—and for the usual reasons. At age 13, I watched Olivia de Havilland as Maid Marian (playing against Errol Flynn's Robin Hood, of course)—and the sight of her breasts through white satins focused and hardened my dawning recognition of fundamental change. (In one of the great fulfillments of my life, I met Miss de Havilland a few years ago, looking as beautiful as ever in the different manner of older age. I told her my story—although I confess that I left out the part about the breasts, rather than

*This article was inspired by Leigh's film at its opening in December 1999. But the film is rarely mentioned, and this article is in no way a review (the most ephemeral and unrepublishable of all literary genres). Rather, I shamelessly used Leigh's wonderful film to write the essay I always meant to compose on my heretofore private passion. This piece originally appeared in *The American Scholar.*

the whole persona, as my inspiration. She was very gracious; I was simply awestruck.) So I abandoned my intensity for Gilbert and Sullivan, but I have never lost my affection, or forgotten a jot or tittle of the writ.

I have thus lived, for several decades, in the ambiguity of many Savoyards: feeling a bit sheepish, even apologetic, given my vaunted status as a card-carrying intellectual, about my infatuation for this prototypical vestige of low-brow entertainment imbued, at best, with middlebrow pretensions. Two major themes fueled my fear that this continued affection could only represent a misplaced fealty to my own youth, based on a refusal to acknowledge that such unworthy flowers could only seem glorious before the realities of life spread like crabgrass through the splendor of adolescent turf.

First, some of Gilbert's texts do strike a modern audience as silly and forced—as in the seemingly endless punning on "orphan" versus "often" in *The Pirates of Penzance,* though no worse, really, or longer for that matter, than the opening dialogue about cobbling in *Julius Caesar,* with its similar takes on "soul" versus "sole" and "awl" versus "all." Perhaps the whole corpus never rises above such textual juvenility. Creeping doubt might then generate a corresponding fear that most of Sullivan's music, however witty and affecting to the Victorian ear, must now be downgraded as either mawkish (as in his "Lost Chord," once widely regarded as the greatest song ever written, but now forgotten), or pompous (as in his "Onward Christian Soldiers").

Second, evidence of declining public attention might fuel these fears about quality. While scarcely extinct, or even moribund, Gilbert and Sullivan's works have surely retreated to a periphery of largely amateur performance, spiced now and again by an acclaimed, but transient, professional foray, often in highly altered form (Joseph Papp's rock version of *The Pirates of Penzance,* or the revival of a favorite from the 1930s, *The Hot Mikado,* in the ultimate historical venue for terminations, Washington, D.C.'s Ford's Theater). England's D'Oyly Carte Opera, Gilbert and Sullivan's original company, performed for more than one hundred years, but expired about a decade ago from a synergistic mixture of public indifference and embarrassingly poor performances. And America's finest professional troupe, the New York Gilbert and Sullivan Players, seems to mock this acknowledged decline in status by their own featured acronym of GASP.

And yet, despite occasional frissons of doubt, I have never credited these negative assessments, and I continue to believe that the comic operas of Gilbert and Sullivan prevailed over a vast graveyard of contemporary (and later) works—including the efforts of such truly talented composers as Victor Herbert and Sigmund Romberg—because they embody the elusive quality of absolute excellence, the goal of all our creative work, and the hardest of human attributes to nurture, or even to define.

Peter Rainer, reviewing *Topsy Turvy* for *New York* magazine, intended only praise in writing, "The beauty of Gilbert and Sullivan's art, which is also its mystery, is that, gloriously minor, it's more redolent and lasting than many works regarded as major." But we will never gain a decent understanding of excellence if we continue to use this standard distinction between major and minor forms of art as our primary taxonomic device.

We live in a fractal world, where scales that we choose to designate as major (a great tenor at the Metropolitan Opera, for example) hold no intrinsically higher merit than styles traditionally judged as vernacular or minor (a self-taught banjo player on a country porch, for example). Each scale builds a corral of exactly the same shape to hold all progeny of its genre; and each corral reserves a tiny corner for its few products of absolute excellence. To continue this metaphor, a magnified photo of this corner for a "minor" art cannot be distinguished from the same corner for a "major" art seen through the wrong end of a pair of binoculars. And if these two photos, mounted on a wall at the same scale, cannot be told apart, then we must seek a different criterion for judgment based on the common morphology of scale-independent excellence.

As a man with Darwinian training, I do admit a bias toward accepting long survival as the first rough criterion in our guidebook for identifying species of true excellence. I continue to regard as sagacious a childhood bet that I made with my brother, although he never paid up—that Beethoven would outlast the rock hit of that particular moment, "Roll Over Beethoven." But if we must set aside the spurious correlation of minor with transient and major with enduring, and then take the more radical step of rejecting the taxonomy of major versus minor altogether, then the mystery surrounding the survival of Gilbert and Sullivan cannot be ascribed to the minor status of their chosen genre. And yet the mystery remains, and even intensifies, when thus stripped of its customary context. Why their work, and no others of the time? And if excellence be the common substrate of such endurance, how can we recognize this most elusive quality *before* the test of time provides a merely empirical confirmation?

I have no original insight to propose on this question of questions, but I can offer a quirky little personal testimony about Gilbert and Sullivan that might, at best, focus some useful discussion. Unless a creative person entirely abjures any goal or desire to communicate his efforts to fellow human beings, then I suspect that all truly excellent works must exist simultaneously on two planes—and must be constructed (whether consciously or not) in such duality. I also confess to the elitist view that the novel and distinguishing aspect of excellent works will be fully accessible to very few consumers—initially, perhaps, to none at all.

Two reasons regulate this "higher" plane: the motivating concept may be novel beyond any power of contemporary comprehension ("ahead of its time," in a common but misleading phrase, for time marks no necessary incrementation of quality, and the first recorded human art, the 35,000-year-old cave paintings of Chauvet, matches Picasso at its best); or the virtuosity of execution may extend beyond the discriminatory powers of all but a very few viewers or listeners.

But, on the second, vernacular plane, excellent works must exude the potential to be felt as superior (albeit not fully known) to consumers with sufficient dedication and experience to merit the accolade of "fan." A serious but untutored devotee of music in mid-eighteenth-century Leipzig should have been able to attend services at the Thomaskirche, hear the works of J. S. Bach, and recognize them as something weirdly, excruciatingly, and fascinatingly different from anything ever heard before. For the second criterion of virtuosity, a modern fan of opera might hear (as I did) Domingo, Voigt, and Salminan, under Levine and the Metropolitan Opera Orchestra, in the first act of *Die Walküre*—and just know, without being able to say why, that he had heard something surpassingly rare and transcendent.

I think that all geniuses "out of time"—if they do not go mad—consciously construct their works on these two planes, one accessible to reasonably skilled aficionados of the moment, and one for Plato's realm (and for possible future comprehension). Thus, on the vernacular plane, Bach had to accept whatever reputation (and salary) he could muster in his own time and parochial place by becoming the premier organ virtuoso of his age (a vain annoyance to many who hated the bravado and loathed the encrusting of beloved tunes with such frills of improvisation, but a source of awe and respect for the relative cognoscenti). Moreover, since Bach composed in an age that had not yet formulated our modern concept of the individual genius as an innovator, we cannot even know how he understood his own uniqueness—beyond observing that this superiority had not escaped his notice. Then, on Plato's plane, Bach could write for the angels. And Darwin had to content himself with making an explicit division of his life's work into a comprehensible vernacular plane for all educated people (the factual basis of evolution) and a second plane that even his most dedicated supporters could not grasp in all its subtle complexity (the theory of natural selection, based on a radical philosophy that inverted all previous notions about organic history and design).

This problem of composition on two planes becomes even more explicit and severe for any artist working in a genre designated as "popular" in our peculiar taxonomies of human creativity. The master of an elite genre does not aim for widespread appeal in any case—so the vernacular plane of his duality already permits a great deal of rarefied complexity (and his second plane can be as personal and as arcane as he desires). But the equal master of a popular genre must build his vernacular plane in a far more accessible place of significantly lesser complexity. How much higher, then, can his second plane ascend before the two levels lose all potential contact, and the work dissolves into incoherence?

Finally, I must preface my thoughts about Gilbert and Sullivan with one further and vital caveat. My confessed elitism remains entirely democratic—for structural rather than ethical reasons. (I happen to embrace the ethical reasons as well, but the structural claim embodies the premise that excellence cannot be achieved without such respect for the sensibility of consumers.) No person can achieve excellence in a popular genre without a rigorous and undeviating commitment to providing a personal best at all times. At the first moment of compromise—the first "dumbing down" for "easier" or "wider" acceptability, the first boilerplating for reasons of simple weariness or an overcommitted schedule—one simply falls into the abyss. (I rarely speak so harshly, but I do believe that this particular gate remains so strait, and this special path so narrow.) The difference between elite and popular genres bears no relationship whatever to any notion of absolute quality. This common distinction between genres is purely sociological. Excellence remains as rare and as precious in either category. The pinnacle of supernal achievement holds no more DiMaggios to play center field than Domingos to inquire about the location of Wälse's sword.

And so, I simply submit that Gilbert and Sullivan have survived for the best and most defendable of all reasons: their work bears a unique stamp of excellence, best illustrated by its optimal and simultaneous functioning on both levels—on the vernacular plane of accessibility to all people who like the genre (for these works tickle the funnybone, delight the muse of melody, and expose, in a gentle but incisive way, the conceits and foibles of all people and cultures); and on Plato's plane, by the most fiendishly clever union of music and versification ever accomplished in the English language. Moreover, since excellence demands both full respect and undivided attention to consumers at both levels, the uncanny genius of Gilbert and Sullivan rests largely upon their skill in serving these two audiences with, I must assume, a genuine affection for the different and equal merits of their product in both realms.

Sullivan has often been depicted as a man yearning to fulfill his supposed destiny as England's greatest classical composer since the immigrant Handel, or the earlier native Purcell—a higher spirit tethered to a more earthly Gilbert only for practical and pecuniary reasons generated by the tables of Monte Carlo and his other expensive vices. Part of his character (a rather small part, I suspect), abetted by the sanctimonious urgings of proper society (especially after Victoria dubbed him Sir Arthur, but left his partner as mere Mr. Gilbert), pulled him toward "serious" composition, but love beyond need, and a good nose for the locus of his own superior skills, led him to resist the blandishments of "higher" callings unsuited to his special gifts.

Gilbert, though often, and wrongly, cast as an acerbic martinet, came to better personal terms with the genre favored by his own muse. He supervised every detail of staging, and rehearsed his performers to the point of exhaustion. But they honored Gilbert's fierce commitment, and gave him all their loyalty, because they also knew his unfailing respect for their professionalism. Flaccid concord does not build an optimal foundation for surpassing achievement in any case.

The particular intensity of this "two level" problem for creating excellence in mass entertainment has always infused the genre—surely more so today (when the least common denominator for general appeal stands so low) than in Gilbert and Sullivan's time, when Shakespearean references, and even a Latin quip or two, might work on the vernacular plane. Chuck Jones, with Bugs Bunny and his pals at Looney Tunes, holds first prize for the twentieth century, but Disney, at his best, set the standard for works that function, without any contradiction or compromise, both as unoffensive sweetness for innocent children at one level, and as mordant and sardonic commentaries, informed by immense technical skill in animation, for sophisticated adults.

Gilbert's drawing of his Sorceror, John Wellington Wells.

Pinocchio (1940) must rank as the first masterpiece in this antic style of dual entertainment. But Disney then lost the way (probably in a conscious and politically motivated decision), as saccharine commercialism, replete with pandering that precluded excellence at either level, enveloped most of his studio's work. But a recent film rediscovered this wonderful and elusive path: *Toy Story II,* a sweet fable with brilliant animation for kids, and a rather dark, though still comic, tale for adults about life in a world of Scylla and Charybdis, with no real pathway between (leading to the ultimate existential message of "just keep truckin'." Our hero, a "collectible" cowboy modeled on an early TV star, must either stay with the boy who now loves him, and resign himself to eventual residence and dismemberment on the scrap heap, or go with his fellow TV buddies, whom he has just met in joyous discovery of his own origins, to permanent display in Japan—that is, to immortality in the antisepsis of a glass display case).

I can only claim amateur status as an exegete of Gilbert and Sullivan, but I can offer a personal testimony that may help to elucidate these two necessary planes of excellence. My first book, *Ontogeny and Phylogeny,* traced the history of biological views on the relationship between embryological development and evolutionary change—and I remain committed to the principle that systematic alterations during a life span often mirror either a historical sequence or a stable hierarchy of fully developed forms at rising levels of complexity in our current world.

As stated at the outset, I fell in love with Gilbert and Sullivan at a tender age, and imbibed all the words and music before I could possibly understand their full context and meaning. I therefore, and invariably, enjoy a bizarre and exhilarating, if mildly unsettling, experience every time I attend a performance today. An old joke, based on an ethnic stereotype that may pass muster as a mock on a privileged group in an age of political correctness, asks why the dour citizens of Switzerland often burst into inappropriate laughter during solemn moments at Sunday church services. "When they get the jokes they heard at Saturday night's party." Similar experiences attend my current Sundays with Gilbert and Sullivan, but my delays between hearing and comprehension extend to forty years or more!

I know all the words by childhood rote, but I couldn't comprehend their full cleverness at this time of implantation. Thus I carry the text like an idiot savant—with full accuracy and limited understanding because, although now fully capable, I make no conscious effort to ponder or analyze the sacred writ during my daily life. But whenever I attend a performance, I enjoy at least one "Swiss moment" when I listen with an adult ear and suddenly experience a jolt that can only induce an enormous grin for exposing human folly by personal example: "Oh yes, of course, how stupid of me. So that's what those words, which I have known and recited nearly all my conscious life, actually mean."

At this point I must highlight an aspect of this argument that I had hoped to elucidate *en passant,* and without such overt pedantry. But I have failed in all efforts to achieve this end, and had best bite the bullet of embarrassing explicitness. I do, indeed, intend this primary comparison between the vernacular versus Plato's plane as two necessary forms of excellence, and between my differing styles of childhood and adult affection for Gilbert and Sullivan. But I am, most emphatically, not arguing that the vernacular plane bears any legitimate analogy to any of the conventional descriptions of childhood as primitive, undeveloped, lesser, unformed, or even unsophisticated.

(If I really thought, in any conscious part of my being, that the vernacular character of my "popular" writings on science implied any disrespect for my audience or any adulteration of my content, I could not proceed because excellence would then lie beyond my grasp by inherent definition—and my personal quest for the two planes of this goal provides my strongest conscious motivation for this aspect of my career. Good popular writing in science builds an honored branch of our humanistic tradition, extending back to Galileo's composition of both his great books as accessible and witty dialogues in Italian, not as abstract treatises in Latin, and to Darwin's presentation of the *Origin of Species* as a book for all educated readers.)

I only compare childhood's love with the vernacular plane of excellence because both base their accurate perception and discernment upon the immediacy of unanalyzed attraction—and many forms of our highest achievement do lie beyond words, or even beyond conscious formulation, in a realm of "knowing that" rather than "stating why." For example, I first saw Joe DiMaggio play when I was eight years old, but already a reasonably knowledgeable fan of the game. I knew, with certainty, that his play and presence surpassed all others. But I had probably never even heard the word *grace,* and I surely could not formulate any concept of *excellence.*

Thus, I know that Gilbert and Sullivan fulfill excellence's primary criterion of full and simultaneous operation on two planes because I have experienced them both, and sequentially, during my own life—and this temporal separation permits me to untangle the different appeals. Some intrusions of adult understanding upon childhood's rote strike me as simply funny, not particularly illustrative of anything about excellence, but worth mentioning to set a context and to potentiate my full confession. Failure to understand does not inhibit—and may actually abet—rote memorization or unconscious infusion.

As one silly example, I recently attended a performance of *H.M.S. Pinafore,* and had my Swiss moment during Josephine's aria, as she wrestles with the dilemma of following her true love for a poor sailor or making an advantageous union with Sir Joseph Porter, "the ruler of the Queen's navy." She says of her true love, Ralph Rackstraw, "No golden rank can he impart. No wealth of house or land. No fortune save his trusty heart. . . ." And the proverbial lightbulb finally illuminated my brain. I did not know the meaning of "save" as "except" at age eleven. At that time I remember wondering how fortune could "rescue" Ralph's admirable ticker—but I never resolved the line, and didn't revisit the matter for forty-five years.

Some little examples in this mode even prove embarrassing, and therefore ever so salutary in the service of humility for arrogant intellectuals. In *Iolanthe,* for example, the Lord Chancellor berates himself for mistaking a powerful fairy queen for an insignificant schoolmarm:

> A plague on this vagary,
> I'm in a nice quandary!
> Of hasty tone with dames unknown
> I ought to be more chary;
> It seems that she's a fairy
> From Andersen's library,
> And I took her for the proprietor
> Of a Ladies' Seminary!

Now, and obviously on the second plane, part of Gilbert's literary joke lies in his conscious distortion of words to force rhymes with others that we properly stress on the penultimate syllable—especially "fairy," the key to the entire verse. But I didn't know the correct versions behind many of these distortions—and I pronounced "vagary" on the second syllable, thus exposing my pretentiousness, until a bit of auditory dissonance led me to a dictionary only about ten years ago!

As operative examples of the two planes—and of Gilbert and Sullivan's achievement of excellence through their unparalleled success in both domains, without disrespect for the vernacular, or preciousness on the upper level—consider these Swiss moments of my sequential experiences in music and text. I have relished my unanalyzed vernacular pleasure all my life, but I now appreciate the surpassingly rare quality of these works all the more because, in my maturity, I have added a few glimpses upon the depth and uniqueness of their dual representation.

Sullivan, to cite some examples in just one aspect of his efforts on the second plane, had mastered all major forms of the classical repertory. He especially appreciated the English roots and versions of certain styles. Nothing can surpass his elegant Handelian parody in *Princess Ida,* when the three irredeemably stupid sons of King Gama seek freedom from mechanical restriction before a battle by removing their armor, piece by piece, as Arac intones his formal melody to the graceful accompaniment of strings alone ("This helmet, I suppose, was meant to ward off blows"). And, still speaking of *Princess Ida,* Sullivan took an ultimate risk, and showed his genuine mastery, when he wrote a truly operatic aria of real quality ("Oh goddess wise") to accompany Gilbert's delicious spoof of Ida's serious pretensions. Durward Lely, who played the tenor lead of Cyril in the original production, said of this aria, "As an example of mock heroics it seems to me unsurpassable." The composer, in modern parlance, really had balls.

Whenever Sullivan wrote his parodies of classical forms, he did so for a wickedly funny, and devastatingly appropriate, dramatic reason. But his device must remain inaccessible unless one knows the musical style behind Sullivan's tread on the higher plane. Still, the songs work wonderfully on the vernacular plane, even when a listener cannot grasp the intended musical joke because he does not know the classical form under parody. This I can assert with certainty, albeit for a limited domain, because I loved and somehow caught the "specialness" of the following two songs in my youth, but didn't learn about the musical forms (and thus recognize the parody) until my adult years, and even then didn't understand the intended contextual joke until a much later Swiss moment.

Even at age ten, I would have identified the trio ("A British tar is a soaring soul") as my favorite song from Act I of *Pinafore.* I knew that Gilbert's text described this song as a "glee," but I knew no meaning for the word besides "mirth," and therefore did not recognize his citation of a genre of unaccompanied part songs for three or more male voices, especially popular in eighteenth-century England (and the source of the term "glee club," still used to describe some amateur singing groups).

At an intermediary stage toward the second plane, I did recognize the song as a perfectly composed glee in the old style, including its a cappella setting of the two stanzas for Ralph Rackstraw, the Boatswain, and the Boatswain's Mate. But I only got the joke a few years ago. Sir Joseph Porter gives three copies of the song to Ralph, claiming that he wrote the ditty himself "to encourage independence of thought and action in the lower branches of the service." The three men start to perform the unfamiliar piece by singing at sight from Sir Joseph's score. They begin, and manage to continue for an entire verse of four lines, in

Gilbert's drawing (signed with his pseudonym "Bab") of a British tar as "a soaring soul . . . whose energetic fist is ready to resist a dictatorial word."

perfect homophony—Sir Joseph's obvious intention for the entire piece. But, as simple sailors after all, these men can boast little experience in sight-singing, and the poor Boatswain's Mate soon falls behind the others. The second verse of four lines therefore "devolves" into an elegant, if mock, polyphonic texture—that is, a part song, a perfect and literal glee (made all the more absurd in fulfilling, by this "imperfection," Sir Joseph's stated intention to encourage independence among his men).

(I am confident that I have caught Sullivan's intent aright, but I remain surprised that modern performers generally don't seem to get the point, or don't choose to honor the potential understanding of at least some folks in their audience. Only twice, in about ten performances, have I seen the song performed by sailors holding the scores, with the Boatswain's Mate becoming frustrated as he falls behind. Why not honor Sullivan's theatrical instincts, and his lovely, if relatively sophisticated, musical joke? The piece, and the humor in the polyphonic discombobulation, works perfectly well even for listeners who only experience the vernacular plane, whereas the dual operation, with full respect accorded to both planes, expresses the *sine qua non* of excellence.)

The madrigal (so identified) from *The Mikado* ("Brightly dawns our wedding day") exploits a similar device. Yum-Yum prepares for her wedding with Nanki-Poo, but under the slight "drawback" that her husband must be beheaded in a month. The four singers, including the bride and bridegroom, try to cheer themselves up through the two verses of this perfect madrigal, or four-part song in alternating homophonic and polyphonic sections. They start optimistically ("joyous hour we give thee greeting") but cannot sustain the mood ("all must sip the cup of sorrow"), despite the words that close both verses ("sing a merry madrigal, fal-la").

To "get" the full joke, one must recognize the rigidly formal and unvarying texture of the musical part—thus setting a contrast with the singers' progressive decline into textual gloom and actual tears. Gilbert made his quartet stand rigidly still throughout the piece, as if declaiming their ancient madrigal in a formal concert hall—thus emphasizing the almost antic contrast of constant and stylish form with growing despair of feeling. In this case, I do sense that most modern directors grasp the intended point, but suspect that so few listeners now know the musical language and definition of a madrigal that the contrast between form and feeling must be reinforced by some other device. I do not object to such modernization, which clearly respects authorial intent, but I feel that directors have acquitted their proper intuition with highly varying degrees of success.

My vote for last place goes to Jonathan Miller's otherwise wonderful resetting at an English seaside resort. (As Gilbert's central joke, after all, *The Mikado*'s Japanese setting is a sham, for each character plays an English stock figure thinly disguised in oriental garb—the ineffably polite upper-class Mikado, who boils his victims in oil but invites them to tea before their execution; his wastrel son; the phony noble pretender who will do any job for a price, and who gave his name, Pooh-Bah, to men of such multiple employments and salaries.) Miller's quartet sings the madrigal stock-still, in true Gilbertian fashion, while workmen ply their tools and move their ladders about the set to emphasize the key contrast. A decent concept, perhaps, but Sullivan never wrote a more beautiful song, and you can't hear a note amid the distractions and actual clatter of the workmen's movements. Pure kudos, on the other end, to Canada's Stratford Company. They recognized that most of the audience would not draw, from the song itself, the essential contrast between structured form and dissolving emotion—so they set the singers in the midst of a formal Japanese tea ceremony to transfer the musical point to visual imagery that no one could miss, and without disrupting the music either visually or aurally.

I have thus grown to appreciate Sullivan more and more over the years, but I still lean to the view, probably held by a majority of Savoyards, that Gilbert's text must be counted as more essential to their joint survival. (Gilbert might have found another collaborator of passable talent, though no one could match Sullivan, but Sir Arthur would surely have gone the way of Victor Herbert, despite his maximal musical quality, without Gilbert's texts.) My Swiss moments with Gilbert therefore bring me even greater pleasure, make me

Jack Point, Gilbert's jester, showing "a grain or two of truth among the chaff" of his humor.

wonder how much we have all missed, and lead me to honor, even more, the talent of a man who could so love his vernacular art, while not debasing or demeaning these accomplishments one whit by festooning the texts with so many gestures and complexities that few may grasp, but that never interfere with the vernacular flow. Gilbert thus surpasses Jack Point, his own jester in *Yeomen of the Guard,* by simultaneous expression of his two levels:

> I've jibe and joke
> And quip and crank
> For lowly folk
> And men of rank. . . .
> I can teach you with a quip, if I've a mind
> I can trick you into learning with a laugh;
> Oh, winnow all my folly, and you'll find
> A grain or two of truth among the chaff!

As a small example of Gilbert's structural care, I don't know how many hundred times I have heard the duet between Captain Corcoran and Little Buttercup at the beginning of *Pinafore*'s Act II. Buttercup hints at a secret that, when revealed, will fundamentally change the Captain's status. (We later learn, in the classic example of Gilbertian topsy-turvy, that Buttercup, as the captain's nursemaid "a many years ago," accidentally switched him with another baby in her charge—who just happens to be none other than Ralph Rackstraw, now revealed as nobly born, and therefore fit to marry the Captain's [now demoted] daughter after all.) The song, as I knew even from day one at age ten, presents a list of rhyming proverbs shrouded in mystery. But I never discerned the textual structure until my Swiss moment just a few months ago. (I do not, by the way, claim that this text displays any great profundity. I merely point out that Gilbert consistently applies such keen and careful intelligence to his efforts, even when most consumers may not be listening. One must work, after all, primarily for one's own self-respect.)

I suppose I always recognized that Buttercup's proverbs all refer to masquerades and changes of status, as she hints to the Captain:

> Black sheep dwell in every fold;
> All that glitters is not gold;
> Storks turn out to be but logs;
> Bulls are but inflated frogs.

But I never grasped the structure of Captain Corcoran's response. He admits right up front that Buttercup's warnings confuse him: "I don't see at what you're driving, mystic lady." He then parries her proverbizing, supposedly in kind, at least according to his understanding:

> Though I'm anything but clever,
> I could talk like that forever.

Gilbert's simple device expresses the consummate craftsmanship, and the unvarying intelligence, that pervades the details of all his texts. The captain also pairs his proverbs in rhymed couplets. But each of his pairs includes two proverbs of opposite import, thus intensifying, by the song's structure, the character of the captain's befuddlement:

> Once a cat was killed by care;
> Only brave deserve the fair.
> Wink is often good as nod;
> Spoils the child who spares the rod.

(Just to be sure that I had not experienced unique opacity during all these years of incomprehension, I asked three other friends of high intelligence and good Gilbertian knowledge—and not a one had ever recognized the structure behind the Captain's puzzlement.)

The Captain of the Pinafore *hears Little Buttercup's rhymed warnings with puzzlement.*

To cite one more example of Gilbert's structural care and complexity, the finale to Act I of *Ruddigore* begins with three verses, each sung by a different group of characters with varied feelings about unfolding events. (In this Gilbertian parody of theatrical melodrama, Robin Oakapple has just been exposed as the infamous Ruthven Murgatroyd, Baronet of Ruddigore, condemned by a witch's curse to commit at least one crime each day. This exposure frees his younger brother, Despard, who had borne the title and curse while his big brother lived on the lam, but who may now reform to a desired life of calm virtue. Meanwhile, Richard Dauntless, Ruthven's foster brother and betrayer, wins the object of his dastardly deed, the jointly beloved Rose Maybud.)

Each of the three groups proclaims happiness surpassing a set of metaphorical comparisons. But I had never dissected the structure of Gilbert's imagery, and one verse had left me entirely confused—until another Swiss moment at a recent performance. The eager young lovers, Richard and Rose, just united by Richard's betrayal, begin by proclaiming that their own joy surpasses several organic images of immediate fulfillment:

> Oh happy the lily when kissed by the bee . . .
> And happy the filly that neighs in her pride.

Despard and his lover, Margaret, now free to pursue their future virtue, then confess to happiness surpassing several natural examples of delayed gratification:

> Oh, happy the flowers that blossom in June,
> And happy the bowers that gain by the boon.

But the lines of the final group had simply puzzled me:

> Oh, happy the blossom that blooms on the lea,
> Likewise the opossum that sits on a tree.

Why cite a peculiar marsupial living far away in the Americas, except to cadge a rhyme? And what relevance to Ruddigore can a flower on a meadow represent? I resolved Gilbert's intent only when I looked up the technical definition of a lea—what do we city boys know?—and learned its application to an arable field temporarily sown with grass for grazing. Thus the verse cites two natural images of items far and uncomfortably out of place—an irrelevant opossum in distant America, and a lone flower in a sea of grass. The three singers of this verse have all been cast adrift, or made superfluous, by actions beyond their control—for Hannah, Rose's aunt and caretaker, must now live alone; Zorah, who also fancied Richard, has lost her love; and Adam, Robin's good and faithful manservant, must now become the scheming accomplice of the criminalized and rechristened Ruthven.

Finally I recognized the structure of the entire finale as a progressive descent from immediate and sensual joy into misery—from Rose and Richard's current raptures; to Despard and Margaret's future pleasures; to the revised status of Hannah, Zorah, and Adam as mere onlookers; to the final verse, sung by Ruthven alone, and proclaiming the wretchedness of his new criminal role—not better than the stated images (as in all other verses), but worse:

> Oh, wretched the debtor who's signing a deed!
> And wretched the letter that no one can read!

For Ruthven only cites images of impotence, but he must become an *active* criminal:

> But very much better their lot it must be
> Than that of the person
> I'm making this verse on
> Whose head there's a curse on—
> Alluding to me!

Again, I claim no great profundity for Gilbert, but only the constant intelligence of his craftsmanship, and his consequent ability to surprise us with deft touches, previously unappreciated. In this case, moreover, we must compare Gilbert with the medieval cathedral builders who placed statues in rooftop positions visible only to God, and therefore not subject to any human discernment or approbation. For these verses rush by with such speed, within so dense an orchestral accompaniment, and surrounded by so much verbiage for the full chorus, that no one can hear all the words in any case. Hadn't Gilbert, after all, ended the patter trio in *Ruddigore*'s second act by recognizing that all artists must, on occasion, puncture their own bubbles or perish in a bloat of pride?

> This particularly rapid, unintelligible patter
> Isn't generally heard, and if it is it doesn't matter.

Mr. Justice Stewart famously claimed, in true (if unintended) Gilbertian fashion, that he might not be able to define pornography, but that he surely knew the elusive product when he saw it. I don't know that we can reach any better consensus about the orthogonal, if not opposite, phenomenon of excellence, but my vernacular plane knows, while my limited insight into Plato's plane can only glimpse from afar (and through a dark glass), that Gilbert and Sullivan have passed through this particular gate of our mental construction, and that their current enthusiasts need never apologize for loving such apparent trifles, truly constructed as pearls beyond price.

Ko-Ko, reciting the tale of Titwillow to win Katisha and (literally) save his neck.

Ko-Ko, to win Katisha's hand and save his own neck, weaved a tragically affecting tale about a little bird who drowned himself because blighted love has broken his heart. But Ko-Ko spun his entire story—the point of Gilbert's humor in Sullivan's wonderfully sentimental song—on the slightest of evidentiary webs, for the bird spoke but a single mysterious word, reiterated in each verse: "tit willow." Even the most trifling materials in our panoply of forms can, in the right hands, be clothed in the magic of excellence.

5

Art Meets Science in *The Heart of the Andes:* Church Paints, Humboldt Dies, Darwin Writes, and Nature Blinks in the Fateful Year of 1859

THE INTENSE EXCITEMENT AND FASCINATION THAT Frederic Edwin Church's *Heart of the Andes* solicited when first exhibited in New York in 1859 may be attributed to the odd mixture of apparent opposites so characteristic of our distinctive American style of showmanship—commercialism and excellence, hoopla and incisive analysis. The large canvas, more than ten by five feet, and set in a massive frame, stood alone in a darkened room, with carefully controlled lighting and the walls draped in black.

Dried plants and other souvenirs that Church had collected in South America probably graced the room as well. Visitors marveled at the magisterial composition, with a background of the high Andes, blanketed in snow, and a foreground of detail so intricate and microscopically correct that Church might well be regarded as the Van Eyck of botany.

But public interest also veered from the sublime to the merely quantitative, as rumors circulated that an unprecedented sum of twenty thousand dollars had been paid for the painting (the actual figure of ten thousand dollars was impressive enough for the time). This tension of reasons for interest in Church's great canvases has never ceased. A catalog written to accompany a museum show of his great Arctic landscape, *The Icebergs,* contains, in order, as its first three pictures, a reproduction of the painting, a portrait of Church, and a photo of the auctioneer at Sotheby's gaveling the sale at $2.5 million as "the audience cheered at what is [or was in 1980, at the time of this sale] the highest figure ever registered at an art auction in the United States."

A far more important, but basically ill-founded, tension—the supposed conflict between art and science—dominates our current scholarly discussion of Church and his views about nature and painting. This tension, however, can only be deemed retrospective, a product of divisions that have appeared in our society since Church painted his most famous canvases. Church did not doubt that his concern with scientific accuracy proceeded hand in hand with his drive to depict beauty and meaning in nature. His faith in this fruitful union stemmed from the views of his intellectual mentor Alexander von Humboldt, a great scientist who had ranked landscape painting among the three highest expressions of our love of nature.

Church sent *The Heart of the Andes* to Europe after its great American success in 1859. He wanted, above all, to show the painting to Humboldt, then ninety years old, and who, sixty years before, had begun the great South American journey that would become the source of his renown. Church wrote to Bayard Taylor on May 9, 1859:

> The "Andes" will probably be on its way to Europe before your return to the City. . . . [The] principal motive in taking the picture to Berlin is to have the satisfaction of placing before Humboldt a transcript of the scenery which delighted his eyes sixty years ago—and which he had pronounced to be the finest in the world.

But Humboldt died before the painting arrived, and Church's act of homage never bore fruit. Later in 1859, as *The Heart of the Andes* enjoyed another tri-

Frederic Edwin Church's great landscape painting, The Heart of the Andes.

umph of display in the British Isles, Charles Darwin published his epochal book, *The Origin of Species,* in London. These three events, linked by their combined occurrence in 1859—the first exhibition of *The Heart of the Andes,* the death of Alexander von Humboldt, and the publication of *The Origin of Species*—set the core of this essay. They present, in my view, a basis for understanding the central role of science in Church's career and for considering the larger issue of relationships between art and the natural world.

As a professional scientist, I hold no credentials for judging or interpreting Church's paintings. I can only say that I have been powerfully intrigued (stunned would not be too strong a word) by his major canvases throughout my life, beginning with childhood visits to the Metropolitan Museum of Art in my native New York City, when *The Heart of the Andes,* medieval armor, and Egyptian mummies grabbed my awe and attention in that order.*

But if I have no license to discourse on Church, at least I inhabit the world of Humboldt and Darwin, and I can perhaps clarify why Humboldt became such a powerful intellectual guru for Church and an entire generation of artists and scholars, and why Darwin pulled this vision of nature up from its roots, substituting another that could and should have been read as equally ennobling, but that plunged many votaries of the old order into permanent despair.

When Church began to paint his great canvases, Alexander von Humboldt may well have been the world's most famous and influential intellectual. If his name has faded from such prominence today, this slippage only records a curiosity and basic unfairness of historical judgment. The history of ideas emphasizes innovation and downgrades popularization. The great teachers of any time exert enormous influence over the lives and thoughts of entire generations, but their legacy fades as the hagiographic tradition exalts novel thoughts and discards context. No one did more to change and enhance science in the first half of the nineteenth century than Alexander von Humboldt, the cardinal inspiration for men as diverse as Charles Darwin, Alfred Russel Wallace, Louis Agassiz (whom Humboldt financed at a crucial time), and Frederic Edwin Church.

Humboldt (1769–1859) studied geology in his native Germany with another

---

*I greatly amused my Hungarian grandmother (see essay 1) on my first visit at about age five—when I asked her if she had worn such armor as a girl in that far-off land. My mother, after all, had told me that my grandmother was "middle aged."

The Icebergs, *by Frederic Edwin Church.*

great teacher, A. G. Werner. Following Werner's interest in mining, Humboldt invented a new form of safety lamp and a device for rescuing trapped miners. Early in his career, Humboldt developed a deep friendship with Goethe, a more uncertain relationship with Schiller, and a passion to combine personal adventure with the precise measurements and observations necessary to develop a science of global physical geography. Consequently, recognizing that the greatest diversity of life and terrain would be found in mountainous and tropical regions, he embarked on a five-year journey to South America in 1799, accompanied by the French botanist Aimé Bonpland. During this greatest of scientific adventures, Humboldt collected sixty thousand plant specimens, drew countless maps of great accuracy, wrote some of the most moving passages ever penned against the slave trade, proved the connection between the Orinoco and Amazon rivers, and established a mountaineering record (at least among westerners inclined to measure such things) by climbing to nineteen thousand feet (though not reaching the summit) on Chimborazo. On his way home in 1804, Humboldt visited the United States and had several long meetings with Thomas Jefferson. Back in Europe, he met and befriended Simon Bolívar, becoming a lifelong adviser to the great liberator.

Humboldt's professional life continued to revolve around his voyage and the meticulous records and diaries that he had kept. Over the next twenty-five years he published thirty-four volumes of his travel journal illustrated with 1,200 copper plates, but never finished the project. His large and beautiful maps became the envy of the cartographic world. Most important (in influencing Church and Humboldt's other disciples), Humboldt conceived, in 1827–28, a plan for a multivolume popular work on, to put the matter succinctly, everything. The first two volumes of *Kosmos* appeared in 1845 and 1847, the last three in the 1850s. *Kosmos,* immediately translated into all major Western languages, might well be ranked as the most important work of popular science ever published.

Humboldt's primary influence on Church can scarcely be doubted. Church owned, read, and reread both Humboldt's travel narratives and *Kosmos.* In an age when most painters aspired to a European grand tour to set the course of their work and inspiration, Church followed a reverse route, taking his cue from Humboldt. After his apprenticeship with Thomas Cole, Church first traveled, at Humboldt's direct inspiration, to the high tropics of South America, in 1853 and 1857. In Quito, he sought out and occupied the house that Humboldt had inhabited nearly sixty years before. He painted the great canvases of his most fruitful decade (1855–65) as embodiments of Humboldt's aesthetic philosophy and convictions about the unity of art and science. Even subjects maximally distant from the tropics bear Humboldt's mark of influence. *The Icebergs* and Church's general fascination with polar regions closely parallel Humboldt's second major expedition, his Siberian sojourn of 1829. Church did not visit Europe until 1867, and this cradle of most Western painting did not provoke a new flood of great creativity.

We can best grasp Humboldt's vision by examining the plan of *Kosmos.* On the first page of his preface, Humboldt states the grand aim of his entire work:

> The principal impulse by which I was directed was the earnest endeavor to comprehend the phenomena of physical objects in their general connection, and to represent nature as one great whole, moved and animated by internal forces.

"Nature," he adds later, "is a unity in diversity of phenomena; a harmony, blending together all created things, however dissimilar in form and attributes; one great whole animated by the breath of life." This twofold idea of natural unity forged by a harmony of internal laws and forces represented no mere rhapsodizing on Humboldt's part; for this vision expressed his view of natural causation. This view of life and geology also embodied the guiding principles that animated Church and that Darwin would tear down with a theory of conflict and balance between internal and external (largely random) forces.

Volume one of *Kosmos* covers, on the grandest possible scale, the science that we would call physical geography today. Humboldt ranges from the most distant stars to minor differences in soil and climate that govern the distribution of vegetation. (*Kosmos* is fundamentally a work in geography, a treatise about the natural forms and places of things. Thus, Humboldt includes little conventional biology in his treatise and discusses organisms primarily in terms of their geographic distribution and appropriate fit to environments.)

Kosmos takes seriously, and to the fullest possible extent, Humboldt's motivating theme of unity. If volume one presents a physical description of the universe, then volume two—an astounding tour de force that reads with as much beauty and relevance today as in Church's era—treats the history and forms of human sensibility toward nature. (The last three volumes of *Kosmos,* published many years later, present case studies of the physical world; these volumes never became as popular as the first two.) Humboldt wrote of his overall design:

> I have considered Nature in a two-fold point of view. In the first place, I have endeavored to present her in the pure objectiveness of external phenomena; and secondly, as the reflection of the image impressed by the senses upon the inner man, that is, upon his ideas and feelings.

Humboldt begins volume two with a discussion of the three principal modes (in his view) for expressing our love of nature—poetic description, landscape painting (need I say more for the influence upon Church?), and cultivation of exotic plants (Church made a large collection of dried and pressed tropical plants). The rest of the volume treats, with stunning erudition and encyclopedic footnotes, the history of human attitudes toward the natural world.

Humboldt embodied the ideals of the Enlightenment as well and as forcefully as any great intellectual—as Voltaire, or Goya, or Condorcet. If he lived so long, and past the hour of maximal flourishing for this philosophy, he remained firm in his convictions, a beacon of hope in a disillusioned world. Humboldt conveyed the Enlightenment's faith that human history moved toward progress and harmony based on the increasing spread of intellect. People may differ in current accomplishments, but all races are equally subject to similar improvement. In the most famous nineteenth-century statement of equality made by a scientist (see also essay 27), Humboldt wrote:

> While we maintain the unity of the human species, we at the same time repel the depressing assumption of superior and inferior races of men. There are nations more susceptible of cultivation, more highly civilized, more ennobled by mental cultivation than others, but none in themselves nobler than others. All are in like degree designed for freedom.

In expressing his liberal belief in progress, Humboldt contrasts his perception of unity with the standard views, based on division and separation, of such social conservatives as Edmund Burke. For Burke and other leaders of the reaction against liberalism, feeling and intellect must be treated as separate domains; emotion, the chief mode of the masses, leads to danger and destruction. The masses must therefore be restrained and ruled by an elite capable of mastering the constructive and empowering force of intellect.

Humboldt's vision, in direct contrast, emphasizes the union and positive interaction between feeling and analysis, sentiment and observation. Sentiment, properly channeled, will not operate as a dangerous force of ignorance, but as a prerequisite to any deep appreciation of nature:

> The vault of heaven, studded with nebulae and stars, and the rich vegetable mantle that covers the soil in the climate of palms, cannot surely fail to produce on the minds of these laborious observers of nature an impression more imposing and more worthy of the majesty of creation than on those who are unaccustomed to investigate the great mutual relations of phenomena. I cannot, therefore, agree with Burke when he says, "it is our ignorance of natural things that causes all our admiration, and chiefly excites our passions."

Romantic nonsense might proclaim a superiority of untrammeled feeling over the dryness of accurate observation and measurement, but the Enlightenment's faith in rationality located highest truth in the mutual reinforcement of feeling and intellect:

> It is almost with reluctance that I am about to speak of a sentiment, which appears to arise from narrow-minded views, or from a certain weak and morbid sentimentality—I allude to the fear entertained by some persons, that nature may by degrees lose a portion of the charm and magic of her power, as we learn more and more how to unveil her secrets, comprehend the mechanism of the movements of the heavenly bodies, and estimate numerically the intensity of natural forces. . . . Those who still cherish such erroneous views in the present age, and amid the progress of public opinion, and the advancement of all branches of knowledge, fail in duly appreciating the value of every enlargement of the sphere of intellect, and the importance of the detail of isolated facts in leading us on to general results.

Humboldt viewed the interaction of feeling and intellect as an upwardly spiraling system, moving progressively toward deep understanding. Feeling excites our interest and leads us to a passionate desire for scientific knowledge of details and causes. This knowledge in turn enhances our appreciation of natural beauty. Feeling and intellect become complementary sources of understanding; knowing the causes of natural phenomena leads us to even greater awe and wonder.

> Thus do the spontaneous impressions of the untutored mind lead, like the laborious deductions of cultivated intellect, to the same intimate persuasion, that one sole and indissoluble chain binds together all nature. . . . Every imposing scene in nature depends so materially upon the mutual relation of the ideas and sentiments simultaneously excited in the mind of the observer.

Humboldt rooted his theory of aesthetics in this idea of mutual reinforcement. A great painter must also be a scientist, or at least committed to the detailed and accurate observation, and to the knowledge of causes, that motivate a professional scientist. For the visual arts, landscape painting becomes the principal mode of expressing the unity of knowledge (as poetry serves the literary arts and cultivation of exotic plants the practical arts). A great landscape painter is the highest servant of both nature and the human mind.

Church accepted Humboldt's aesthetic theory as his own guide (and why not, for I think that no one has ever improved upon this primary statement of humanism). Church achieved primary recognition and respect as the most scientific of painters (when such a designation implied admiration, not belittlement). Critics and connoisseurs viewed his penchant for accuracy in observation and rendering, both for intricate botanical details in his foregrounds and for geological forms in his backgrounds, as a primary source of quality in his art and as a key to his success in awakening feelings of awe and sublimity in his viewers.

I do not, of course, say that Church attempted, or that Humboldt advocated, a slavish rendering of particular places with snapshot accuracy. Humboldt did stress the value of colored sketches from nature, even of photographs (though he felt, in the nascent years of this art, that photography could only capture the basic forms of a landscape, not the important details). But Humboldt realized that any fine canvas must be conceived and executed as an imaginative reconstruction, accurate in all details of geology and vegetation, but not a re-creation of a particular spot:

> A distinction must be made in landscape painting, as in every other branch of art, between the elements generated by the more limited field of contemplation and direct observation, and those which spring from the boundless depth and feeling and from the force of idealizing mental power.

None of Church's great tropical paintings represent particular places. He often constructed idealized vantage points so that he could encompass all life zones, from the vegetation of lush lowlands to the snow-clad Andean peaks, in a single composition. (For example, although Church's most famous painting of Cotopaxi includes no lowland plants, most of his other canvases of this great volcano feature palm trees and other luxuriant plants that do not grow in such proximity to the mountain.) Moreover, though likely with no conscious intent, Church did not always depict his geological background accurately. Volcanologist Richard S. Fiske discovered that Church painted the symmetrical cone of Cotopaxi with steeper sides than the actual mountain possesses. We may, however, view this "license" as a veering toward accuracy, for Humboldt himself had drawn Cotopaxi with even steeper slopes!

Humboldt's influence over Church extended well beyond general aesthetic philosophy and the value of science and accurate observation. One may identify landscape painting as the principal mode of glorifying nature in the visual arts, but which among the infinitude of earthly landscapes best captures the essence of wonder? Humboldt replied with the aesthetic conviction that still motivates such modern ecological movements as the battle to save the rain forests of the Amazon. Maximal diversity of life and landscape defines the *summum bonum* of aesthetic joy and intellectual wonder. This maximal diversity thrives in two circumstances that enjoy their greatest confluence in the High Andes of South America. First, the vastly greater diversity of vegetation in tropical regions marks the equatorial zone as immensely more varied than temperate areas inhabited by most Western peoples. Second, diversity will be greatly enhanced by a range of altitudes, for the sequence of lowland to mountaintop in a single district may span the entire panoply of lowland environments from equator to pole, with an equatorial mountaintop acting as a surrogate for the Arctic. Thus, the higher the mountains, the wider the range of diversity. The Himalayas might win our preference, but they lie too far north of the equator and do not include zones of tropical lowland vegetation. The Andes of South America became the premier spot on earth for landscape painting, for only here does the full luxuriance of the lowland jungle stand in the shadow of such a massive range of snow-clad peaks. Humboldt therefore chose South America, as did Darwin, Wallace, and Frederic Edwin Church, much to the benefit of art and history. Humboldt wrote:

> Are we not justified in hoping that landscape painting will flourish with a new and hitherto unknown brilliancy when artists

One of Church's versions of Cotopaxi, *showing the full range of environments from tropical lowland vegetation to the snow-clad volcanic peak.*

> of merit shall more frequently pass the narrow limits of the Mediterranean, and when they shall be enabled, far in the interior of continents, in the humid mountain valleys of the tropical world, to seize, with the genuine freshness of a pure and youthful spirit, on the true image of the varied forms of nature?

When Church was still a small boy, Humboldt's travel writings also played a major role in setting the life course of a young English graduate who planned to become a country parson (not from any particular zeal for religion, and probably to maximize time for avocational interests in natural history). But Charles Darwin veered down a different course to become one of history's most important intellectuals—and Humboldt served as his primary influence. Darwin read two books that focused his interests upon natural history in a more serious and professional way: J. F. W. Herschel's *Preliminary Discourse on the Study of Natural History* and Humboldt's *Personal Narrative* of the South American voyages (1814–29). As an old man, Darwin reminisced in his autobiography:

> [These books] stirred up in me a burning zeal to add even the most humble contribution to the noble structure of Natural Science. No one or a dozen other books influenced me nearly so much as these two.

Moreover, directly inspired by Humboldt's views on the importance of tropical travel, Darwin hatched a plot to visit the Canary Islands with some entomologist friends. Darwin involved his mentor, botanist J. S. Henslow, in the plan, and this decision led, clearly if indirectly, to Darwin's invitation to sail on the *Beagle,* the beginning and *sine qua non* of his rendezvous with history. Mathematician George Peacock asked Henslow to recommend a keen young naturalist to Captain FitzRoy, and Henslow, impressed with Darwin's general zeal and desire for tropical travel, suggested his young protégé for the job. The *Beagle* spent five years circumnavigating the globe, but the trip had been conceived primarily as a surveying voyage to South America, and Darwin spent the bulk of his time in and around Humboldt's favorite places. More than mere accident underlies the fact that the twin discoverers of natural selection, Darwin and Alfred Russel Wallace, both cited Humboldt as their inspiration, and both made their most extensive, youthful trips to South America. On April 28, 1831, as Darwin prepared for the *Beagle* voyage, he wrote to his sister Caroline:

> My head is running about the tropics: in the morning I go and gaze at Palm trees in the hot-house and come home and read Humboldt; my enthusiasm is so great that I can hardly sit still on my chair.

Darwin's first view of the richness of tropical life led him to rhapsody, for the real objects even exceeded Humboldt's descriptions. In Brazil, Darwin wrote in his diary for February 28, 1832:

> Humboldt's glorious descriptions are and will for ever be unparalleled; but even he with his dark blue skies and the rare union of poetry with science which he so strongly displays when writing on tropical scenery, with all this falls far short of the truth. The delight one experiences in such times bewilders the mind; if the eye attempts to follow the flights of a gaudy butterfly, it is arrested by some strange tree or fruit; if watching an insect one forgets it in the stranger flower it is crawling over; if turning to admire the splendor of the scenery, the individual character of the foreground fixes the attention. The mind is a chaos of delight, out of which a world of future and more quiet pleasure will arise. I am at present fit only to read Humboldt; he like another sun illuminates everything I behold.

And, more succinctly, in a letter to his mentor Henslow a few months later, on May 18: "I never experienced such intense delight. I formerly admired Humboldt, I now almost adore him."

Darwin did not read Humboldt only for visceral wonder; he evidently studied Humboldt's aesthetic theories with some care as well, as several entries in the *Beagle* diary testify. Consider this comment from Rio de Janeiro in 1832:

> During the day I was particularly struck with a remark of Humboldt's who often alludes to "the thin vapor which without changing the transparency of the air, renders its tints more harmonious, softens its effects," etc. This is an appearance which I have never observed in the temperate zones. The atmosphere, seen through a short space of half or three-quarters of a mile, was perfectly lucid, but at a greater distance all colors were blended into a most beautiful haze.

Or this passage, from his summary comments upon returning in 1836:

> I am strongly induced to believe that, as in Music, the person who understands every note, will, if he also has true taste, more thoroughly enjoy the whole; so he who examines each part of a fine view, may also thoroughly comprehend the full and combined effect. Hence a traveler should be a botanist, for in all views plants form the chief embellishment. Group masses of naked rocks, even in the wildest forms. For a time they may afford a sublime spectacle, but they will soon grow monotonous; paint them with bright and varied colors, they will become fantastick [sic]; clothe them with vegetation, and they must form at least a decent, if not a most beautiful picture.

Humboldt himself could not have written a better passage on the value of diversity and his favorite theme of aesthetic appreciation enhanced by detailed knowledge of individual parts—the union of artistic pleasure and scientific understanding.

So we reach the pivotal year of our drama, 1859. Humboldt lies dying in Berlin, while two powerful and influential men, half a world apart in geography and profession, reach an apex of fame founded on Humboldt's inspiration: Frederic Edwin Church displays *The Heart of the Andes,* and Charles Darwin publishes *The Origin of Species.*

And we encounter a precious irony, an almost painfully poignant outcome. Humboldt himself, in the preface to volume one of *Kosmos,* had noted the paradox that great works of science condemn themselves to oblivion as they open floodgates to reforming knowledge, while classics of literature can never lose relevance:

> It has frequently been regarded as a subject of discouraging consideration, that while purely literary products of intellectual activity are rooted in the depths of feeling, and interwoven with the creative force of imagination, all works treating of empirical knowledge, and of the connection of natural phenomena and physical laws, are subject to the most marked modifications of form in the lapse of short periods of time. . . . Those scientific works which have, to use a common expression, become antiquated by the acquisition of new funds of knowledge, are thus continually being consigned to oblivion as unreadable.

By Darwin's hand, Humboldt's vision suffered this fate of superannuation in 1859. The exterminating angel cannot be equated with the fact of evolution itself, for some versions of evolution as necessarily progressive and internally driven fit quite well with Humboldt's notion of pervasive harmony. Rather, Darwin's particular theory, natural selection, and the radical philosophical context of its presentation, drove Humboldt's pleasant image to oblivion. Frederic Edwin Church, alas, felt even more committed than Humboldt to the philosophical comfort of their shared vision, for Church, unlike Humboldt, had rooted a good portion of his Christian faith—for him a most important source of inspiration and equanimity—in a view of nature as essential harmony in unity.

Consider just three aspects of the new Darwinian worldview, all confuting central aspects of Humboldt's vision.

1. Nature must be reconfigured as a scene of competition and struggle, not higher and ineffable harmony. Order and good design arise by natural selection, and only as a side consequence of struggle. Hobbes's "war of all against all" denotes the causal reality of most daily interactions in nature. The struggle should be viewed as metaphorical and need not involve bloody battle (a plant, Darwin tells us, may be said to struggle against an inclement environment at the edge of a desert). But, more often than not, competition proceeds by the sword, and some die that others may live. The struggle, moreover, operates for the reproductive success of individual organisms, not directly in the service of any higher harmony. Darwin, in one of his most trenchant metaphors, seems to tear right through Humboldt's faith and Church's canvases in depicting apparent harmony as dangerously misleading:

> We behold the face of nature bright with gladness, we often see superabundance of food; we do not see, or we forget, that the birds which are idly singing round us mostly live on insects or seeds, and are thus constantly destroying life; or we forget how largely these songsters, or their eggs, or their nestlings, are destroyed by birds and beasts of prey.

2. Evolutionary lineages follow no intrinsic direction toward higher states or greater unification. Natural selection only yields local adaptation, as organisms alter in response to modifications in their environment. The geological and climatological causes of environmental change impose no inherent direction either. Evolution is opportunistic.

3. Evolutionary changes do not arise by an internal and harmonious force. Evolution expresses a balance between the internal characteristics of organisms and the external vector of environmental change. These internal and external forces both include strong random components, further obviating any notion of impulse toward union and harmony. The internal force of genetic mutation, the ultimate source of evolutionary variation, works randomly with respect to the direction of natural selection. The external force of environmental change alters capriciously with respect to the progress and complexity of organisms.

Many other humanists joined Frederic Edwin Church in feeling crushed by this new and apparently heartless view of nature. Few themes, in fact, reverberate more strongly through late nineteenth- and early twentieth-century literature than the distress and sadness provoked by losing the comfort of a world lovingly constructed with intrinsic harmony among all its constituent parts. Thomas Hardy, in a striking poem titled "Nature's Questioning," lets the natural objects and organisms of Darwin's new world express their despair through stunned silence:

> When I look forth at dawning, pool,
> Field, flock, and lonely tree,
> All seem to gaze at me
> Like chastened children sitting silent in a school.
>
> Upon them stirs in lippings mere
> (As if once clear in call,
> But now scarce breathed at all)—
> "We wonder, ever wonder, why we find us here!"

I am no devotee of psychobiography or psychohistory, and I will not indulge in speculative details about the impact of Darwin's revolution on Church's painting. But we cannot ignore the coincidences of 1859, and their impact upon the last thirty years of Church's life. When I began this project,* I was shocked to learn that Church had lived until 1900. His work and its meaning had been so firmly fixed, in my eyes, into the world just before Darwin's watershed, that I had trouble imagining his corporeal self peering into the twentieth century. (Church reminds me of Rossini, living into Wagner's era, but with all his work done thirty

---

*I originally wrote this article for the catalog of a retrospective exhibit of Church's paintings, displayed in Washington, D.C., at the National Gallery of Art in 1989.

years before in a different age of *bel canto;* or of Kerensky, deposed by Lenin, but then living for more than fifty years as an aging exile in New York.)

My impression of surprise arose in part from the facts of Church's output. He continued to produce some canvases into the 1890s, but he painted no more great landscapes after the 1860s. I know that several non-ideological reasons help to explain Church's withdrawal. For one, he became very wealthy from his painting (contrary to the stereotype of struggling artists) and spent much of his later life designing and furnishing his remarkable home, Olana, on the Hudson River in upstate New York. For another factor (and one could hardly state a better reason), he experienced severe health problems with inflammatory rheumatism and eventually lost the use of his painting arm. Still, I wonder if the collapse of his vision of nature, wrought by Darwin's revolution, also played a major role in destroying both his enthusiasm for painting such landscapes ever again. If an uplifting harmony turns into a scene of bloody battle, does not the joke become too bitter to bear?

Several scholars have claimed that the large number of books about science in Church's library at Olana prove his continuing concern for keeping up with the latest ideas in natural history. But this argument cannot be sustained, and the list, in my judgment as a historian of the sciences of natural history, actually implies an opposite conclusion. Yes, Church owned many books about science, but as Sherlock Holmes once recognized the *absence* of a bark as the crucial evidence for the nonexistence of a dog, the key to Church's collection lies in the books he did *not* own. Church maintained a good collection of Humboldt; he bought Wallace's books on the geographic distribution of animals and on tropical biology, Darwin on the *Beagle* voyage and the *Expression of the Emotions in Man and Animals* (1872). He purchased the major works by Christian evolutionists who continued to espouse the idea of necessary progress mediated by internal forces of vital matter—H. F. Osborn and N. S. Shaler, for example. But Church did not own either of Darwin's evolutionary treatises, *The Origin of Species* (1859) or *The Descent of Man* (1871). More important, he apparently collected not a single work of a mechanistic or materialist bent—not a word of E. H. Haeckel and only a text on religion by T. H. Huxley, though sales of their books far exceeded all others among late-nineteenth-century popularizations of evolution. I think that Frederic Edwin Church probably did undergo a crisis of confidence akin to the pain and bewilderment suffered by the organisms of Hardy's poem—and that he could not bear to face the consequences of Darwin's world.

I do not wish to end this article on a somber note—not only because I try to maintain a general cheerfulness of temperament, but also because such a termination would not provide a factually correct or aesthetically honorable end for my story. I want to finish by affirming an aspect of Humboldt's vision that I regard as more important than his falsified view of natural harmony and, therefore, as upholding the continuing power and beauty of Church's great paintings. I also want to suggest that Hardy's sadness and Church's silence may not represent the most fruitful or appropriate responses of humanists to Darwin's new world—an initial reaction of shock and dismay, perhaps, but not the considered conclusion of more reflection and understanding from both sides.

First of all, Humboldt correctly argued, as quoted earlier, that great works of science supersede themselves by sowing seeds for further advances. This, Humboldt adds, marks an aspect of science's joy, not its distress:

> However discouraging such a prospect must be, no one who is animated by a genuine love of nature, and by a sense of the dignity attached to its study, can view with regret anything which promises future additions and a greater degree of perfection to general knowledge.

Second, and of far more importance for this essay, Humboldt rightly emphasized the interaction of art and science in any deep appreciation of nature. Therefore Church advanced a grand vision, as right and as relevant today as in his own time, in his fidelity to the principle and actuality of natural observation combined with the shaping genius of imagination. Indeed, I would go further and argue that this vision may now be even more important and relevant today than in the era of Humboldt and Church. For never before have we been surrounded with such confusion, such a drive to narrow specialization, and such indifference to the striving for connection and integration that defines the best in the humanist tradition. Artists dare not hold science in contempt, and scientists will work in a moral and aesthetic desert—a most dangerous place in our age of potentially instant destruction—without art. Yet integration becomes more difficult to achieve than ever before, as jargons divide us and anti-intellectual movements sap our strength. Can we not still find inspiration in the integrative visions of Humboldt and Church?

I will not deny that such integration becomes more difficult in Darwin's world—a bleaker place, no doubt, than Humboldt's. But in another sense, the very bleakness of Darwin's world points to the right solution, a viewpoint perceived with crystal clarity by Darwin himself. Nature simply is what she is; nature does not exist for our delectation, our moral instruction, or our pleasure. Therefore, nature will not always (or even preferentially) match our hopes.

Humboldt asked too much of nature, and pinned too much of his philosophy on a particular outcome. He therefore chose a dubious, even a dangerous, tactic—for indifferent nature may not supply the answers that our souls seek.

Darwin grasped the philosophical bleakness with his characteristic courage. He argued that hope and morality cannot, and should not, be passively read in the construction of nature. Aesthetic and moral truths, as *human* concepts, must be shaped in human terms, not "discovered" in nature. We must formulate these answers for ourselves and then approach nature as a partner who can answer other kinds of questions for us—questions about the factual state of the universe, not about the meaning of human life. If we grant nature the independence of her own domain—her answers unframed in human terms—then we can grasp her exquisite beauty in a free and humble way. For then we become liberated to approach nature without the burden of an inappropriate and impossible quest for moral messages to assuage our hopes and fears. We can pay our proper respect to nature's independence and read her own ways as beauty or inspiration in our different terms. I therefore give the last word to Darwin (diary entry of January 16, 1832), who could not deny the apparent truth of natural selection as a mechanism of change, but who never lost his sense of beauty or his childlike wonder. Darwin stood in the heart of the Andes as he wrote:

> It has been for me a glorious day, like giving to a blind man eyes, he is overwhelmed by what he sees and cannot justly comprehend it. Such are my feelings, and such may they remain.

III

Darwinian Prequels and Fallout

6

The Darwinian Gentleman at Marx's Funeral: Resolving Evolution's Oddest Coupling

WHAT COULD POSSIBLY BE DEEMED INCONGRUOUS ON A shelf of Victorian bric-a-brac, the ultimate anglophonic symbol for miscellany? What, to illustrate the same principle on a larger scale, could possibly seem out of place in London's Highgate Cemetery—the world's most fantastic funerary park of overgrown vegetation and overblown statuary, described as a "Victorian Valhalla . . . a maze of rising terraces, winding paths, tombs and catacombs . . . a monument to the Victorian age and to the Victorian attitude to death . . . containing some of the most celebrated—and often most eccentric—funerary architecture to be found anywhere" (from *Highgate Cemetery* by F. Barker and J. Gay, published in 1984 by

E. Ray Lankester. Why did such a conservative gent attend the funeral of Karl Marx?

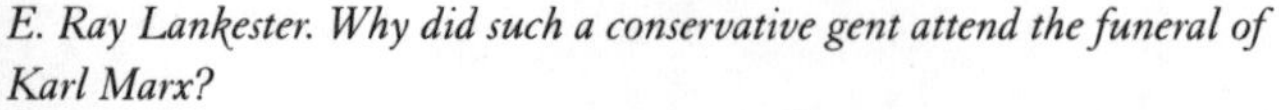

John Murray in London, the same firm that printed all Darwin's major books—score one for British continuity!).

Highgate holds a maximal variety of mortal remains from Victoria's era—from eminent scientists like Michael Faraday, to literary figures like George Eliot, to premier pundits like Herbert Spencer, to idols of popular culture like Tom Sayers (one of the last champions of bare-knuckle boxing), to the poignancy of early death for ordinary folks—the young Hampstead girl "who was burned to death when her dress caught fire," or "Little Jack," described as "the boy missionary," who died at age seven on the shores of Lake Tanganyika in 1899.

But one monument in Highgate Cemetery might seem conspicuously out of place, to people who have forgotten an odd fact from their high-school course in European history. The grave of Karl Marx stands almost adjacent to the tomb of his rival and arch opponent of all state intervention (even for street lighting and sewage systems), Herbert Spencer. The apparent anomaly only becomes exacerbated by the maximal height of Marx's monument, capped by an outsized bust. (Marx had originally been buried in an inconspicuous spot adorned by a humble marker, but visitors complained that they could not find the site, so in 1954, with funds raised by the British Communist Party, Marx's gravesite reached higher and more conspicuous ground. To highlight the anomaly of his presence, this monument, until the past few years at least, attracted a constant stream of the most dour, identically suited groups of Russian or Chinese pilgrims, all snapping their cameras, or laying their "fraternal" wreaths.)

Marx's monument may be out of scale, but his presence could not be more appropriate. Marx lived most of his life in London, following exile from Belgium, Germany, and France for his activity in the revolution of 1848 (and for general political troublemaking: he and Engels had just published the *Communist Manifesto*). Marx arrived in London in August 1848, at age thirty-one, and lived there until his death in 1883. He wrote all his mature works as an expatriate in England; and the great (and free) library of the British Museum served as his research base for *Das Kapital*.

Let me now introduce another anomaly, not so easily resolved this time, about the death of Karl Marx in London. This item, in fact, ranks as my all-time-favorite, niggling little incongruity from the history of my profession of evolutionary biology. I have been living with this bothersome fact for twenty-five years, and I made a pledge to myself long ago that I would try to discover some resolution before ending this series of essays. Let us, then, return to Highgate Cemetery, and to Karl Marx's burial on March 17, 1883.

Friedrich Engels, Marx's lifelong friend and collaborator (also his financial "angel," thanks to a family textile business in Manchester), reported the short, small, and modest proceedings (see Philip S. Foner, ed., *Karl Marx Remembered: Comments at the Time of His Death* [San Francisco: Synthesis Publications, 1983]). Engels himself gave a brief speech in English that included the following widely quoted comment: "Just as Darwin discovered the law of evolution in organic nature, so Marx discovered the law of evolution in human history." Contemporary reports vary somewhat, but the most generous count places only nine mourners at the gravesite—a disconnect between immediate notice and later influence exceeded, perhaps, only by Mozart's burial in a pauper's grave.

(I exclude, of course, famous men like Bruno and Lavoisier, executed by state power and therefore officially denied any funerary rite.)

The list, not even a minyan in length, makes sense (with one exception): Marx's wife and daughter (another daughter had died recently, thus increasing Marx's depression and probably hastening his end); his two French socialist sons-in-law, Charles Longuet and Paul Lafargue; and four nonrelatives with long-standing ties to Marx, and impeccable socialist and activist credentials: Wilhelm Liebknecht, a founder and leader of the German Social-Democratic Party (who gave a rousing speech in German, which, together with Engels's English oration, a short statement in French by Longuet, and the reading of two telegrams from workers' parties in France and Spain, constituted the entire program of the burial); Friedrich Lessner, sentenced to three years in prison at the Cologne Communist trial of 1852; G. Lochner, described by Engels as "an old member of the Communist League"; and Carl Schorlemmer, a professor of chemistry in Manchester, but also an old communist associate of Marx and Engels, and a fighter at Baden in the last uprising of the 1848 revolution.

But the ninth and last mourner seems to fit about as well as that proverbial snowball in hell or that square peg trying to squeeze into a round hole: E. Ray Lankester (1847–1929), already a prominent young British evolutionary biologist and leading disciple of Darwin, but later to become—as Professor Sir E. Ray Lankester K.C.B. (Knight, Order of the Bath), M.A. (the "earned" degree of Oxford or Cambridge), D.Sc. (a later honorary degree as doctor of science), F.R.S. (Fellow of the Royal Society, the leading honorary academy of British science)—just about the most celebrated, and the stuffiest, of conventional and socially prominent British scientists. Lankester moved up the academic ladder from exemplary beginnings to a maximally prominent finale, serving as professor of zoology at University College London, then as Fullerian Professor of Physiology at the Royal Institution, and finally as Linacre Professor of Comparative Anatomy at Oxford University. Lankester then capped his career by serving as director (from 1898 to 1907) of the British Museum (Natural History), the most powerful and prestigious post in his field. Why, in heaven's name, was this exemplar of British respectability, this basically conservative scientist's scientist, hanging out with a group of old (and mostly German) communists at the funeral of a man described by Engels, in his graveside oration, as "the best hated and most calumniated man of his times"?

Even Engels seemed to sense the anomaly, when he ended his official report of the funeral, published in *Der Sozialdemokrat* of Zürich on March 22, 1883, by writing: "The natural sciences were represented by two celebrities of the first rank, the zoology Professor Ray Lankester and the chemistry Professor

Schorlemmer, both members of the Royal Society of London." Yes, but Schorlemmer was a countryman, a lifelong associate, and a political ally. Lankester did not meet Marx until 1880, and could not, by any stretch of imagination, be called a political supporter, or even a sympathizer (beyond a very general, shared belief in human improvement through education and social progress). As I shall discuss in detail later in this essay, Marx first sought Lankester's advice in recommending a doctor for his ailing wife and daughter, and later for himself. This professional connection evidently developed into a firm friendship. But what could have drawn these maximally disparate people together?

We certainly cannot seek the primary cause for warm sympathy in any radical cast to Lankester's biological work that might have matched the tenor of Marx's efforts in political science. Lankester may rank as the best evolutionary morphologist in the first generation that worked through the implications of Darwin's epochal discovery. T. H. Huxley became Lankester's guide and mentor, while Darwin certainly thought well of his research, writing to Lankester (then a young man of twenty-five) on April 15, 1872: "What grand work you did at Naples! [at the marine research station]. I can clearly see that you will some day become our first star in Natural History." But Lankester's studies now read as little more than an exemplification and application of Darwin's insights to several specific groups of organisms—a "filling in" that often follows a great theoretical advance, and that seems, in retrospect, not overly blessed with originality.

As his most enduring contribution, Lankester proved that the ecologically diverse spiders, scorpions, and horseshoe crabs form a coherent evolutionary group, now called the Chelicerata, within the arthropod phylum. Lankester's

Karl Marx's grave site in Highgate Cemetery, London.

research ranged widely from protozoans to mammals. He systematized the terminology and evolutionary understanding of embryology, and he wrote an important paper on "degeneration," showing that Darwin's mechanism of natural selection led only to local adaptation, not to general progress, and that such immediate improvement will often be gained (in many parasites, for example) by morphological simplification and loss of organs.

In a fair and generous spirit, one might say that Lankester experienced the misfortune of residing in an "in between" generation that had imbibed Darwin's insights for reformulating biology, but did not yet possess the primary tool—an understanding of the mechanism of inheritance—so vitally needed for the next great theoretical step. But then, people make their own opportunities, and Lankester, already in his grumpily conservative maturity, professed little use for Mendel's insights upon their rediscovery at the outset of the twentieth century.

In the first biography ever published—the document that finally provided me with enough information to write this essay after a gestation period of twenty-five years!—Joseph Lester, with editing and additional material by Peter Bowler, assessed Lankester's career in a fair and judicious way (*E. Ray Lankester and the Making of British Biology,* British Society for the History of Science, 1995):

> Evolutionary morphology was one of the great scientific enterprises of the late nineteenth century. By transmuting the experiences gained by their predecessors in the light of the theory of evolution, morphologists such as Lankester threw new light on the nature of organic structures and created an overview of the evolutionary relationships that might exist between different forms. . . . Lankester gained an international reputation as a biologist, but his name is largely forgotten today. He came onto the scene just too late to be involved in the great Darwinian debate, and his creative period was over before the great revolutions of the early twentieth century associated with the advent of Mendelian genetics. He belonged to a generation whose work has been largely dismissed as derivative, a mere filling in of the basic details of how life evolved.

Lankester's conservative stance deepened with the passing years, thus increasing the anomaly of his early friendship with Karl Marx. His imposing figure only enhanced his aura of staid respectability (Lankester stood well over six feet tall, and he became quite stout, in the manner then favored by men of high station). He spent his years of retirement writing popular articles on natural history for newspapers, and collecting them into several successful volumes. But few of these pieces hold up well today, for his writing lacked both the spark and the depth of the great British essayists in natural history: T. H. Huxley, J. B. S. Haldane, J. S. Huxley, and P. B. Medawar.

As the years wore on, Lankester became ever more stuffy and isolated in his elitist attitudes and fealty to a romanticized vision of a more gracious past. He opposed the vote for women, and became increasingly wary of democracy and mass action. He wrote in 1900: "Germany did not acquire its admirable educational system by popular demand . . . the crowd cannot guide itself, cannot help itself in its blind impotence." He excoriated all "modern" trends in the arts, especially cubism in painting and self-expression (rather than old-fashioned storytelling) in literature. He wrote to his friend H. G. Wells in 1919: "The rubbish and self-satisfied bosh which pours out now in magazines and novels is astonishing. The authors are so set upon being 'clever,' 'analytical,' and 'up-to-date,' and are really mere prattling infants."

As a senior statesman of science, Lankester kept his earlier relationship with Marx safely hidden. He confessed to his friend and near contemporary A. Conan Doyle (who had modeled the character of Professor Challenger in *The Lost World* upon Lankester), but he never told the young communist J. B. S. Haldane, whom he befriended late in life and admired greatly, that he had known Karl Marx. When, upon the fiftieth anniversary of the Highgate burial, the Marx-Engels Institute of Moscow tried to obtain reminiscences from all people who had known Karl Marx, Lankester, by then the only living witness of Marx's funeral, replied curtly that he had no letters and would offer no personal comments.

Needless to say, neither the fate of the world nor the continued progress of evolutionary biology depends in the slightest perceptible degree upon a resolution of this strange affinity between two such different people. But little puzzles gnaw at the soul of any scholar, and answers to small problems sometimes lead to larger insights rooted in the principles utilized for explanation. I believe that I have developed a solution, satisfactory (at least) for the dissolution of my own former puzzlement. But, surprisingly to me, I learned no decisive fact from the literature that finally gave me enough information to write this essay—the recent Lankester biography mentioned above, and two excellent articles on the relationship of Marx and Lankester: "The friendship of Edwin Ray Lankester and Karl Marx," by Lewis S. Feuer (*Journal of the History of Ideas* 40 [1979]: 633–48), and "Marx's Darwinism: a historical note," by Diane B. Paul (*Socialist Review* 13 [1983]: 113–20). Rather, my proposed solution invokes a principle that may seem disappointing and entirely uninteresting at first, but that may embody a generality worth discussing, particularly for the analysis of historical sequences—a common form of inquiry in both human biography and evolutionary biology. In short, I finally realized that I had been *asking the wrong question* all along.

A conventional solution would try to dissolve the anomaly by arguing that Marx and Lankester shared far more similarity in belief or personality than appearances would indicate, or at least that each man hoped to gain something direct and practical from the relationship. But I do not think that this ordinary form of argument can possibly prevail in this case.

To be sure, Lankester maintained a highly complex and, in some important ways, almost secretive personality beneath his aura of Establishment respectability. But he displayed no tendencies at all to radicalism in politics, and he surely included no Marxist phase in what he might later have regarded as the folly of youth. But Lankester did manifest a fierce independence of spirit, a kind of dumb courage in the great individualistic British tradition of "I'll do as I see fit, and bugger you or the consequences"—an attitude that inevitably attracted all manner of personal trouble, but that also might have led Lankester to seek interesting friendships that more timid or opportunistic colleagues would have shunned.

Despite his basically conservative views in matters of biological theory, Lankester was a scrappy fighter by nature, an indomitable contrarian who relished professional debate, and never shunned acrimonious controversy. In a remarkable letter, his mentor T. H. Huxley, perhaps the most famous contrarian in the history of British biology, warned his protégé about the dangers of sapping time and strength in unnecessary conflict, particularly in the calmer times that had descended after the triumph of Darwin's revolution. Huxley wrote to Lankester on December 6, 1888:

> Seriously, I wish you would let an old man, who has had his share of fighting, remind you that battles, like hypotheses, are not to be multiplied beyond necessary. . . . You have a fair expectation of ripe vigor for twenty years; just think what may be done with that capital. No use to *tu quoque* me ["thou also"—that is, you did it yourself]. Under the circumstances of the time, warfare has been my business and duty.

To cite the two most public examples of his scrappy defense of science and skepticism, Lankester unmasked the American medium Henry Slade in

September 1876. Slade specialized in séances (at high fees), featuring spirits that wrote messages on a slate. Lankester, recognizing Slade's *modus operandi,* grabbed the slate from the medium's hands just before the spirits should have begun their ghostly composition. The slate already contained the messages supposedly set for later transmission from a higher realm of being. Lankester then sued Slade for conspiracy, but a magistrate found the medium guilty of the lesser charge of vagrancy, and sentenced him to three months at hard labor. Slade appealed and won on a technicality. The dogged Lankester then filed a new summons, but Slade decided to pack up and return to a more gullible America. (As an interesting footnote in the history of evolutionary biology, the spiritualistically inclined Alfred Russel Wallace testified on Slade's behalf, while Darwin, on the opposite side of rational skepticism, quietly contributed funds for Lankester's efforts in prosecution.)

Three years later, in the summer of 1879, Lankester visited the laboratory of the great French physician and neurologist Jean-Martin Charcot. To test his theories on the role of electricity and magnetism in anesthesia, Charcot induced insensitivity by telling a patient to hold an electromagnet, energized by a bichromate battery, in her hand. Charcot then thrust large carpet needles into her affected arm and hand, apparently without causing any pain.

The skeptical Lankester, no doubt remembering the similar and fallacious procedures of Mesmer a century before, suspected psychological suggestion, rather than any physical effect of magnetism, as the cause of anesthesia. When Charcot left the room, Lankester surreptitiously emptied the chemicals out of the battery and replaced the fluid with ordinary water, thus disabling the device. He then urged Charcot to repeat the experiment—with the same result of full anesthesia! Lankester promptly confessed what he had done, and fully expected to be booted out of Charcot's lab *tout de suite.* But the great French scientist grabbed his hand and exclaimed, "Well done, Monsieur," and a close friendship then developed between the two men.

One additional, and more conjectural, matter must be aired as we try to grasp the extent of Lankester's personal unconventionalities (despite his conservative stance in questions of biological theory) for potential insight into his willingness to ignore the social norms of his time. The existing literature maintains a wall of total silence on this issue, but the pattern seems unmistakable. Lankester remained a bachelor, although he often wrote about his loneliness and his desires for family life. He was twice slated for marriage, but both fiancées broke their engagements for mysterious and unstated reasons. He took long European vacations nearly every year, and nearly always to Paris, where he maintained clear distance from his professional colleagues. Late in life,

The famous Vanity Fair *caricature of E. Ray Lankester, drawn by Spy.*

Lankester became an intimate platonic friend and admirer of the great ballerina Anna Pavlova. I can offer no proof, but if these behaviors don't point toward the love that may now be freely discussed, but then dared not speak its name (to paraphrase the one great line written by Oscar Wilde's paramour, Lord Alfred Douglas), well, then, Professor Lankester was far more mysterious and secretive than even I can imagine.

Still, none of these factors, while they may underscore Lankester's general willingness to engage in contentious and unconventional behavior, can explain any special propensity for friendship with a man like Karl Marx. (In particular, orthodox Marxists have always taken a dim view of personal, particularly sexual, idiosyncrasy as a self-centered diversion from the social goal of revolution.) Lankester did rail against the social conservatives of his day, particularly against hidebound preachers who opposed evolution, and university professors who demanded the standard curriculum of Latin and Greek in preference to any newfangled study of natural science.

But Lankester's reforming spirit centered only upon the advance of science—and his social attitudes, insofar as he discussed such issues at all, never transcended the vague argument that increasing scientific knowledge might liberate the human spirit, thus leading to political reform and equality of opportunity. Again, this common attitude of rational scientific skepticism only evoked the disdain of orthodox Marxists, who viewed this position as a bourgeois escape for decent-minded people who lacked the courage to grapple with the true depth of social problems, and the consequent need for political revolution. As Feuer states in his article on Marx and Lankester: "Philosophically, moreover, Lankester stood firmly among the agnostics, the followers of Thomas Henry Huxley, whose standpoint Engels derided as a 'shamefaced materialism.'"

If Lankester showed so little affinity for Marx's worldview, perhaps we should try the opposite route and ask if Marx had any intellectual or philosophical reason to seek Lankester's company. Again, after debunking some persistent mythology, we can find no evident basis for their friendship.

The mythology centers upon a notorious, if understandable, scholarly error that once suggested far more affinity between Marx and Darwin (or at least a one-way hero-worshiping of Darwin by Marx) than corrected evidence can validate. Marx did admire Darwin, and he did send an autographed copy of *Das Kapital* to the great naturalist. Darwin, in the only recorded contact between the two men, sent a short, polite, and basically contentless letter of thanks. We do know that Darwin (who read German poorly and professed little interest in political science) never spent much time with Marx's *magnum opus.* All but the first 105 pages in Darwin's copy of Marx's 822-page book remain uncut (as does the table of contents), and Darwin, contrary to his custom when reading books carefully, made no marginal annotations. In fact, we have no evidence that Darwin ever read a word of *Das Kapital.*

The legend of greater contact began with one of the few errors ever made by one of the finest scholars of this, or any other, century—Isaiah Berlin in his 1939 biography of Marx. Based on a dubious inference from Darwin's short letter of thanks to Marx, Berlin inferred that Marx had offered to dedicate volume two of *Das Kapital* to Darwin, and that Darwin had politely refused. This tale of Marx's proffered dedication then gained credence when a second letter, ostensibly from Darwin to Marx, but addressed only to "Dear Sir," turned up among Marx's papers in the International Institute of Social History in Amsterdam. This letter, written on October 13, 1880, does politely decline a suggested dedication: "I Shd. prefer the Part or Volume not be dedicated to me (though I thank you for the intended honor) as it implies to a certain extent my approval of the general publication, about which I know nothing." This second document seemed to seal Isaiah Berlin's case, and the story achieved general currency.

To shorten a long story, two scholars, working independently and simultaneously in the mid-1970s, discovered the almost comical basis of the error—see Margaret A. Fay, "Did Marx offer to dedicate *Capital* to Darwin" (*Journal of the History of Ideas* 39 [1978]: 133–46); and Lewis S. Feuer, "Is the 'Darwin-Marx correspondence' authentic?" (*Annals of Science* 32: 1–12). Marx's daughter Eleanor became the common-law wife of the British socialist Edward Aveling. The couple safeguarded Marx's papers for several years, and the 1880 letter, evidently sent by Darwin to Aveling himself, must have strayed into the Marxian collection.

Aveling belonged to a group of radical atheists. He sought Darwin's official approval, and status as dedicatee, for a volume he had edited on Darwin's work and his (that is Aveling's, not necessarily Darwin's) view of its broader social meaning (published in 1881 as *The Student's Darwin,* volume two in the International Library of Science and Freethought). Darwin, who understood Aveling's opportunism and cared little for his antireligious militancy, refused with his customary politeness, but with no lack of firmness. Darwin ended his letter to Aveling (not to Marx, who did not treat religion as a primary subject in *Das Kapital*) by writing:

> It appears to me (whether rightly or wrongly) that direct arguments against christianity and theism produce hardly any effect on the public; and freedom of thought is best promoted by the gradual illumination of men's minds which follows from the advance of science. It has, therefore, been always my object to avoid writing on religion, and I have confined myself to science.

Nonetheless, despite this correction, Marx might still have regarded himself as a disciple of Darwin, and might have sought the company of a key Darwinian in the younger generation—a position rendered more plausible by Engels's famous comparison (quoted earlier) in his funerary oration. But this interpretation must also be rejected. Engels maintained far more interest in the natural sciences than did Marx (as best expressed in two books by Engels, *Anti-Dühring* and *The Dialectics of Nature*). Marx, as stated above, certainly admired Darwin as a liberator of knowledge from social prejudice, and as a useful ally, at least by analogy. In a famous letter of 1869, Marx wrote to Engels about Darwin's *Origin of Species:* "Although it is developed in the crude English style, this is the book which contains the basis in natural history for our view."

But Marx also criticized the social biases in Darwin's formulation, again writing to Engels, and with keen insight:

> It is remarkable how Darwin recognizes among beasts and plants his English society with its division of labor, competition, opening up of new markets, "invention," and the Malthusian "struggle for existence." It is Hobbes's *bellum omnium contra omnes* [the war of all against all].

Marx remained a committed evolutionist, of course, but his interest in Darwin clearly diminished through the years. An extensive scholarly literature treats this subject, and I think that Margaret Fay speaks for a consensus when she writes (in her article previously cited):

> Marx . . . though he was initially excited by the publication of Darwin's *Origin* . . . developed a much more critical stance towards Darwinism, and in his private correspondence of the 1860s poked gentle fun at Darwin's ideological biases. Marx's Ethnological Notebooks, compiled circa 1879–1881, in which Darwin is cited only once, provide no evidence that he reverted to his earlier enthusiasm.

To cite one final anecdote, the scholarly literature frequently cites Marx's great enthusiasm (until the more scientifically savvy Engels set him straight) for a curious book published in 1865 by the now (and deservedly) unknown French explorer and ethnologist P. Trémaux, *Origine et transformations de l'homme et des autres êtres* (Origin and Transformation of Man and Other Beings). Marx professed ardent admiration for this work, proclaiming it *einen Fortschritt über Darwin* (an advance over Darwin). The more sober Engels bought the book at Marx's urging, but then dampened his friend's ardor by writing: "I have arrived at the conclusion that there is nothing to his theory if for no other reason than because he neither understands geology nor is capable of the most ordinary literary historical criticism."

I had long been curious about Trémaux and sought a copy of his book for many years. I finally purchased one a while ago—and I must say that I have never read a more absurd or more poorly documented thesis. Basically, Trémaux argues that the nature of the soil determines national characteristics, and that higher civilizations tend to arise on more-complex soils formed in later geological periods. If Marx really believed that such unsupported nonsense could exceed the *Origin of Species* in importance, then he could not have properly understood or appreciated the power of Darwin's facts and ideas.

We must therefore conclude that Lankester harbored no secret sympathy for Marxism, and that Marx sought no Darwinian inspiration in courting Lankester's friendship. Our puzzle only deepens: What brought these disparate men together; what kind of bond could have nurtured their friendship? The first question, at least, can be answered, and may even suggest a route toward resolving the second, and central, conundrum of this essay.

Four short letters from Lankester remain among Marx's papers. (Marx probably wrote to Lankester as well, but no evidence of such reciprocity has surfaced.) These letters clearly indicate that Marx first approached Lankester for medical advice in the treatment of his wife, who was dying, slowly and painfully, of breast cancer. Lankester suggested that Marx consult his dear friend (and co-conspirator in both the Slade and Charcot incidents), the physician H. B. Donkin. Marx took Lankester's advice, and proclaimed himself well satisfied with the result, as Donkin, whom Marx described as "a bright and intelligent man," cared, with great sensitivity, both for Marx's wife and then for Marx himself in their final illnesses.

We do not know for sure how Marx and Lankester first met, but Feuer develops an eminently plausible hypothesis in his article cited previously—one, moreover, that may finally lead us to understand the basis of this maximally incongruous pairing. The intermediary may well have been Charles Waldstein, born in New York in 1856, the son of a German Jewish immigrant. Waldstein, who later served as professor of classical archaeology at Cambridge, knew Lankester well when they both lived in London during the late 1870s. Waldstein became an intimate friend of Karl Marx, an experience that he remembered warmly in an autobiographical work written in 1917 (when he had attained eminence and respectability under the slightly, but portentously, altered name of Sir Charles Walston):

> In my young days, when I was little more than a boy, about 1877, the eminent Russian legal and political writer . . . Professor Kovalevsky, whom I had met at one of G. H. Lewes and George Eliot's Sunday afternoon parties in London, had introduced me to Karl Marx, then living in Hampstead. I had seen very much of this founder of modern theoretic socialism, as well as of his most refined wife; and, though he had never succeeded in persuading me to adopt socialist views, we often discussed the most varied topics of politics, science, literature, and art. Besides learning much from this great man, who was a mine of deep and accurate knowledge in every sphere, I learnt to hold him in high respect and to love the purity, gentleness, and refinement of his big heart. He

> seemed to find so much pleasure in the mere freshness of my youthful enthusiasm and took so great an interest in my own life and welfare, that one day he proposed that we should become *Dutz-freunde.*

The last comment is particularly revealing. Modern English has lost its previous distinction (*thou* versus *you*) between intimate and formal address, a difference that remains crucially important—a matter not to be taken lightly—in most European languages. In German, *Dutz-freunde* address each other with the intimate *Du,* rather than the formal *Sie* (just as the verb *tutoyer,* in French, means to use the intimate *tu* rather than the formal *vous*). In both nations, especially in the far more conservative social modes of nineteenth-century life, permission to switch from formal to intimate address marked a rare and precious privilege reserved only for one's family, one's God, one's pets, and one's absolutely dearest friends. If an older and established intellectual like Marx suggested such a change of address to a young man in his early twenties, he must have felt especially close to Charles Waldstein.

Lankester's first letter to Marx, written on September 19, 1880, mentions Waldstein, thus supporting Feuer's conjecture: "I shall be very glad to see you at Wellington Mansions. I had been intending to return to you the book you kindly lent to me—but had mislaid your address and could not hear from Waldstein who is away from England." Lankester and Waldstein remained close friends throughout their lives. Waldstein's son responded to Feuer's inquiry about his father's relationship with Lankester by writing, in 1978, that he retained a clear childhood memory of "Ray Lankester . . . coming to dinner from time to time at my home—a very fat man with a face like a frog."

Waldstein's memories of Marx as a kind man and a brilliant intellectual mentor suggest an evident solution to the enigma of Marx and Lankester—once we recognize that we had been asking the wrong question all along. No error of historical inquiry can match the anachronistic fallacy of using a known present to misread a past circumstance that could not possibly have been defined or influenced by events yet to happen. When we ask why a basically conservative biologist like Lankester could have respected and valued the company of an aging agitator like Karl Marx, we can hardly help viewing Marx through the glasses of later human catastrophes perpetrated in his name—from Stalin to Pol Pot. Even if we choose to blame Marx, in part, for not foreseeing these possible consequences of his own doctrines, we must still allow that when he died in 1883, these tragedies only resided in an unknowable future. Karl Marx, the man who met Lankester in 1880, must not be confused with Karl

Marx, the posthumous standard-bearer for some of the worst crimes in human history. We err when we pose E. Ray Lankester, the stout and imposing relic of Victorian and Edwardian biology, with Karl Marx, cited as the rationale for Stalin's murderous career—and then wonder how two such different men could inhabit the same room, much less feel warm ties of friendship.

In 1880, Lankester was a young biologist with a broad view of life and intellect, and an independent mind that cared not a fig for conventional notions of political respectability, whatever his own basically conservative convictions. Showing a rare range of interest among professional scientists, he also loved art and literature, and had developed fluency in both German and French. Moreover, he particularly admired the German system of university education, then a proud model of innovation, especially in contrast with the hidebound classicism of Oxford and Cambridge, so often the object of his greatest scorn and frustration.

Why should Lankester not have enjoyed, even cherished, the attention of such a remarkable intellect as Karl Marx—for that he was, whatever you may think of his doctrines and their consequences? What could possibly have delighted Lankester more than the friendship of such a brilliant older man, who knew art, philosophy, and the classics so well, and who represented the epitome of German intellectual excellence, the object of Lankester's highest admiration? As for the ill, aging, and severely depressed Karl Marx, what could bring more solace in the shadow of death than the company of bright, enthusiastic, optimistic young men in the flower of their intellectual development?

Waldstein's memories clearly capture, in an evocative and moving way, this aspect of Marx's persona and final days. Many scholars have emphasized this feature of Marx's later life. Diane Paul, for example, states that "Marx had a number of much younger friends. . . . The aging Marx became increasingly difficult in his personal relationships, easily offended and irritated by the behavior of old friends, but he was a gracious mentor to younger colleagues who sought his advice and support." Seen in this appropriate light of their own time, and not with anachronistic distortion of later events that we can't escape but they couldn't know, Marx and Lankester seem ideally suited, indeed almost destined, for the warm friendship that actually developed.

All historical studies—whether of human biography or of evolutionary lineages in biology—potentially suffer from this "presentist" fallacy. Modern chroniclers know the outcomes that actually unfolded as unpredictable consequences of past events—and they often, and inappropriately, judge the motives and actions of their subjects in terms of futures unknowable at the time. Thus, and far too frequently, evolutionists view a small and marginal lineage of pond-dwelling Devonian fishes as higher in the scale of being and destined for success because we know, but only in retrospect, that these organisms spawned all modern terrestrial vertebrates, including our exalted selves. And we overly honor a peculiar species of African primates as central to the forward thrust of evolution because our unique brand of consciousness arose, by contingent good fortune, from such a precarious stock. And as we northerners once reviled Robert E. Lee as a traitor, we now tend to view him, in a more distant and benevolent light, as a man of principle and a great military leader—though neither extreme position can match or explain this fascinating man in the more appropriate context of his own time.

A little humility before the luck of our present circumstances might serve us well. A little more fascination for past realities, freed from judgment by later outcomes that only we can know, might help us to understand our history, the primary source for our present condition. Perhaps we might borrow a famous line from a broken man, who died in sorrow, still a stranger in a strange land, in 1883—but who at least enjoyed the solace of young companions like E. Ray Lankester, a loyal friend who did not shun the funeral of such an unpopular and rejected expatriate.

History reveals patterns and regularities that enhance our potential for understanding. But history also expresses the unpredictable foibles of human passion, ignorance, and dreams of transcendence. We can only understand the meaning of past events in their own terms and circumstances, however legitimately we may choose to judge the motives and intentions of our forebears. Karl Marx began his most famous historical treatise, his study of Napoleon III's rise to power, by writing, "Men make their own history, but they do not make it just as they please."

7

The Pre-Adamite in a Nutshell

WINSTON CHURCHILL FAMOUSLY DESCRIBED THE SOVIET Union as "a riddle wrapped in a mystery inside an enigma." This essay, impoverished by contrast, features only two levels of puzzlement—the tale of an anonymous author defending an odd theory that only becomes, in Alice's immortal words, curiouser and curiouser as one reads. However, in a fractal universe, a single mote can mirror the cosmos, giving literal meaning to Blake's famous image of the "world in a grain of sand, and a heaven in a wild flower." Forgotten documents that now seem ridiculous can offer us maximal instruction in human foibles and in the history of our attempts to make sense of a complex natural world—the enterprise that we call "science."

In his important book on the development of conventions for illustrating extinct faunas of the geological past, *Scenes from Deep*

Time (University of Chicago Press, 1992), the British historian of geology (and former paleontologist) M. J. S. Rudwick reproduced a figure from 1860 that, in his words, "broke the standard mold by suggesting a sequence in deep time." Previously, most authors had presented only one or two reconstructions of particular past moments or intervals—with Mesozoic dinosaurs and large Cenozoic mammals already emerging as "industry standards."

Few charts, however, had attempted to depict the flow of change through successive faunas. But this elaborate version from 1860, inserted as a large foldout (eight by fourteen inches in my copy) into an ordinary octavo book, shows the history of life in three layers, with dinosaurs and their allies on the bottom, large mammals in the middle (including the nearly obligatory giant ground sloth, mammoth, and Irish elk), and modern creatures in a top layer that also affirms a human presence by including the great pyramids of Egypt in the upper right.

Although authors usually describe such pictures as supposedly unbiased factual summaries, nearly all complex representations must, whether consciously or not, express favored theories about patterns and causes for the history of life. This author, at least, makes no bones (bad pun) about the didactic character of her chart. In particular, she depicts the first two layers as a continuous development, even though the basic faunas of the two stages differ so profoundly. In a dramatic middle location, to reinforce this central claim, *Iguanodon* (then falsely depicted in a crocodilian pose, but now recognized as a bipedal duck-billed dinosaur) slithers up a slope connecting the two layers.

But the top layer of modern life has been completely severed from all that came before by an intervening lifeless interval, depicted as a forbidding, ice-covered world. The original chart intensifies the theme by printing this lifeless layer in stark white, thus establishing a marked contrast with the three faunas, all overprinted in a soothing yellowish pink. Obviously the author believed that life's history includes two distinct phases: a long older period of continuity and at least occasional change, abruptly terminated by a cold and lifeless world and then followed by the much shorter interval of modern organisms.

To state the first conundrum: The book containing this plate appeared in 1860 under anonymous authorship, and with the title *Pre-Adamite Man: The Story of Our Old Planet and Its Inhabitants.* I have never before been so unsuccessful in searching through the literature, but I have found absolutely nothing about the life or other works of the author beyond her name: Isabelle Duncan. I have frequently written about the sadness and anger of women naturalists in these Victorian times. They often published anonymously in highly restricted formats of sentimentalized effusions and versifications for children or dilet-

A highly original history of life shown in three successive scenes, bottom to top: dinosaurs, large mammals of the Ice Age, and modern creatures. By Isabelle Duncan, 1860.

---

tantes, although several of these women developed fully professionalized skills and longed for equal participation with male colleagues. But one can usually find *some* documentation in standard bibliographic listings, or in the scholarship of modern feminist historians, who have taken the rehabilitation, or at least the identification, of these forgotten women as a mission and solemn duty.

Nonetheless, I have located nary an informative word about Isabelle Duncan. I may well have missed something, and I would be grateful for any assistance from readers (whose commentaries, expansions, and corrections have given me so much pleasure, and granted me so much enlightenment, over the years).* I can at least say that other scholars who devoted far more time to the search came up equally empty-handed. Rudwick himself simply writes, "The author is now obscure, but in its time the book was evidently not: it had reached a fourth edition within a couple of years." The most important scholarly work on the Pre-Adamite theory, R. H. Popkin's biography of the movement's founder (*Isaac La Peyrère (1596–1676): His Life, Work and Influence* [E. J. Brill, 1987]), gives a short summary of the contents, but only refers to the author as "one Isabella Duncan"—the conventional way to say "I know her name, but otherwise zilch, zippo, nada, goose eggs, and bugger-all about her."

Given the barely post-Darwinian date of Duncan's book (*The Origin of Species* had appeared the year before, in 1859), one might assume that Duncan

---

*And again, the universal company and republic of intellectuals did not fail me. Although Isabelle Duncan has not been mentioned in print (except as the otherwise anonymous author of *Pre-Adamite Man*) since 1915, and never in more than a cursory line or two, several scholars knew her identity from two sources: from letters of Jane Carlyle, wife of Thomas Carlyle, who both admired the book and knew the Duncan family, and through her more famous father-in-law, Henry Duncan (1774–1846), a Scottish minister and social reformer, now best remembered as "the father of savings banks" (at least according to the Savings Banks Museum in Ruthwell, Scotland.) But I then hit pay dirt when Stephen D. Snobelen, a young historian of science at Cambridge University, sent me his excellent dissertation on Isabelle Duncan, her evangelical religious views, her commitment to the concordist tradition with scientific findings and her remarkable book, which went through six editions between 1860 and 1866 (probably with a total press run of some six thousand copies—a very respectable sale for the times). Snobelen's superb detective work in this former terra incognita (including a location of her portrait and the correct spelling of her name—all previous sources had called her Isabella rather than her preferred Isabelle) has since been published as a first scholarly account of this remarkable woman's life and identity: "Of stones, men and angels: the competing myth of Isabelle Duncan's *Pre-Adamite Man* (1860)," *Studies in the History and Philosophy of Biology and the Biomedical Sciences* 32 (2001): 59–104.

wrote to introduce lay readers to new discoveries in geology and anthropology, and to prepare them for assessing (whether pro or con) the onslaught of Darwin's revisions to traditional views about human history. But her motivation arose from an entirely different, and now nearly invisible, source—the subject of our second riddle: an old theory of biblical exegesis that can be traced to some of the early church fathers, but that arose in explicit form amid the millenarian movements of the mid–seventeenth century, and then kicked around (attracting at least the passing interest of such famous characters as Spinoza, Voltaire, Napoleon, and Goethe) until geological discoveries about the earth's great age, combined with anthropological findings of prehistoric human artifacts, pushed the subject into a renewed light of Darwinian debate.

The problem that inspired this so-called Pre-Adamite theory—the claim that humans existed before Adam, and that the man described in the first chapters of Genesis only denotes God's later creation of the Jews and other allied peoples—must have occurred to anyone who ever read the Bible with a critical eye. Several passages in Genesis, if taken at face value, seem to imply the existence of pre-Adamites. The subject may never come up in polite company, but if Adam and Eve mark a unique creation as a single pair, then whom did their son Cain marry? His unnamed sister, we must assume (at least if we wish to avoid the even more repugnant Oedipal alternative). But can we accept such incest at our very roots (although, admittedly, the story of Lot's daughters indicates some affinity for the theme within the Book of Genesis).

More explicitly, why did God need to place a mark upon Cain after he killed his brother Abel? God punishes Cain, who had been "a tiller of the ground," by ordaining that the soil, having imbibed his brother's blood, will never produce a crop for him again—and that he must therefore become "a fugitive and a vagabond . . . in the earth." But Cain begs for mitigation by pleading "every one that findeth me shall slay me." So God relents and places the famous mark upon him: "Therefore whosoever slayeth Cain, vengeance shall be taken on him sevenfold. And the Lord set a mark upon Cain, lest any finding him should kill him" (Genesis 4:15). But why did Cain need such a fancy ID, if no one else (except, perhaps, a few unnamed sibs) then inhabited the earth?

Moreover, what interpretation should we give to the two famously ambiguous passages of Genesis 6? First, the statement in verse 2 "that the sons of God saw the daughters of men . . . [and] took them wives." (Now, are the sons members of Adam's line and the daughters of some other stock, or vice versa? In either case, the comment seems to designate two distinct lineages, one perhaps pre-Adamic.) Second, the initial phrase of verse 4: "There were giants in the earth in those days." The Hebrew word *Nephilim* may be ambiguous, and King

James's translators may have goofed in imputing gigantic proportions, but one can easily read the comment as a reference to some kind of pre-Adamic stock.

At an opportune time in the midst of millennial anxieties sweeping Europe after the end of the Thirty Years War in 1648, Isaac La Peyrère, a French Protestant intellectual in the employ of the powerful Prince of Condé, joined these traditional doubts with a new exegetical argument to launch the pre-Adamite theory in 1655. His book, published in liberal Amsterdam and translated into English the next year (under the title *Men Before Adam*), created a goodly stir and caused its author a peck of trouble. La Peyrère was arrested and severely questioned. He finally accepted what we would call a plea bargain today—namely, that all would be forgiven if he converted to Catholicism, repented his pre-Adamite heresy, and apologized personally to the Pope, which he did (to Alexander VII) early in 1657.

The put-up nature of the job stands out in two tales, probably true according to Popkin, about La Peyrère's meeting with Alexander VII—that the General of the Jesuit order told La Peyrère how he and the Pope had "laughed delightedly" when they read the book, and that the Pope apparently said, upon meeting this "dangerous" heretic: "Let us embrace this man who is before Adam." In any case, La Peyrère never abandoned his theory, lived another twenty years, and died near Paris as a lay member of the Seminary of Oratorians. (In his later defenses of the pre-Adamite theory, La Peyrère followed the cautious route that Galileo had rejected thirty years before. He simply stated that the idea held great interest, made eminent sense of all the evidence, but could not be true because the Church had so decreed.)

La Peyrère proposed his theory from a millenarian and universalistic perspective. If God had created Adam as the progenitor of Jewish history, and if other races had previously existed, then the Jews must lead all people to a final redemption. La Peyrère focused upon the traditional Christian belief that a Jewish conversion would herald the blessed millennium. This time must now be at hand. The Jewish Messiah would soon appear and, in league with the King of France, return to Jerusalem in triumph, where both men would reign over a unified and fully Christian world. France, as a tolerant nation, must seek and welcome a Jewish influx, for if these favored people could be gathered together without constraint or persecution, then their Messiah would surely come.

Ironically, in stark contrast with La Peyrère's messianic view that all people, Adamite and pre-Adamite alike, would be equally redeemed, most subsequent (especially nineteenth-century) invocations of pre-Adamism used the doctrine to support racism—particularly the "polygenist" theory that each major human race arose as a separately created species, with Adam as the final progenitor of superior white folks (the latest with the mostest, so to speak), and several earlier Pre-Adams, one for each lower race (the firstest with the leastest). In other words, whites are Adamites; all other people are inferior pre-Adamites.

My interest in the history of pre-Adamism stems from its status as such a radically different way to treat a subject so central to the sciences—the origin and history of human diversity. In our secular age, we feel convinced (and rightly so, I would argue) that the empirical methods of science must be employed to answer such basically factual questions (while religion takes legitimate interest in entirely different spiritual questions about the meaning of life, and ethical inquiries about the proper conduct of life). But pre-Adamism represents, for someone trained in science, a "weirdly different" approach—one that should be called basically literary or hermeneutic rather than religious per se, even though analysis focuses upon a religious text, namely the Bible.

Pre-Adamite theorists formulated and justified their arguments by interpretation of scripture, rather than by appeal to factual information from the burgeoning sciences of anthropology and geology (although supporters of pre-Adamism also used the data, first of voyages of exploration leading to contact with diverse peoples throughout the world, and then of the fossil record and the discovery of deep time, to buttress their fundamentally textual thesis). I find the notion of such a parallel tradition fascinating, primarily for what it reveals about the diversity of human approaches—some ultimately fruitful, others doomed by false premises from the start—to difficult common problems. Pre-Adamism and science run on two parallel tracks, working by entirely different initial assumptions, methods of argument, and standards of proof. The two approaches also span roughly the same period, for La Peyrère's founding document appeared at the dawn of Newton's generation, the traditional origin for modern science as a dominant worldview; while the triumph of an evolutionary account by the end of the nineteenth century removed the underpinnings of exegetical pre-Adamism as a theory about actual datings and timings for the origin of various human groups.

Thus we must understand La Peyrère's theory not as a kooky exercise in biblical apologetics, but as a courageous and radical claim within the conventional theology of his day (in both Catholic and Protestant circles). For, by arguing that the Pentateuch—the Bible's first five books, traditionally attributed to Moses as sole and divinely inspired author—only described the local history of the Jewish people, and not the entire chronology of all humans, La Peyrère challenged a precept that almost no scholar had ever dared to question in public (though private doubts and thoughts had always been rife): the conviction that the Bible, as the inspired word of God, means exactly what it says. In so doing, La Peyrère helped to open the floodgates to a major theological movement that has swept the field of religious studies ever since, again in parallel with science (by using rational literary, rather than rational empirical, techniques): "higher criticism" and other exegetical approaches dedicated to interpreting the Bible as a fallible document, cobbled together from numerous sources of varying reliability, but subject to deeper understanding when all questions may be asked (and answers sought without fear), and no dogmas need be obeyed *a priori.*

We sense the difference and distance between these parallel roads of exegetical and scientific approaches to human prehistory when we describe the foundation of La Peyrère's argument—a claim that scientists might read as almost laughably arcane and irrelevant to any "real" inquiry about human origins, but that played an important role in this different hermeneutical tradition of literary interpretation. La Peyrère cited the usual arguments from Genesis, as noted above, but he centered his theory upon a novel, if peculiar, interpretation of a single passage in St. Paul's Epistle to the Romans (5:12–14):

> Wherefore, as by one man sin entered into the world, and death by sin; and so death passed upon all men, for that all have sinned. For until the law sin was in the world: but sin is not imputed when there is no law. Nevertheless, death reigned from Adam to Moses, even over them that had not sinned after the similitude of Adam's transgression. . . .

Now, here's the rub: traditional (and probably accurate) readings interpret "the law" as referring to Moses' receipt of God's word in the Pentateuch. The passage then states that although Adam had sinned, his sin could not be "officially" imputed to people until Moses received the divine word that specified the nature of Adam's transgression, and the price that all subsequent people must pay (the doctrine of original sin). Nonetheless, all pre-Mosaic people had to suffer death as a result of Adam's sin, even those who lived righteously and never disobeyed God in "the similitude of Adam's transgression"—and even though they had not yet received the law from Moses, and therefore could not fully understand why they must die.

As we say in the modern vernacular, "I got no problem with that"—but La Peyrère did, and for a definite (if entirely idiosyncratic) reason. La Peyrère insisted that Paul's invocation of "the law" referred to God's instructions to Adam, not to his dispensation to Moses. Now, if "sin was in the world" before the law, and Adam first received the law, and if only people could sin—well then, *ipso facto* and QED, people must have existed before Adam's creation. Most folks today, scientists and theologians alike, would consider this reading as a woefully thin foundation for such a radical chain of consequences. But so be it; *autres temps, autres moeurs* (other times, other customs), as they say in La Peyrère's country.

We need this background, and this concept of science and literary exegesis as parallel tracks in the exploration of human prehistory, to understand Isabelle Duncan's book and theory with any sympathy—for her ideas sound kooky even beyond La Peyrère's vision about the King of France all buddy-buddy with a Jewish Messiah in a blissful world to come. Yet, when we understand that her argument emerges not from science, but from a tradition of biblical exegesis trying to harmonize itself with science, then her mode of reasoning becomes clearer (even though her particular claims don't, and can't, improve in plausibility). Rudwick states this important point in writing of Duncan's book: "Such a theory may now seem bizarre, but it belongs . . . to a flourishing Anglo-American subculture of biblically based cosmological speculations, often with powerful social and racial implications."

Duncan, facing the new Darwinian world of 1860, should not be compared too strictly with La Peyrère, plunged into the millenarian fervors of mid-seventeenth-century Europe. But both do belong to a biblical tradition larger than their own particular pre-Adamite conclusions—a reconciliationist rather than a contrarian approach to science and other secular studies based on empirical evidence. Reconciliationists accept the Bible as the truthful and inspired word of God, but also insist that the factual discoveries of science must be respected. Since truth and inspiration don't require a literal reading, the biblical text may be interpreted, but never denied or controverted, to harmonize with scientific conclusions. Contrarians, with American "young earth creationism" as the most prominent modern version, simply know what the Bible says. If science disagrees, then science must be wrong. Case closed.

The classic problem for reconciliationists interested in the emerging sciences of geology and paleontology has always centered upon the apparent claim in Genesis 1 that God created both the cosmos and all living creatures in a sequence of six days, and that the earth, as inferred from biblical chronologies of the patriarchs and kings, cannot be much more than five or six thousand years old. Many long books have been written on this complex subject, but a simplified summary might identify three major traditions in reconciliationist arguments on this crucial topic.

First, the "gap" theory argues that Genesis must be read literally, but that an unspecified amount of time—sufficient to fit anything that geology might discover about the earth's age—intervened between verse 1 ("In the beginning God created the heaven and the earth"), and the particular descriptions that begin with verse 2.

Second, the "day-age" theory argues that Genesis 1 got the sequence right, but that the Hebrew word *yom,* translated as "day" in the King James version, may refer to intervals of unspecified duration. So each biblical "day" may represent as long a period of time as geological discovery requires. Third, the "only local" theory holds that Genesis intends to describe only the particular origin of the Jewish people in the Near East, not the full chronology of all earthly time. Adam's children therefore marry the offspring of earlier people from different regions, and Noah's flood may be read as a local inundation, therefore avoiding such crushing problems as whether the progenitors of all living species could fit into one boat. Nearly all pre-Adamite theories belong primarily to this third tradition.

We must grant Isabelle Duncan at least one general point, however harshly we judge the quality of her particular argument, either in the context of her own times or ours: she developed a novel version of conciliationist pre-Adamism by following La Peyrère's idiosyncratic procedure of building a case upon a single biblical text—in her version a much more sensible analysis of Genesis 1 and 2 than La Peyrère ever applied to Romans 5. Almost all previous versions of pre-Adamism had invoked the theory to explain our *current* racial diversity, usually to the detriment of people outside the European cultural context of the theory itself. But Duncan employed the literary and exegetical traditions of pre-Adamism to explain the geological antiquity of humans on earth, while affirming the unity of all living people by descent from a single and recent Adam.

In short, Duncan argued for two entirely distinct and separate creations, both featuring humans. God created pre-Adamites near the end of the first creation; but he then destroyed all life before unleashing a second creation, this time beginning with Adam, the progenitor of all living humans. Thus, pre-Adamites left human artifacts in late geological sediments, but all modern humans are Adamites of the second creation.

La Peyrère's reading of Romans could claim no basis beyond his personal idiosyncrasy. But Duncan's interpretation of the first two chapters of Genesis—a strikingly novel analysis within the history of pre-Adamite thought—represented a false solution to a genuine insight. I have often been amazed at how few people, including creationists who swear that the Bible must be read literally, even remember that the creation stories of Genesis 1 and 2 tell entirely different tales, when read at face value. Genesis 1 presents the traditional sequence of creation in six days, proceeding from the earth itself, to light, to plants, to the sun and moon, to animals in a "rising" series from fish to mammals, and, finally on the sixth day, to human beings—with male and female created together: "So God created man in his own image, in the image of God created he him; male and female created he them" (Genesis 1:27).

But the tale of Genesis 2 could hardly be more different. God makes Adam at the outset, a single male on a lifeless planet: "And the Lord God formed man of the dust of the ground, and breathed into his nostrils the breath of life; and man became a living soul" (Genesis 2:7). God then places Adam in the Garden of Eden, and subsequently creates plants, and then animals, to assuage the isolation of his first creature: "It is not good that the man should be alone" (2:18). God then brings all the animals before Adam, giving his first man the privilege of assigning their names.

But Adam remains lonely, so God makes "an help meet for him" (2:20) from one of his ribs: "And the Lord God . . . took one of his ribs, and closed up the flesh instead thereof; And the rib, which the Lord God had taken from man, made he a woman, and brought her unto the man. And Adam said, This is now bone of my bones, and flesh of my flesh; she shall be called Woman" (2:21–23).

I suspect that we generally forget these striking differences because we cobble the two stories together into the combined vernacular version that pleases us most. We borrow the six-day sequence from Genesis 1, but we love the story of Eve's manufacture from Adam's rib, and of the initial situation in Eden—so we graft these "plot devices" from Genesis 2 onto the different resolution of Genesis 1 (simultaneous creation of male and female).

No scholarly debate or serious theological objection now attends the obvious and well-documented explanation of these discrepancies. The two stories differ because they derive from two prominent texts among the many separate sources that ancient compilers used to construct the Bible. Modern critics call these texts the "E" and "J" documents to note the different designations of God in the two sources—Elohim versus Yahweh, with the latter title conventionally transliterated as Jehovah in European Christian traditions. (Written Hebrew uses no explicit vowels, so early Christians had to make inferences from the tetragrammaton, or four-lettered sequence, of the Hebrew text: YHWH. Since the alphabet of Latin, the common tongue of early Christian writers, includes neither *Y* nor *W,* the necessary substitutions, plus the inferred vowels, yielded Jehovah.) The Pentateuch cannot, therefore, represent Moses' unique composition as dictated directly by God. The contradictions within Genesis and other books arise from the amalgamation of inevitably different texts. No religious belief should be threatened thereby. The Bible is not, in any case, a factual treatise about natural history.

But Isabelle Duncan did not work within this scholarly tradition. In her conventional piety, she stuck rigidly to the old belief in an inerrant and coherent text, subject to interpretation of course, but necessarily true at face value. In her conciliationist respect for new discoveries of science, she also believed that this inerrant text, when properly read, could not contradict any genuine empirical discovery. Her unique version of pre-Adamism arose from these twinned convictions.

The two creation stories, she acknowledged, must be read as genuinely different in content. But if the biblical text must also be inerrant, what can these successive and disparate tales mean? Duncan must solve her problem by exegetical analysis, not by empirical evidence from science—and she must do so, according to her own lights, "with unshaken submission to the testimony of Scripture." But how can this double reconciliation (of Genesis 1 with Genesis 2, and of the entire biblical account with scientific evidence) be accomplished if the two creation stories truly conflict?

Duncan begins her book by exposing the paradox within her assumption that the biblical text may be metaphorical, but not factually false:

> In the first and second chapters of the Book of Genesis, we find two distinct accounts of the Creation of Man, materially differing from each other, yet generally interpreted as referring to the same event. To my mind, this interpretation has long presented serious difficulties.

She then locates the main problem in Adam's different position within the two tales—created after the other animals in chapter 1, but before them in chapter 2:

> While in the first chapter, these and many other tribes of the lower animals come into existence on the fifth day, and therefore before man, in the other, man is made and placed in Eden before the creation of these humble races, which were formed by a special act of God, intended to minister to a felt necessity of his newly-created child.

She then summarizes attempts by religious scholars to reconcile the two texts as consonant accounts of the same event, for God may surely choose redundancy as a literary strategy! "I do not affirm that Moses, as an inspired writer, was precluded from giving a second account of the same transaction." But she cannot escape the plain textual evidence, discussed just above, of a contradictory sequence between the two stories. "If we are to look upon the second chapter as standing in this relation to the first [as a second telling of the same event], we must at least expect that they will not be found contradictory to one another. . . . There shall be no irreconcilable difference between them."

Duncan then devises the ingenious solution that inspired her novel version of the old pre-Adamite theory. *Both* texts are true, but they tell two stories in proper temporal order, about two distinct events of creation in the history of life on earth. (The Hebrew word *Adam* may be read as generic, rather than the proper name of a particular fellow, so the stories may designate different progenitors.) Duncan then summarizes her entire thesis:

> I was thus led, with a conviction which has become always stronger by reading and reflection [note her two explicit literary criteria, with no reference to the empirical data of science], to perceive that the true way of explaining these passages is to refer them to two distinct creations, belonging respectively to periods far removed from one another, and occurring under conditions extremely different.

To explain the long duration, revealed for the first creation by geology and paleontology, Duncan adopts the traditional "day-age" theory of reconciliation: "I hope . . . to give sufficient reasons for adopting the belief now so generally received by thinking persons, that the six days of creation were in fact six ages, or cycles of ages." The separate and second creation of Adam, the progenitor of all living people, then suffers no challenge from the geological discovery of deep time, for any needed length can be absorbed by the long history of the first creation.

So far, so good (and not so wacky). But Duncan's model of sequential creations then leads us to ask a difficult question about the once extensive but now extinct race of pre-Adamites. Where are they? Archaeological evidence had, after much debate, finally established the contemporaneity of human artifacts with bones of large extinct mammals (mammoths, cave bears, woolly rhinos), indubitably assignable to Duncan's first, or pre-Adamite, creation. But no unambiguous evidence of human bodies—the bones of my bones, if not the unfossilizable flesh of my flesh—had been recovered (and none would be located until the 1890s, when Eugen du Bois discovered the remains of *Homo erectus* in Java). So if arrowheads and axes testify to a pre-Adamite existence, but no pre-Adamite bodies ever make their way into the fossil record, what happened to the physical evidence?

> Does nothing remain to indicate what he was, or how he spent his time, or what was his character? The birds and beasts of these ages, their plants and trees, their flowers and fruits have left distinct traces in every part of the world. Have none survived of man? . . . Where are his remains? We have the bones of the lower animals in abundance in the rocks of their respective eras, where are those of the Pre-Adamites?

Only at this point does Duncan fall into what scientists might label as a realm of folly wrought by overcommitment to a theory, but that Duncan no doubt regarded as a simple extension in the logic of a developing argument. A venerable scientific motto proclaims that "absence of evidence is not evidence of absence." In the early and exploratory days of a theory, failure to confirm will spur a search for evidence, whereas positive disproof will always refute a hypothesis. If this failure continues as the theory develops, and finally persists beyond a reasonable hope for future affirmation, then of course the theory must be dropped. In the case of human evolution, where sturdy flint tools greatly exceed fragile bones in capacity for preservation in the geological record, evidence of artifacts without bodies only spurred the search for bones—an expectation fulfilled within thirty years of Duncan's publication. (If science had still not found any bones, 140 years later, then we would be considering alternatives—but not Duncan's scenario.)

Duncan followed the logic of her exegesis instead. If the Bible promises eventual *bodily* resurrection to all sons of Adam in the second creation, then God probably redeemed the offspring of pre-Adam as well, and at the catastrophic termination of the first creation—hence, only tools, but no bones for pre-Adamites. But where, then, did the resurrected pre-Adamites go?

In a stunning solution to her greatest conundrum, Duncan proposes that the resurrected pre-Adamites must now be the angels of our legends and purported visitations:

> I venture to suggest that the Angel Host, whose mysterious visits to our world are so often recorded in the Bible—whose origin

> is so obscure—whose relations to Adam's family are so close, yet so unexplained . . . were in their original this very pre-Adamite race, holy, pure and like their Maker so long as they kept their first estate.

But one hypothetical solution engenders other collateral problems that must then also be encompassed within a logic already severely stressed and stretched. If these pre-Adamites were good enough to become angels in our eyes after their resurrection, why did God exterminate them in the first place, while dooming all other, and presumably innocent, plants and animals to a common grave (with no subsequent resurrection for these lower unfortunates)? The instigating event must have been something truly awful to contemplate; what could have distressed God so deeply that global destruction represented his only reasonable option?

To complete her argument, Duncan resolves this last puzzle. A wayward group of pre-Adamites rebelled against God, and the entire creation had to suffer for their transgression. As for these miscreants, they remain among us as the fallen angels—Satan and his devil host. Moreover, in destroying the earth (and leaving signs for us to recognize this event as an ice age, as recently discovered by the Swiss naturalist Louis Agassiz), and then resurrecting the bad with the good, God gave two warnings to his subsequent Adamites that they too should fear the wrath to come if they transgressed and followed Satan:

> Lucifer was the tempter, a Pre-Adamite of mortal mould, ambitious, enterprising, proud and able. His victims too were men, who yielding an ear, more or less willingly, to his falsehoods, subjected themselves to the same condemnation. The divine anger involved the ruin of the rebels. . . . God left on our globe, everywhere, the unmistakable evidence of the stupendous power he wields when he comes forth in His Majesty to shake terribly the earth.

We now understand that Duncan constructed her fascinating and groundbreaking chart for the full pageant of life through time (discussed in the introduction to this essay) not as a scientific innovation, but as a theological scenario for the earth's history, as constructed within the pre-Adamite tradition of textual analysis. The key white strip of Agassiz's ice-covered world may be validated by geological science, but this catastrophe represents, for Duncan, the agent of God's wrath after the satanic group of pre-Adamites fell from grace, and "God's great plough" (Agassiz's own description of the glacial age, by the way, but for different purposes and intentions) swept the planet, destroying the work of the first creation, and preparing a furrow to welcome the new race of Adamites, our own puny selves, to a humbled planet.

What, ultimately, can we say for Isabelle Duncan's theory, beyond noting an entertainment value far in excess of most incorrect proposals about human prehistory? A scientist might be tempted simply to dismiss her view as a disproven conjecture: she invented an elaborate theory to explain why prehistoric artifacts, but not prehistoric bones, had been preserved in the geological record—and she was dead wrong because we have since found bones aplenty.

But if we dig a bit deeper and ask *why* she developed such a peculiar explanation (bizarre to a scientist, to be sure, but even a bit odd for most theologians of her time), then we need to consider the more general theme of restriction. We can then learn something important from Isabelle Duncan because her blazingly obvious restrictions may help us to analyze our own, more subtle, limitations—for we always view the natural world within a blinkered mental compass, and we usually don't know how to see beyond our presuppositions (the reason, of course, why many false views of indubitable past geniuses seem so strange to us today).

Duncan operated within the limited procedures of literary exegesis upon a document that she did not permit herself to view as potentially inaccurate. Such a conviction does not leave much maneuvering room for the broad range of hypotheses that we must allow ourselves to entertain if we wish to resolve truly difficult questions about the natural world. This perspective inevitably leads us to ask whether the more obvious limitation imposed upon Isabelle Duncan—the peripheral space granted to intellectual women in her time—also contributed to her overly narrow focus. Did she accept her limited lot, or did she long to rebel? In only one passage of her otherwise impersonal (however passionate) book does she lift the veil of her frustration and allow her readers a brief peek beneath. She needs to refute a potential objection to her claim that God resurrected the pre-Adamites as angels. These people must have included both males and females, but our literature only mentions male angels. So where did the female pre-Adamites go? Duncan answers that they also became angels, but invisible angels because our literary biases place them beyond notice, just as our social biases often relegate contemporary women and children to a similar fate:

> There are many other indubitable truths on which for ages the Bible has been silent. The very existence of women on the earth

> during centuries, might be questioned were it allowed to be necessary that the Bible should assert it, and there are long ages during which we have no notice of little children.

In other words, absence of evidence is not evidence of absence. Or, as Hamlet said in the same scene that includes his sardonic commentary, "What a piece of work is man":

> O God I could be bounded in a nutshell, and could count myself a king of infinite space, were it not that I have bad dreams.

I'm afraid that we must pay the price in scary thoughts if we wish to fracture the confines of our mental comfort.

8

Freud's Evolutionary Fantasy

In 1897, the public schools of Detroit carried out an extensive experiment with a new and supposedly ideal curriculum. In the first grade, children would read *The Song of Hiawatha* because, at this age, they recapitulated the "nomadic" and "savage" stages of their evolutionary past and would therefore appreciate such a like-minded hero. During the same years, Rudyard Kipling wrote poetry's greatest paean to imperialism, "The White Man's Burden." Kipling admonished his countrymen to shoulder the arduous responsibility of serving these "new-caught, sullen people, half-devil and half-child." Teddy Roosevelt, who knew the value of a good line, wrote to Henry Cabot Lodge that Kipling's effort "was very poor poetry but made good sense from the expansion point of view."

These disparate incidents record the enormous influence upon popular culture of an evolutionary idea that ranks second only to natural selection itself for impact beyond biology. This theory held, mellifluously and perhaps with a tad of obfuscation in terminology, that "ontogeny recapitulates phylogeny," or that an organism, during the course of its embryonic growth, passes through a series of stages representing adult ancestors in their proper historical order. The gill slits of a human embryo record our distant past as a fish, while our later embryonic tail (subsequently resorbed) represents the reptilian stage of our ancestry.

Biology abandoned this idea some fifty years ago, for a variety of reasons chronicled in my book *Ontogeny and Phylogeny* (Harvard University Press, 1977), but not before the theory of recapitulation—to cite just three examples of its widespread practical influence—had served as the basis for an influential proposal that "born criminals" acted by necessity as the unfortunate result of a poor genetic shake as manifested by their retention of apish features successfully transcended in the ontogeny of normal people; buttressed a variety of racist claims by depicting adults in "primitive" cultures as analogs of Caucasian children in need of both discipline and domination; and structured the primary-school curricula of many cities by treating young children as equivalent to grown men and women of a simpler past.

The theory of recapitulation also played a profound, but almost completely unrecognized, role in the formulation of one of the half-dozen most influential movements of the twentieth century: Freudian psychoanalysis. Although the legend surrounding Freud tends to downplay the continuity of his ideas with preexisting theories, and to view psychoanalysis as an abrupt and entirely novel contribution to human thought, Freud trained as a biologist in the heyday of evolution's first discovery, and his theory sank several deep roots in the leading ideas of Darwin's world. (See Frank J. Sulloway's biography *Freud, Biologist of the Mind* [Basic Books, 1979], with its argument that nearly all creative geniuses become surrounded by a mythology of absolute originality.)

The "threefold parallelism" of classical recapitulation theory in biology equated the child of an advanced species both with an adult ancestor and with adults of any "primitive" lineages that still survived (the human embryo with gill slits, for example, represents both an actual ancestral fish that lived some 300 million years ago and all surviving fishes as well; similarly, in a racist extension, white children might be compared both with fossils of adult *Homo erectus* and with modern adult Africans). Freud added a fourth parallel: the neurotic adult who, in important respects, represents a normal child, an adult ancestor, or a normal modern adult from a primitive culture. This fourth term for adult pathologies did not originate with Freud, but arose within many theories of the time—as in Lombroso's notion of *l'uomo delinquente* (criminal man), and in various interpretations of neonatal deformity or mental retardation as the retention of an embryonic stage once normal in adult ancestors.

Freud often expressed his convictions about recapitulation. He wrote in his *Introductory Lectures on Psychoanalysis* (1916), "Each individual somehow recapitulates in an abbreviated form the entire development of the human race." In a note penned in 1938, he evoked a graphic image for his fourth term: "With neurotics it is as though we were in a prehistoric landscape—for instance in the Jurassic. The great saurians are still running around; the horsetails grow as high as palms."

Moreover, these statements do not represent merely a passing fancy or a peripheral concern. Recapitulation occupied a central and pervasive place in Freud's intellectual development. Early in his career, before he formulated the theory of psychosexual stages (anal, oral, and genital), he wrote to Wilhelm Fliess, his chief friend and collaborator, that sexual repression of olfactory stimuli represented our phyletic transition to upright posture: "Upright carriage was adopted, the nose was raised from the ground, and at the same time a number of what had formerly been interesting sensations connected with the earth became repellent" (letter of 1897). Freud based his later theory of psychosexual stages explicitly upon recapitulation: the anal and oral stages of childhood sexuality represent our quadrupedal past, when senses of taste, touch, and smell predominated. When we evolved upright posture, vision became our primary sense and reoriented sexual stimuli to the genital stage. Freud wrote in 1905 that oral and anal stages "almost seem as though they were harking back to early animal forms of life."

In his later career, Freud used recapitulation as the centerpiece for two major books. In *Totem and Taboo* (1913), subtitled *Some Points of Agreement Between the Mental Life of Savages and Neurotics,* Freud inferred a complex phyletic past from the existence of the Oedipus complex in modern children and its persistence in adult neurotics, and from the operation, in primitive cultures, of incest taboos and totemism (identification of a clan with a sacred animal that must be protected, but may be eaten once a year in a great totemic feast). Freud argued that early human society must have been organized as a patriarchal horde, ruled by a dominant father who excluded his sons from sexual contact with women of the clan. In frustration, the sons killed their domineering father, but then, in their guilt, could not possess the women (incest taboo). They expiated their remorse by identifying their slain father with a totemic animal, but celebrated their triumph by reenactment during the annual totemic feast. Modern children relive this act of primal parricide in the Oedipus complex. Freud's last book, *Moses and Monotheism* (1939), reiterates the same theme in a particular context.

Moses, Freud argues, was an Egyptian who cast his lot with the Jews. Eventually his adopted people killed him and, in their overwhelming guilt, recast him as a prophet of a single, all-powerful God, thus also creating the ethical ideals that lie at the heart of Judeo-Christian civilization.

A new discovery, hailed as the most significant in many years by Freudian scholars, has now demonstrated an even more central role for recapitulation in Freud's theory than anyone had ever imagined or been willing to allow—although, again, almost every commentator has missed the connection because Freud's biological influences have been slighted by a taxonomy that locates him in another discipline, and because the eclipse of recapitulation has placed this formerly dominant theory outside the consciousness of most modern scholars. In 1915, in the shadow of war and as he began his sixtieth year, Freud labored with great enthusiasm on a book that would set forth the theoretical underpinnings of all his work—the "metapsychology," as he called the project. He wrote twelve papers for his work, but later abandoned his plans for unknown reasons much discussed by scholars. Five of the twelve papers were eventually published (with *Mourning and Melancholia* as the best known), but the other seven were presumed lost or destroyed. In 1983, Ilse Grubrich-Simitis discovered a copy, in Freud's hand, of the twelfth and most general paper. The document had resided in a trunk, formerly the property of Freud's daughter Anna (who died in 1983), and otherwise filled with the papers of Freud's Hungarian collaborator Sándor Ferenczi. Harvard University Press published this document in 1987 under the title *A Phylogenetic Fantasy* (translated by Axel and Peter T. Hoffer and edited and explicated by Dr. Grubrich-Simitis).

The connection with Ferenczi reinforces the importance of recapitulation as a centerpiece of Freud's psychological theory. Freud had been deeply hurt by the estrangement and opposition of his leading associates, Alfred Adler and Carl Jung. But Ferenczi remained loyal, and Freud strengthened both personal and intellectual ties with him during this time of stress. "You are now really the only one who still works beside me," Freud wrote to Ferenczi on July 31, 1915. In preparing the metapsychological papers, Freud's interchange with Ferenczi became so intense that these works might almost be viewed as a joint effort. The twelfth paper, the phylogenetic fantasy, survived only because Freud sent a draft to Ferenczi for his criticism. Ferenczi had received the most extensive biological training of all Freud's associates, and no one else in the history of psychoanalysis maintained so strong a commitment to recapitulation. When Freud sent his phylogenetic fantasy to Ferenczi on July 12, 1915, he ended his letter by stating, "Your priority in all this is evident."

Ferenczi wrote a remarkable work titled *Thalassa: A Theory of Genitality*

(1924), perhaps best known today in mild ridicule for claiming that much of human psychology records our unrecognized yearning to return to the comforting confines of the womb, "where there is no such painful disharmony between ego and environment that characterizes existence in the external world." By his own admission, Ferenczi wrote *Thalassa* "as an adherent of Haeckel's recapitulation theory."

Ferenczi viewed sexual intercourse as an act of reversion toward a phyletic past in the tranquillity of a timeless ocean—a "thalassal regressive trend . . . striving towards the aquatic mode of existence abandoned in primeval time." He interpreted the weariness of postcoital repose as symbolic of oceanic tranquillity. He also viewed the penis as a symbolic fish, so to speak, reaching toward the womb of the primeval ocean. Moreover, he pointed out, the fetus that arises from this union passes its embryonic life in an amniotic fluid, thus recalling the aquatic environment of our ancestors.

Ferenczi tried to locate even earlier events in our modern psychic lives. He also likened the repose following coitus to a striving further back toward the ultimate tranquillity of a Precambrian world before the origin of life. Ferenczi viewed the full sequence of a human life—from the coitus of parents to the final death of their offspring—as a recapitulation of the gigantic tableau of our entire evolutionary past (Freud would not proceed nearly this far into such a realm of conflating possible symbol with reality). Coitus, in the repose that strives for death, represents the early earth before life, while impregnation recapitulates the dawn of life. The fetus, in the womb of its symbolic ocean, then passes through all ancestral stages from the primal amoeba to a fully formed human. Birth recapitulates the colonization of land by reptiles and amphibians, while the period of latency, following youthful sexuality and before full maturation, repeats the torpor induced by ice ages.

With this recollection of human life during the ice ages, we can connect Ferenczi's thoughts with Freud's phylogenetic fantasy—for Freud, eschewing Ferenczi's overblown, if colorful, inferences about an earlier past, begins with the glacial epoch in trying to reconstruct human history from current psychic life. The basis for Freud's theory lay in his attempt to classify neuroses according to their order of appearance during human growth.

Theories inevitably impose themselves upon our perceptions; no exclusive, objective, or obvious way exists for describing nature. Why should we classify neuroses primarily by their *time* of appearance? Neuroses might be described and ordered in a hundred other ways (by social effect, by common actions or structure, by emotional impacts upon the psyche, by chemical changes that might cause or accompany them). Freud's decision stemmed directly from his commitment to an evolutionary explanation of neurosis—a scheme, moreover, that Freud chose to base upon the theory of recapitulation. In this view, sequential events of human history set the neuroses—for neurotic people become fixated at a stage of growth that normal people transcend. Since each stage of growth recapitulates a past episode in our evolutionary history, each neurosis fixates on a particular prehistoric stage in our ancestry. These behaviors may have been appropriate and adaptive then, but they now produce neuroses in our vastly different modern world. Therefore, if neuroses can be ordered by time of appearance, we will obtain a guide to their evolutionary meaning (and causation) as a series of major events in our phyletic history. Freud wrote to Ferenczi on July 12, 1915, "What are now neuroses were once phases of human conditions." In the *Phylogenetic Fantasy,* Freud asserts that "the neuroses must also bear witness to the history of the mental development of mankind."

Freud begins by acknowledging that his own theory of psychosexual stages, combined with Ferenczi's speculations, may capture some truly distant aspects of phylogeny by their appearance in the development of very young children. For the phylogenetic fantasy, however, he confines himself to more definite (and less symbolic) parts of history that lie recorded in two sets of neuroses developing later in growth—the transference neuroses and the narcissistic neuroses of his terminology. As the centerpiece of the phylogenetic fantasy, Freud orders these neuroses in six successive stages: the three transference neuroses (anxiety hysteria, conversion hysteria, and obsessional neurosis), followed by the three narcissistic neuroses (dementia praecox [schizophrenia], paranoia, and melancholia-mania [depression]).

> There exists a series to which one can attach various far-reaching ideas. It originates when one arranges the . . . neuroses . . . according to the point in time at which they customarily appear in the life of the individual. . . . Anxiety hysteria . . . is the earliest, closely followed by conversion hysteria (from about the fourth year); somewhat later in prepuberty (9–10) obsessional neuroses appear in children. The narcissistic neuroses are absent in childhood. Of these, dementia praecox in classic form is an illness of the puberty years, paranoia approaches the mature years, and melancholia-mania the same time period, otherwise not specifiable.

Freud interprets the transference neuroses as recapitulations of behaviors that we developed to cope with difficulties of human life during the ice ages: "The temptation is very great to recognize in the three dispositions to anxiety hysteria, conversion hysteria, and obsessional neurosis regressions to phases that the whole human race had to go through at some time from the beginning to the end of the Ice Age, so that at that time all human beings were the way only some of them are today." Anxiety hysteria represents our first reaction to these difficult times: "Mankind, under the influence of the privations that the encroaching Ice Age imposed upon it, has become generally anxious. The hitherto predominantly friendly outside word, which bestowed every satisfaction, transformed itself into a mass of threatening perils."

In these parlous times, large populations could not be supported, and limits to procreation became necessary. In a process adaptive for the time, humans learned to redirect their libidinal urges to other objects, and thereby to limit reproduction. The same behavior today, expressed as a phyletic memory, has become inappropriate and therefore represents the second neurosis—conversion hysteria: "It became a social obligation to limit reproduction. Perverse satisfactions that did not lead to the propagation of children avoided this prohibition. . . . The whole situation obviously corresponds to the conditions of conversion hysteria."

The third neurosis, obsession, records our mastery over these difficult conditions of the Ice Age. We needed to devote enormous resources of energy and thought to ordering our lives and overcoming the hostilities of the environment. This same intensely directed energy may now be expressed neurotically in obsessions to follow rules and to focus on meaningless details. This behavior, once so necessary, now "leaves as compulsion, only the impulses that have been displaced to trivialities."

Freud then locates the narcissistic neuroses of later life in the subsequent, postglacial events of human history that he had already identified in *Totem and Taboo.* Schizophrenia records the father's revenge as he castrates his challenging sons:

> We may imagine the effect of castration in that primeval time as an extinguishing of the libido and a standstill in individual development. Such a state seems to be recapitulated by dementia praecox which . . . leads to giving up every love-object, degeneration of all sublimations, and return to auto-erotism. The youthful individual behaves as though he had undergone castration.

(In *Totem and Taboo,* Freud had only charged the father with expelling his sons from the clan; now he opts for the harsher punishment of castration. Commentators have attributed this change to Freud's own anger at his "sons"

Adler and Jung for their break with his theories, and their foundation of rival schools. By castration, Freud could preclude the possibility of their future success. I am not much attracted to psychoanalytic speculations of this genre. Freud was, of course, not unaware that a charge of castration posed difficulty for his evolutionary explanation—for the mutilated sons could leave no offspring to remember the event in heredity. Freud speculates that younger sons were spared, thanks to the mother's intercession; these sons lived to reproduce but were psychically scarred by the fate that had befallen their brothers.)

The next neurosis, paranoia, records the struggle of exiled sons against the homosexual inclinations that must inevitably arise within their bonded and exiled group: "It is very possible that the long-sought hereditary disposition of homosexuality can be glimpsed in the inheritance of this phase of the human condition. . . . Paranoia tries to ward off homosexuality, which was the basis for the organization of brothers, and in so doing must drive the victim out of society and destroy his social sublimations."

The last neurosis of depression then records the murder of the father by his triumphant sons. The extreme swings in mood of the manic-depressive record both the exultation and the guilt of parricide: "Triumph over his death, then mourning over the fact they all still revered him as a model."

From our current standpoint, these speculations seem so farfetched that we may be tempted simply to dismiss them as absurd, even though they emanate from such a distinguished source. Freud's claims are, to be sure, quite wrong, based on knowledge gained in the past half-century. (In particular, Freud's theory is fatally and falsely Eurocentric. Human evolution was not shaped near the ice sheets of northern Europe, but in Africa. We can also cite no reason for supposing that European Neanderthals, who were probably not our ancestors in any case, suffered unduly during glacial times with their abundant game for hunting. Finally, we can offer not a shred of evidence that human social organizations once matched Freud's notion of a domineering father who castrated his sons and drove them away—an awfully precarious way to assure one's Darwinian patrimony.)

But the main reason that we must not dismiss Freud's theory as absurd lies in its consonance with biological ideas then current. Science has since abandoned the biological linchpins of Freud's theory, and most commentators don't know what these concepts entailed or that they ever even existed. Freud's theory therefore strikes us as a crazy speculation that makes absolutely no sense according to modern ideas of evolution. Well, Freud's phylogenetic fantasy *is* bold, wildly beyond data, speculative in the extreme, idiosyncratic—and wrong. But Freud's speculation does become comprehensible once one recognizes the two formerly respectable biological theories underpinning the argument.

The first theory, of course, is recapitulation itself, as discussed throughout this essay. Recapitulation must provide the primary warrant for Freud's fantasy, for recapitulation allowed Freud to interpret a normal feature of childhood (or a neurosis interpreted as fixation to some childhood stage) as necessarily representing an adult phase of our evolutionary past. But recapitulation does not suffice, for one also needs a mechanism to convert the experiences of adults into the heredity of their offspring. Conventional Darwinism could not provide such a mechanism in this case—and Freud understood that his fantasy demanded allegiance to a different version of heredity.

Freud's fantasy requires the passage to modern heredity of events that affected our ancestors only tens of thousands of years ago at most. But such events—anxiety at approaching ice sheets, castration of sons and murder of fathers—have no hereditary impact. However traumatic, such events do not affect the eggs and sperm of parents, and therefore cannot pass into heredity under Mendelian and Darwinian rules.

Freud, therefore, held firmly to his second biological linchpin—the Lamarckian idea, then already unfashionable but still advocated by some prominent biologists, that acquired characters will be inherited. Under Lamarckism, all theoretical problems for Freud's mechanism disappear. Any important and adaptive behavior developed by adult ancestors can pass directly into the heredity of offspring—and quickly. A primal parricide that occurred just ten or twenty thousand years ago may well be encoded as the Oedipal complex of modern children.

I credit Freud for his firm allegiance to the logic of his argument. Unlike Ferenczi, who concocted an untenable melange of symbolism and causality in *Thalassa* (the placenta, for example, as a newly evolved adaptation of mammals, cannot, therefore, enclose a phyletic vestige of the primeval ocean). However, Freud's theory obeyed a rigidly consistent biological logic rooted in two notions since discredited—recapitulation and Lamarckian inheritance.

Freud understood that his theory depended upon the validity of Lamarckian inheritance. He wrote in the *Phylogenetic Fantasy,* "One can justifiably claim that the inherited dispositions are residues of the acquisition of our ancestors." He also recognized that Lamarckism had been falling from fashion since the rediscovery of Mendel's laws in 1900. In their collaboration, Freud and Ferenczi dwelt increasingly upon the necessary role of Lamarckism in psychoanalysis. They planned a joint book on the subject, and Freud dug in with enthusiasm, reading Lamarck's works in late 1916 and writing a paper on the subject (unfortunately never published and apparently not preserved) that he sent to Ferenczi in early 1917. But the project never came to fruition, as the privations of World War I made research and communication increasingly difficult. When Ferenczi nudged Freud one last time in 1918, Freud responded, "Not disposed to work . . . too much interested in the end of the world drama."

Illogic remains slippery and vacuous (*Thalassa* can never be proved or rejected; so the idea has simply been forgotten). But in a logical argument, one must live or die by the validity of required premises. Lamarckism has been firmly rejected, and Freud's evolutionary theory of neurosis falls with the validation of Mendel. Freud himself chronicled with great remorse the slippage of Lamarckism from respectability. In *Moses and Monotheism,* he continued to recognize his need for Lamarckism while acknowledging the usual view of its failure:

> This state of affairs is made more difficult, it is true, by the present attitude of biological science, which rejects the idea of acquired qualities being transmitted to descendants. I admit, in all modesty, that in spite of this I cannot picture biological development proceeding without taking this factor into account.

Since most commentators have not grasped the logic of Freud's theory because they have not recognized the roles of Lamarckism and recapitulation, they fall into a dilemma, particularly if they generally favor Freud. Without these two biological linchpins of recapitulation and Lamarckism, Freud's fantasy sounds crazy. Could Freud really mean that these events of recent history somehow entered the inheritance of modern children and the fixated behavior of neurotics? Consequently, a muted or kindly tradition has arisen for viewing Freud's claims as merely symbolic. He didn't really mean that exiled sons actually killed their father and that Oedipal complexes truly reenacted a specific event of our past. Freud's words should therefore be regarded sympathetically as colorful imagery providing insight into the psychological meaning of neurosis. Daniel Goleman, reporting on the discovery of *A Phylogenetic Fantasy* (in *The New York Times,* 10 February 1987), writes:

> In the manuscript, according to many scholars, Freud appeared to be turning to a literary mechanism he would use often in the explication of his ideas; he put forward a story that might or might not be grounded in reality but whose mythological content revealed what he saw as basic human conflicts.

I strongly reject this "kindly" tradition of watering down Freud's well-formulated mechanism to myth or metaphor. In fact, I don't view this tradition as kindly at all, for in order to make Freud appear cogent in an inappropriate context of modern ideas, such interpretations sacrifice the sharp logic and consistency of Freud's actual argument. Freud's writing gives no indication that he intended his phylogenetic speculation as anything but a potentially true account of actual events. If Freud had meant these ideas only as metaphor, then why did he work out such consistency with biological theory based on Lamarckism and recapitulation? And why did he yearn so strongly for Lamarckism after its popularity had faded?

Freud recognized his fantasy as speculation, of course, but he meant every word as potential reality. In fact, the end of *Totem and Taboo* features an incisive discussion of this very subject, with a firm denial of any metaphorical intent. Freud writes:

> It is not accurate to say that obsessional neurotics, weighed down under the burden of an excessive morality, are defending themselves only against psychical reality and are punishing themselves for impulses which were merely felt. Historical reality has a share in the matter as well.

Freud's closing line then reiterates this argument with a literary fillip. He quotes the famous parody of the first line of John's Gospel ("In the beginning was the Word"), as spoken by Faust in Part 1 of Goethe's drama: *Im Anfang war die Tat* (In the beginning was the Deed).

Finally, in explicating Freud's belief in the reality of his story, and in recognizing the firm logic of his argument, I do not defend his method of speculation devoid of any actual evidence in the historical or archaeological record. I believe that such purely speculative reconstructions of history do more harm than good because they give the study of history a bad name. These speculative reveries often lead students of the "hard" experimental sciences to dismiss the investigation of history as a "soft" enterprise unworthy of the name science. But history, pursued in other ways, includes all the care and rigor of physics or chemistry at its best. I also deplore the overly adaptationist premise that any evolved feature not making sense in our present life must have arisen long ago for a good reason rooted in past conditions now altered. In our tough, complex, and partly random world, many features just don't make functional sense, period. We need not view schizophrenia, paranoia, and depression as postglacial adaptations gone awry: perhaps these illnesses are immediate pathologies, with remediable medical causes, pure and simple.

Freud, of course, recognized the speculative character of his theory. He called his work a phylogenetic "fantasy," and he ultimately abandoned any thought of publication, perhaps because he regarded the work as too outré and unsupported. He even referred playfully to the speculative character of his manuscript, begging that readers "be patient if once in a while criticism retreats in the face of fantasy and unconfirmed things are presented, merely because they are stimulating and open up distant vistas." He then wrote to Ferenczi that scientific creativity must be defined as a "succession of daringly playful fantasy and relentlessly realistic criticism." Perhaps the phase of relentless criticism intruded before Freud dared to publish his phylogenetic fantasy.

We are therefore left with a paradoxical, and at least mildly disturbing, thought. Freud's theory ranks as a wild speculation, based upon false biology and rooted in no direct data at all about phylogenetic history. Yet the manuscript has been published and analyzed with painstaking care more than half a century later. Hundreds of unknown visionaries develop equally farfetched but interesting and coherent speculations every day—but we ignore them or, at best, laugh at such crazy ideas. Rewrite the *Phylogenetic Fantasy* to remove the literary hand of Freud's masterly prose, put Joe Blow's name on the title page, and no one will pay the slightest attention. We live in a world of privilege, and only great thinkers win a public right to fail greatly.

IV

Essays in the Paleontology of Ideas

9

The Jew and the Jewstone

The human mind may love to contemplate exemplary universes of abstract grandeur and idealized perfection, but we can extract equal pleasure from a tiny embodiment of some great thought, or some defining event of a lifetime, in a humble but concrete object that we can hold in our hands and rotate before our eyes. We cherish such explicit reminders—keepsakes, souvenirs, or mementos in our descriptions—for their salience as markers of distinctive moments in our unique trajectory through the general adventure of human life.

For this reason, I have never been able to understand the outright purchase, from catalogs or store shelves, of distinctive items that (I would think) can only have meaning as mementos of particular, preciously individual experiences. I do, for example, cherish a few baseballs signed by personal heroes, but only because they intersected my own life in a meaningful way—the pop foul off DiMaggio's bat that my father caught in 1950 as I sat next to him, and that the great man signed and returned after I mailed him the relic along with a gushing fan letter; the ball signed by Hank Aaron, and presented to me after a talk I gave at Atlanta's Spelman College, as I, nearly speechless for once, could only thank my hosts for the equivalent of an item inscribed by God himself. But what could a ball signed by a Ted Williams or a Pete Rose mean, when ordered from a catalog by anyone willing to fork over a specified sum?

I take special delight in the particular category of things long known and admired in large abstraction, but then seen for the first time in the form of a humble but concrete object. I don't refer to first views of the grand things themselves—the obvious and anticipated thrill of initial contact with the Taj Mahal or the Parthenon—but rather to the sublime surprise of finding my father's card of honorable discharge from the navy after hearing his war stories for so many years, or of seeing my grandfather's name entered on a ship's manifest for his arrival at Ellis Island in 1901 (see essay 1).

As a scholar, most of my thrills in this category arise as unexpected encounters in actual print, in an old book read by real people, of the founding version for stories or concepts once learned in a classroom or textbook, and stored as an important memory implanted by others but never validated by original sources. I get a special jolt when I first see (as my grandmother would have said) *in shvartz*—that is, "in black" ink or printed type—something that had long intrigued my mind, but had never stood right before my eyes in its concrete and original form.

The tale of this essay begins with such an experience of transfer from vague abstraction to factual immediacy. I do not remember where I first heard the story—perhaps in a guest lecture from a distinguished visiting luminary, or as a casual comment from a professor in an undergraduate class at Antioch College? I do not even know whether the tale represents a standard example, well known to all historians of early science,* or an original insight from one teacher's personal research. But I do remember the story itself, and the striking epitome thus provided for the revolutionary character (at its codification in the seventeenth century) of the explanatory system now called "science."

The story cited a memorable example to show how respected styles of former explanation became risible and "mystical" in the light of new views about causality and the nature of the material world. The essence of the difference

---

*Since writing this essay, I have asked some professional historians and have learned that, indeed, the tale of the "weapon salve" became a major issue, much discussed at the time itself, and by later historians ever since, in defining the norms and limits of scientific explanation.

between "prescientific" and scientific explanations, my unremembered source stated, could be epitomized in a popular "prescientific" remedy for the healing of wounds inflicted by swords or other weapons. The prescribed salve must be applied to the wound itself—where, by modern sensibilities, the potion might well work as advertised, since early pharmacists and herbalists had, by experience, discovered many useful remedies, even if we now dismiss their theories about modes of action. But, this particular recipe then required that the salve be applied to the weapon that inflicted the wound as well—for healing required a sympathetic treatment, a rebalancing, a "putting right" of both the injurer and the injuree.

The nub of the revolutionary difference between "prescientific" and scientific explanation, my anonymous source continued, lies beautifully exposed in this microcosm—for the Western world's transition to modernity may virtually be defined by the realization that although some material property of the salve may heal the wound by direct contact, the formerly sensible practice of treating the weapon in a similar way must now be scorned as utter nonsense and absurd mysticism.

This tale about treating the weapon as well as the wound has rattled around my head for twenty years or more, with no documentation beyond a dimly remembered lecture. Then, a few months ago, I bought a copy of Johann Schröder's *Pharmacopoeia medicochymica* (1677 edition of a work first printed in Ulm in 1641), perhaps the most widely used handbook of drugs and other curative remedies from the seventeenth century. And in this copy, published right in the midst of the ferment provoked by Newton's generation at the birth of modern science, I found the formula for the salve that must be applied to weapons as well as wounds—*in shvartz* on page 303, and named *Unguentum Sympatheticum Crollii,* or *Croll's sympathetic ointment* (an unguent is an ointment or salve, and we shall learn more about Mr. Croll a bit later).

The formula for this concoction may raise modern hairs and hackles. We must begin, Schröder tells us, with *adip. veteris aprugn.,* or the fat of an old boar, mixed with bear fat as well. Boil them both in red wine; pour the resulting mixture into cold water and collect the fat that accumulates at the top. Then add a motley assortment of pulverized worms; the brain of a boar (presumably from the same creature that supplied the initial fat); some sandalwood; haematite (a rock containing iron); mumia, or the dust from a corpse; and, to top it all off, *usneae e cranio hominis interempti,* or scrapings from the skull of a man who has been killed.

Schröder then appends a series of notes with permitted variation, including a most welcome statement (to modern sensibilities): *sunt qui omittunt usneam et mumiam* (some people omit the corpse dust and cranial scrapings). But another note then warns us that the cranial scrapings, if included, must be gathered while the moon waxes (that is, during its increase from a thin crescent to a full circle), and under a good astrological sign, preferably with Venus in conjunction (that is, within the reigning zodiacal constellation), but certainly not Mars or Saturn.

Under the next heading, *Usus* (use), Schröder first tells us that this ointment will cure all wounds, unless the nerves or arteries have been severed. The next line—*ungatur telum, quo vulnus inflictum* (the weapon that inflicted the wound is to be anointed)—then advances the argument that so impressed me as a succinct example of the dramatically different account of nature and causality, so soon to be superseded by the rise of modern science.

Schröder's description ends with a set of instructions for proper use of the ointment. We learn that the relevant part must be wrapped in linen, and kept out of the wind in a place neither too hot nor too cold, *ne damnum inferatur Patienti* (so that no damage shall be inflicted upon the patient). Moreover, no dust may fall upon the part, *alioquin Patiens mire affligeretur* (for otherwise the patient will be sorely injured). However, the relevant part, in each case, describes the care and handling of the weapon, not the wound! The two final items on the list also refer exclusively to the anointed weapon: if the wound was inflicted by the point of a sword, then the ointment must be applied from the hilt downward toward the tip; if the weapon cannot be found, a stick of wood, dipped in the victim's blood, may suffice.

In a final paragraph, Schröder offers his only statement about why such a procedure might enjoy success. The cure occurs because the same balsamic (comforting) spirit inheres in both the patient and the blood of his wound—and both must be fortified by the ointment. The weapon, I presume, must be treated because some of the patient's blood still remains (or perhaps merely because the weapon drew the patient's blood, and must therefore be cleansed along with the patient himself, in order to bring both parts of this drama back into harmony).

Oswald Croll (1560–1609), the inventor of this ointment, followed the theories of Paracelsus (1493–1541) in proposing external sources of disease, in opposition to the old Galenic theory of humors. In this major debate of premodern medicine, the externalists believed that "outside" forces or agents entered the human body to cause illness, and that healing substances from the three worldly kingdoms (animal, vegetable, and mineral, but primarily vegetable, as plants provided most drugs and potions) could rid the body of these invaders. The humoral theory, on the other hand, viewed disease as an imbalance among the body's four basic principles: blood (the sanguine, or wet-hot humor), phlegm (the wet-cold humor), choler (dry-hot), and melancholy (dry-cold). Treatment must therefore be directed not toward the expulsion of foreign elements, but to the restoration of internal balance among the humors (bloodletting, for example, when the sanguine humor rises too high; sweating, purging, and vomiting as other devices for setting the humors back into order).

In Paracelsian medicine, by contrast, treatment must be directed against the external agent of disease, rather than toward the restoration of internal harmony by raising or lowering the concentration of improperly balanced humors. How, then, can the proper agents of potential cure be recognized among plants, rocks, or animal parts that might neutralize or destroy the body's invaders? In his article on Paracelsus for the *Dictionary of Scientific Biography,* Walter Pagel summarizes this argument from an age before the rise of modern science:

> Paracelsus . . . reversed this concept of disease as an upset of humoral balance, emphasizing [instead] the external cause of disease. . . . He sought and found the causes of disease chiefly in the mineral world (notably in salts) and in the atmosphere, carrier of star-born "poisons." He considered each of these agents to be a real *ens,* a substance in its own right (as opposed to humors, or temperaments, which he regarded as fictitious). He thus interpreted disease itself as an *ens,* determined by a specific agent foreign to the body, which takes possession of one of its parts. . . . He directed his treatment specifically against the agent of the disease, rather than resorting to the general anti-humoral measures . . . that had been paramount in ancient therapy for "removal of excess and addition of the deficient." . . . Here his notion of "signatures" came into play, in the selection of herbs that in color and shape resembled the affected organ (as, for example, a yellow plant for the liver or an orchid for the testicle). Paracelsus's search for such specific medicines led him to attempt to isolate the efficient kernal (the *quinta essentia*) of each substance. [Our modern word *quintessential* derives from this older usage.]

This doctrine of signatures epitomizes a key difference between modern science and an older view of nature (shared by both the humor theorists and the Paracelsians, despite their major disagreement about the nature of disease). Most scholars of the Renaissance, and of earlier medieval times, viewed the earth and cosmos as a young, static, and harmonious system, created by God, essentially in its current form, just a few thousand years ago, and purposely imbued with pervasive signs of order and harmony among its apparently separate realms—all done to illustrate the glory and subtlety of God's omnipotence, and to emphasize his special focus upon the human species that he had created in his own image.

This essential balance and harmony achieved its most important expression in deep linkages (we would dismiss them today as, at best, loose analogies) between apparently disparate realms. At one level on earth—thus setting the central principle of medicine under the doctrine of signatures—the microcosm of the human body must be linked to the macrocosm of the entire earth. Each part of the human body must therefore be allied to a corresponding manifestation of the same essential form in each of the macrocosmic realms: mineral, vegetable, and animal. Under this conception of nature, so strikingly different from our current views, the idea that a weakened human part might be treated or fortified by its signature from a macrocosmic realm cannot be dismissed as absurd (the orchid flower that looks like the male genitalia, and receives its name from this likeness, as a potential cure for impotence, for example). Oswald Croll, in particular, based his medical views on these linkages of the human microcosm to the earthly macrocosm, and Schröder's *Pharmacopoeia* represents a "last gasp" for this expiring theory, published just as Newton's generation began to establish our present, and clearly more effective, view of nature.

At a second level, the central earth (of this pre-Copernican cosmos) must also remain in harmony with the heavens above. Thus each remedy on earth must correspond with its proper configuration of planets as they move through the constellations of the zodiac. These astronomical considerations regulated when plants and animals (and even rocks), used as remedies for human ills, should be collected, and how they should be treated—hence, in Croll's ointment for wounds and swords, the requirement that cranial scrapings be collected under a good constellation, with the loving Venus, but not the warlike Mars or Saturn, in conjunction.

As a memorable example of this approach to healing under the doctrine of signatures and stable harmony among the realms of nature, consider the accompanying illustration from the last major work of scholarship in this tradition, so soon to be extinguished by modern science—the *Mundus subterraneus,* published in 1664 by perhaps the most learned scholar of his generation, the Jesuit polymath Athanasius Kircher (who wrote an important ethnography of China, came closer than anyone else to deciphering the hieroglyphs of ancient Egypt, wrote major treatises on music and magnetism, and built, in Rome, one of the finest natural history collections ever assembled). In this figure, titled (in

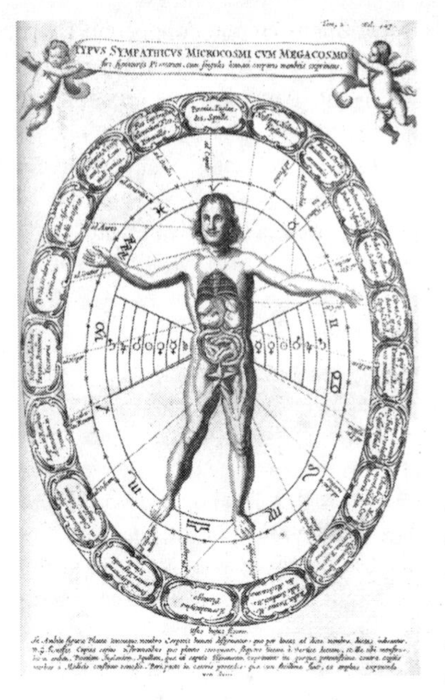

Medicine under the doctrine of signatures as depicted by Athanasius Kircher in 1664. Each ailing part of the body can be linked to a healing plant, constellation of stars, and planet. See text for details.

Crollian fashion) "sympathetic types of the microcosm with the macrocosm," lines radiate from each part of the human body to the names of plants (in the outer rim) that will cure the afflicted organs. To complete the analogies and harmonizations, the inner circle (resting on the man's head and supporting his feet) presents astronomical signs for the zodiacal constellations, while the symmetrical triangles, radiating like wings from the man's sides just below his arms, include similar signs for the planets.

We rightly reject this system today as a false theory based on an incorrect view of the nature of the material world. And we properly embrace modern science as both a more accurate account and a more effective approach to such practical issues as healing the human body from weakening and disease. Viagra does work better than crushed orchid flowers as a therapy for male impotence (though we should not doubt the power of the placebo effect in granting some, albeit indirect, oomph to the old remedy in some cases). And if I get badly cut when slicing a bagel with my kitchen knife, and the wound becomes infected, I much prefer an antibiotic to a salve made of boar fat and skull scrapings that must then be carefully applied both to the knife and to my injury.

Nonetheless, I question our usual dismissal of this older approach as absurd, mystical, or even "prescientific" (in any more than a purely chronological sense). Yes, anointing the wound as well as the weapon makes no sense and sounds like "primitive" mumbo-jumbo in the light of later scientific knowledge. But how can we blame our forebears for not knowing what later generations would discover? We might as well despise ourselves because our grandchildren will, no doubt, understand the world in a different way.

We may surely brand Croll's sympathetic ointment, and its application to the weapon as well as the wound, as ineffective, but Croll's remedy cannot be called either mystical or stupid under the theory of nature that inspired its development—the doctrine of signatures, and of harmony among the realms of nature. To unravel the archaeology of human knowledge, we must treat former systems of belief as valuable intellectual "fossils," offering insight about the human past, and providing precious access to a wider range of human theorizing only partly realized today. If we dismiss such former systems as absurd because later discoveries superseded them, or as mystical in the light of causal systems revealed by these later discoveries, then we will never understand the antecedents of modern views with the same sympathy that Croll sought between weapon and wound, and Kircher proposed between human organs and healing plants.

[In this light, for example, the conventional image of Paracelsus himself as the ultimate mystic who sought to transmute base metals into gold, and to create human homunculi from chemical potions, must be reevaluated. I do not challenge the usual description of Paracelsus as, in the modern vernacular, "one weird dude"—a restless and driven man, subject to fits of rage and howling, to outrageous acts of defiance, and to drinking any local peasant under the table in late-night tavern sessions. But, as a physician, Paracelsus won fame (and substantial success) for his cautious procedures based on minimal treatment and tiny doses of potentially effective remedies (in happy contrast to the massive purging and bloodletting of Galenic doctors), and for a general approach to disease—as an incursion of foreign agents that might be expelled with healing substances—that provided generally more effective remedies than any rebalancing of humors could achieve. Even his chosen moniker of Paracelsus—he was christened Philippus Aureolus Theophrastus Bombastus von Hohenheim—may not bear the arcane and mystical interpretation usually presented in modern accounts. Perhaps he did mean to highlight the boastful claim that he had advanced beyond *(para)* the great Roman physician Celsus. But I rather suspect, as do many other scholars, that Paracelsus may simply represent a Latinized version of his birth name, Hohenheim, or "high country"—for *celsus,* in Latin, means "towering" or "lofty." Many medieval and Renaissance intellectuals converted their vernacular names to classical equivalents—as did Georg Bauer (literally "George the farmer," in German) who became the world's greatest sixteenth-century geologist as the more mellifluous and Latinized Georgius Agricola, with the same meaning; or Luther's leading supporter, Philip Schwartzerd (Mr. "Black Earth" in his native lingo), who adopted a Greek version of the same name as Melanchthon.]

However, while we should heed the scholar's plea for sympathetic study of older systems, we must also celebrate the increase and improvement, achieved by science, in our understanding of nature through time. We must also acknowledge that these ancient and superseded systems, however revealing and fascinating, did impede better resolutions (and practical cures of illness) by channeling thought and interpretation in unfruitful and incorrect directions.

I therefore searched through Schröder's *Pharmacopoeia* to learn how he treated the objects of my own expertise—fossils of ancient organisms. The search for signatures to heal afflicted human parts yielded more potential remedies from the plant and animal kingdoms, but fossils from the mineral realm also played a significant role in the full list of medicines. Mineral remedies discussed by Schröder, and made of rocks in shapes and forms that suggested curative powers over human ailments, included, in terminology that prevailed among students of fossils until the late eighteenth century, the following items in alphabetical order:

1. AETITES, or "pregnant stones" found in eagles' nests, and useful in a suggestive manner: *partum promovet*— "it aids (a woman in giving) birth."
2. CERAUNIA, or "thunder stones," useful in stimulating the flow of milk or blood when rubbed on breasts or knees.
3. GLOSSOPETRA, or "tongue stones," an antidote to poisons of animal wounds or bites.
4. HAEMATITES, or "bloodstones," for stanching the flow of blood: *refrigerat, exiccat, adstringit, glutinat*— "it cools, it dries out, it contracts, and it coagulates."
5. LAPIS LYNCIS, the "lynx stone" or belemnite, helpful in breaking kidney stones and perhaps against nightmares and bewitchings. Many scholars viewed these smooth cylindrical fossils as coagulated lynx urine, but Schröder dismisses this interpretation as an old fable, while supplying no clear alternative.
6. OSTIOCOLLA, or "bone stones," shaped like human bones and useful in helping fractures to mend.

But among all stones that advertise their power of healing by resembling a human organ or the form of an afflication, Schröder seems most confident about the curative powers of the *Lapis judaicus,* or "jew stone" (named for its abundance near Palestine and not, in this case, for its shape or form). The accompanying illustration, included in the most beautiful set of fossil engravings from the early years of paleontology (done by Michael Mercati, curator of the Vatican collections in the mid–sixteenth century, but not published until 1717), mixes true jew stones (bottom two rows, and center figure in top row) with crinoid stem plates in the top row, each separately labeled as *Entrochus,* or "wheel stones" (for these stem plates, of flat and circular form, even bear a central hole for the passage of an imaginary axle).

In Schröder's world, jew stones provide the mineralogical remedy par excellence for one of the most feared and painful of human ailments: kidney stones and other hard growths in body organs and vessels. Schröder tells us that jew stones may be male or female, for sexual distinctions must pervade all kingdoms to validate the full analogy of human microcosm to earthly macrocosm. The smaller, female, jew stones should be used *ad vesicae lapidem* (for stones in the bladder), while the large, male, versions can be employed *ad renum lapidem expellendum* (for the expulsion of stones from the kidneys). The doctrine of signatures suggested at least two reasons for ingesting powdered jew stones to fight kidney stones: first, their overall form resembled the human affliction, and powdering one might help to disaggregate the other; second, the precise system of parallel grooves covering the surface of jew stones might encourage a directed and outward flow (from the body) for the disaggregated particles of kidney stones.

With proper respect for the internal coherence of such strikingly different theories about the natural world, we can gain valuable insight into modes and ranges of human thought. But our fascination for the peculiarity of such differences should not lead us into a fashionable relativism of "I'm okay, you're (or, in this case, you *were*) okay," so what's the difference, so long as our disparate views each express an interesting concept of reality, and also do no harm? A real world regulated by genuine causes exists "out there" in nature, independent of our perceptions (even though we can only access this external reality through our senses and mental operations). And the system of modern science that replaced the doctrine of signatures—despite its frequent and continuing errors and arrogances (as all institutions operated by human beings must experience in spades)—has provided an increasingly more accurate account of this surrounding complexity. Factual truth and causal understanding also correlate, in general, with effectiveness—and incorrect theories do pro-

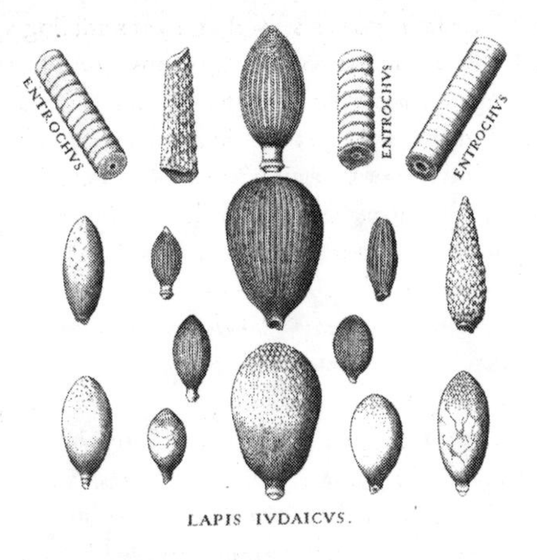

Mercati's mid-sixteenth-century engravings of "jew stones"—fossil sea urchin spines, but then regarded as medicinal rocks useful in curing bladder and kidney stones.

mote harm by their impotence. Snake venoms cannot be neutralized by powdering and ingesting stones that look like serpents' tongues (the *glossopetra* of the preceding list)—for the doctrine of signatures holds no validity in general, and these fossils happen to be shark's teeth in this particular case. So long as we favor an impotent remedy dictated by a false theory, we will not find truly effective cures.

Schröder's quaintly incorrect seventeenth-century accounts of fossils do not only reflect a simple ignorance of facts learned later. Better understanding may also be impeded by false theories that lead scholars to conceptualize fossils in a manner guaranteed to debar more fruitful questions, and to preclude more useful observations. Schröder's theory of fossils as mineral objects, relevant to human knowledge only when their forms suggest sympathy, and therefore potential therapeutic value, to diseased human parts, relegates to virtual inconceivability the key insight that fossils represent the bodies of ancient organisms, entombed within rocks, and often petrified themselves. At least two controlling conclusions of Schröder's theory prevented the establishment of this *sine qua non* for a modern paleontology.

First, the theory of signatures didn't even permit a coherent category for fossils as ancient organisms—and a truth that cannot be characterized, or even named, can hardly be conceptualized at all. In Schröder's taxonomy, fossils belonged, with no separate distinction, to the larger category of "things in rocks that look like objects of other realms." Some of these "things" were organisms—

glossopetra as shark's teeth; *lapis lyncis* as internal shells of an extinct group of cephalopods, the belemnites; *ostiocolla* as true vertebrate bones; and jew stones as the spines of sea urchins. But other "things," placed by Schröder in the same general category, are not organisms—*aetites* as geodes (spherical stones composed of concentric layers, and formed inorganically); *ceraunia* as axes and arrowheads carved by ancient humans (whose former existence could not be conceived on Schröder's earth, created just a few thousand years ago); and *haematites* as a red-hued inorganic mineral made of iron compounds.

Second, how could the status of fossils as remains of ancient organisms even be imagined under a theory that attributed their distinctive forms, not to mention their very existence, to a necessarily created correspondence with similar shapes in the microcosm of the human body? If *aetites* aid human births because they look like eggs within eggs; if *ceraunia* promote the flow of milk because they fell from the sky; if *glossopetra* cure snake bites because they resemble the tongues of serpents; if *haematites,* as red rocks, stanch the flow of blood; if *ostiocolla* mend fractures because they look like bones; and if jew stones expel kidney stones because they resemble the human affliction, but also contain channels to sweep such objects out of our bodies—then how can we ever identify and separate the true fossils among these putative remedies when we conceptualize the entire motley set only as a coherent category of mineral analogs to human parts?

I thus feel caught in the ambivalence of appreciating both the fascinating weirdness and the conceptual coherence of such ancient systems of human thought, while also recognizing that these fundamentally incorrect views stymied a truer and more humanely effective understanding of the natural world. So I ruminated on this ambivalence as I read Schröder's opening dedicatory pages to the citizens of Frankfurt am Main—and the eyes of this virtually unshockable street kid from New York unexpectedly fell upon something *in shvartz* that, while known perfectly well as an abstraction, stunned me in unexpected concrete, and helped me to focus (if not completely to resolve) my ambivalence.

Schröder's preface begins as a charming and benign, if self-serving, defense of medicine and doctors, firmly rooted within his controlling doctrine of signatures and correspondence between the human microcosm and the surrounding macrocosm. God of the Trinity, Schröder asserts, embodies three principles of creation, stability, and restitution. The human analogs of this Trinity must be understood as generation to replenish our populations, good government to ensure stability, and restoration when those systems become weak, diseased, or invaded. The world, and the city of Frankfurt am Main, needs good doctors to assure the last two functions, for medicine keeps us going, and cures us when we falter.

So far, so good. I am a realist, and I can certainly smile at the old human foible of justifying personal existence (and profit) by carving out a place within the higher and general order of things. But Schröder then begins a disquisition on the "diabolical forces" that undermine stability and induce degeneration—the two general ills that good physicians fight. Still fine by me, until I read, *in shvartz,* Schröder's description of the most potent earthly devil of all: *Choraeam in his ducunt Judaei*— the Jews lead this chorus (of devils). Two particularly ugly lines then follow. We first learn of the dastardly deeds done by Jews to the *Gojim*— for Schröder even knows the Hebrew word for gentiles, *goyim* (the plural of *goy,* nation—the Latin alphabet contains no "y" and uses "j" instead).

Schröder writes in full: *His enim regulis suis occultis permissum novit Gojim, id est, Christianos, impune et citra homicidii notam, adeoque citra conscientiam interimere*— "for by his secret rules, he [the Jew] is granted permission to kill the *goyim,* that is the Christians, with impunity, without censure, and without pangs of conscience." But, luckily for the good guys, Schröder tells us in his second statement, these evil Jews can be recognized by their innately depraved appearance—that is, by *nota malitiae a Natura ipse impressa* (marks of evil impressed by Nature herself). These identifying signs include an ugly appearance, garrulous speech, and a lying tongue.

Now, I know perfectly well that such blatant anti-Semitism has pervaded nearly all European history for almost two thousand years at least. I also know that such political and moral evil has been rationalized at each stage within the full panoply of changing views about the nature of reality—with each successive theory pushed, shoved, and distorted to validate this deep and preexisting prejudice. I also know—for who could fail to state this obvious point—that the most tragically effective slaughter ever propagated in the name of anti-Semitism, the Holocaust of recent memory, sought its cruel and phony "natural" rationale not in an ancient doctrine of harmony between microcosm and macrocosm, but in a perverted misreading of modern theories about the evolution of human variation.

Nonetheless, my benign appreciation for the fascinating, but false, doctrine of signatures surely received a jolt when I read, unexpectedly and *in shvartz,* an ugly defense of anti-Semitism rooted (however absurdly) within this very conception of nature. More accurate theories can make us free, but the ironic flip side of this important claim has often allowed evil people to impose a greater weight of suffering upon the world by misusing the technological power that flows from scientific advance.

The improvement of knowledge cannot guarantee a corresponding growth of moral understanding and compassion—but we can never achieve a maximal spread of potential benevolence (in curing disease or in teaching the factual reality of our human brotherhood in biological equality) without nurturing such knowledge. Thus the reinterpretation of jew stones as sea urchin spines (with no effect against human kidney stones) can be correlated with a growing understanding that Jews, and all human groups, share an overwhelmingly common human nature beneath any superficiality of different skin colors or cultural traditions. And yet this advancing human knowledge cannot be directed toward its great capacity for benevolent use, and may actually (and perversely) promote increasing harm in misapplication, if we do not straighten our moral compasses and beat all those swords, once anointed with Croll's sympathetic ointment to assuage their destructive capacity, into plowshares, or whatever corresponding item of the new technology might best speed the gospel of peace and prosperity through better knowledge allied with wise application rooted in basic moral decency.

10

When Fossils Were Young

In his first inaugural address, in 1861, Abraham Lincoln expressed some strong sentiments that later guardians of stable governments would hesitate to recall. "This country, with its institutions," he stated, "belongs to the people who inhabit it. . . . Whenever they shall grow weary of the existing government, they can exercise their constitutional right of amending it, or their revolutionary right to dismember or overthrow it." Compared with these grand (and just) precepts, the tiny little reforms that make life just a tad better may pale into risible insignificance, but I would not disparage their cumulative power to alleviate the weariness of existence and to forestall any consequent movement toward Lincoln's more drastic solutions. Thus, while acknowledging their limits, I do applaud, without cynicism, the introduction of workable air conditioning in the subways of New York City, croissants in our bakeries, goat cheese in our markets (how did we once survive on cheddar and Velveeta?), and supertitles in our opera houses.

Into this category of minuscule but unambiguous improvements I would place a small change that has crept from innovation to near ubiquity on the scheduling boards of our nation's airports. Until this recent reform, lists of departures invariably followed strict temporal order—that is, the 10:15 for Chicago came after the 10:10 for Atlanta (and twenty other flights also leaving at 10:10), with the 10:05 for Chicago listed just above, in the middle of another pack for the same time. Fine and dandy—so long as you knew your exact departure time and didn't mind searching through a long list of different places sharing the same moment. But most of us surely find the difference between Chicago and Atlanta more salient than the distinction between two large flocks of flights separated by a few minutes that all experienced travelers recognize as fictional in any case.

A few years ago, some enlightened soul experienced a flash of insight that should have occurred to myriads of travelers decades ago: Why not list flights by cities of destination rather than times of departure—and then use temporal order as a secondary criterion for all flights to each city? Then travelers need only scan an alphabetical list to find Chicago and its much shorter and far less confusing array of temporal alternatives. Listing by city of destination has now virtually replaced the old criterion of departure time at our nation's airports. The transition, however piecemeal, both within and among airports, took only a few years—and life has become just a bit less stressful as a result. The innovator of this brilliant, if tiny, improvement deserves the "voice of the turtle" medal for easing life's little pains (see the Song of Solomon for a full list of the small blessings that lead us to "rise up . . . and come away" to better places).

I suspect that two important and linked properties of cultural change lead to long persistence of truly outmoded and inconvenient systems, followed by rapid transition to something more sensible. First, the outdated modality arose for a good reason at the time, not by mere caprice. A lingering memory of rationality might help explain the great advantages afforded by simple incumbency in both politics and ideology. (I suspect that listing by time of departure made evident sense when the train, or the stagecoach in earlier times, passed through town along the only road, and all travelers went either one way or t'other: "The southbound stage, did you say, sir? Do you want to take the 10:30, the 3:00, or the 5:15?")

The old way, having worked so well for so long, will persist *faute de mieux* until oozing change finally makes the world sufficiently different and some bright soul gets inspired to say: "We can do much better at essentially no cost or bother." The ease of the change and the obvious character of the improvement then induce the second feature of great rapidity, once thought breaks

A

ABDALA Anarach Medicus Arabs.

Adamus	Lonicerus.
Albertus	Magnus.
Alexander	Maſſaria.
Amatus	Luſitanus.
Ambroſius	Paræus.
Andreas	Alpagius Bellunẽſis.
Andreas	Baccius.
Andreas	Cæſalpinus.
Andreas	Dörerus.
Andreas	Lacuna.
Andreas	Libauius.
Andreas	Theuetus.
Andromachus	
Anshelmus	Boetius.
Antonius	Fornefius.
Antonius	Fumanellus.
Antonius	Guaynerius.
Antonius	Mizaldus.
Antonius	Muſa Braſſauolas.
Antonius	Portus.
Antonius	Schnebergerus,
Ariſtoteles,	
Arnoldus	Manlius,
Arnoldus	Villanouanus.
Auenzoar	Arabs,
Auerroes.	
D. Auguſtinus.	
Auicenna.	

Caspar Bauhin's 1613 bibliography, arranged alphabetically, but by the authors' first names.

through the thick wall of human stodginess. In locating a biological analogy to express the speed of such a transition, I would seek an appropriate metaphor in the concept of infection, not evolution.

I mention this phenomenon because, in an accident potentiated by browsing through old books (a grand and useful pleasure that we must somehow learn to preserve—or, more accurately, keep possible—in the forthcoming world of electronic source materials rather than primary documents), I encountered a remarkably similar example, with an entirely sensible beginning at the birth of modern paleontology in 1546, followed by a troublesome middle period resembling our difficulty in searching through large flocks of flights ordered by time, and ending with a resolution by 1650 that has persisted ever since. How shall the names of authors in scholarly bibliographies be ordered? Alphabetically, of course (and "alphabetical order" already had a long pedigree), but how particularly, since people have more than one name?

Consider the example that first inspired my interest and puzzlement—the bibliography from Caspar Bauhin's 1613 volume on the bezoar stone. (The Swiss brothers Caspar (1560–1624) and Jean (1541–1613) Bauhin ranked among the greatest botanists and natural historians of their time. Bezoars are rounded and layered stones found in the internal organs—stomach, gall bladder, kidneys—of large herbivorous mammals, primarily sheep and goats. In the Bauhins' time, both medical and lay mythology attributed magical and curative powers to these stones.)

In an episode of my favorite television comedy classic, *The Honeymooners,* Ed Norton (played by Art Carney), following his promotion to file clerk from his former job in the sewers, gets into terminal trouble for his overly literal method of filing: he places "the Smith affair" under the letter *T.* Bauhin seems to follow a procedure of only slightly greater utility in alphabetizing his sources by their first names. Note his initial entries for the letter *A.* The system works well enough for a few classical authors—the single-named Aristotle and the greatest scientific philosophers of early Islam, Avicenna and Averroës. Albert the Great (Albertus Magnus), the teacher of Thomas Aquinas, also fares well enough, since Magnus is his honorific, not his last name (although I once received a student term paper that listed him as Mr. Magnus).

But I don't relish the need to search through a long list of Andreases to find my two favorite contemporaries of Bauhin (from a time when the system of first and last names had already been stabilized in Europe), the chemist Andreas Libavius and the anatomist (and geologist) Andreas Caesalpinus. When we encounter the lengthy list of Johns—the most common first name both then and now—the system's utility suffers serious deterioration. I have to remember that Caspar Bauhin's own brother bears this name if I wish to honor (and consult) the family lineage. Even worse, how will I ever find my favorite late-sixteenth-century scholar Giambattista della Porta? I must know his first name, then remember that Giambattista is Italian for "John the Baptist," and finally recognize that, in this Latin listing, he will appear as Ioannus, or just plain John.

(Filing by first name cannot be dismissed as absurd in abstract principle, but only as unworkable in actual practice, because most Europeans share just a few common first names, whereas their far more distinctive last names should be preferred as primary labels for bibliographers. As an interesting modern analog, last names can be quite scarce in Latin countries. Therefore, people who bear one of the more common monikers—the Hernández es or the Guzmáns,

Ioannes	Agricola.
Ioan. Antonius	Sarracenus.
Ioan.	Arculanus.
Io. Baptista	Montanus.
Ioan. Baptist.	Syluaticus.
Ioan.	Bauhinus.
Ioan.	Bodinus.
Ioan.	Caluinus.
Ioan.	Collerus.
Ioan.	Costæus.
Ioan.	Crato.
Ioan.	Fernelius.
Ioan.	Fragosus.
Ioan. Georgius	Agricola.
Ioan Georgius	Schenckius.
Ioan.	Gorræus.
Ioan.	Guinterus Andernacus.
Ioan.	Heurnius.
Ioan.	Hugo à Linscotten.
Ioan.	Kentmannus.
Ioan.	Langius.
Ioan.	Manardus.
Ioan.	Matthæus.
Ioan.	Mesues.
Ioan.	Porta.
Ioan.	Renodæus.
Ioan.	Schenckius.
Ioan.	Weckerus.
Ioan.	Wittichius.

How can I find the authors I seek in Bauhin's 1613 bibliography, when I know them by their last name, but Bauhin lists them by their first name—especially when so many folks bear the most common moniker of "John," as for this entire list!

for example—usually add their mother's last name to their father's primary label—as in González y Ramon—not for reasons of political correctness but rather for greater distinction in identification.)

I would have been less puzzled by this obviously suboptimal system for 1613 if I had bothered to ask myself why such a strategy had probably developed as a sensible option in the first place. Instead I found the answer, again by chance, while browsing through the great foundational work of modern geology: the 1546 treatise by Georgius Agricola.

The extensive bibliography of Agricola's *De Natura Fossilium,* the first major treatise on geology under Gutenberg's regime for producing books, uses first names for ordering and spans a wide range of time (from Homer in the ninth century B.C. to Albertus Magnus in the thirteenth century A.D.) and traditions (from the Greek dramatist Aeschylus to the Persian philosopher Zoroaster). All of Bauhin's ancient *A*'s take their rightful places: Aristotle,

Averroës, and Avicenna. But here the system of starting at the very beginning of each name makes perfect sense, because all of Agricola's sources either bear only one name or use the first word of a compound moniker as their distinctive identification.

Agricola, after all, begins the chronological list of modern scholarship. He therefore devotes his bibliography only to sources from antiquity—both because no two-named contemporary had published anything of note in the field of geology, and especially because scholars of the Renaissance (literally "rebirth") believed that knowledge had deteriorated following the golden ages of Greece and Rome and that their noble task required the rediscovery and emulation of a lost perfection that could not, in principle, be superseded.

But as scholarship exploded, and the assumptions of the Renaissance gave way to a notion (especially in science) that genuinely new knowledge could be discovered by people now living (and, in another cultural innovation, identified by their last names), the old method of bibliographic listing—ordering by the very first letter of each name, no matter which part of a compound name expressed the personal distinctiveness of an individual—ceased to make sense. Almost inevitably, the modern system of listing by the most salient marker—nearly always the "last" name for people of European extraction but (obviously) retaining the old system for single-named ancients—gained quick acceptance and became universal by the century's end. In the bibliography from an important chemical treatise of 1671 (the *Experimentum Chymicum Novum* of J. J. Becher; see page 182), our "modern" system has already prevailed. Agricola now joins Aristotle and Avicenna under the *A*'s, and I can finally find my minor hero Caesalpinus under *C,* even if I can't remember his nondistinctive first name, Andreas.

In short, the original bibliographic system in printed scientific books worked well for Agricola in 1546 because all his sources (authors from antiquity) used only one name. But by the time of Bauhin's bezoar book in 1613, the old system no longer made sense, because a fundamental change in scholarly practice—honoring new knowledge discovered by living people—combined with a social change in naming, filled Bauhin's bibliographic list with modern, two-named folks still listed by the old system as John, Bill, or Mike. Bauhin, in other words, got stuck in a transient intermediary state bound for imminent extinction—still ordering by the time of the afternoon stage for London, when his readers wanted to know the cities first and the times afterward. Within a few decades, some bright soul devised the simple reform (listing by the most salient name) that brought an old institution (the bibliography) into line with a new practice (the relevance and dominance of modern authors).

SCRIPTORES,
QUORUM INVENTIS USUS SUM, ATQUE
EX IPSIS HI, QUI NON EXTANT, AB ALIIS UT
rerum, de quibus ſcribunt, autores citantur.

Ælius Lampridius
Ælius Spartianus
Æſchylus
Aetius Amidenus
Albertus
Alexander Aphrodiſienſis
Alexander Cornelius
Alexander qui ſcripſit res Lyciacas
D. Ambroſius
Antiphon
Apion Pliſtonices
Archelaus
Ariſteas Proconneſius
Ariſtophanes
Ariſtoteles
Aſurabas
Averroes
D. Auguſtinus
Aulus Gellius
Avicenna
Bocchus
C. Plinius Secundus ſenior
Caſsiodorus
Columella
Cornelius Celſus
Cornelius Nepos
Cornelius Tacitus
Cteſias
Diemachus
Democritus qui ſcripſit De lapidibus
Demoſtratus
Diodorus Siculus
Dionyſius Afer
Dioſcorides
Empedocles Agrigentinus
Eratoſthenes
Euripides
Fabius Pictor
Fl. Vopiſcus
Galen. Pergamenus
Græcus ignotus qui ſcripſit De admirandis auditionibus.
Hermogenes
Herodotus
Heſychius
D. Hieronymus
Hierocles
Hippocrates
Homerus
Horus
Jacchus
Iſmenias
Juba
M. Varro
Martialis
Megaſthenes
Metrodorus
Mithridates
Mneſias
Mutianus
Nicanor
Nicias
Oribaſius
Ovidius
Paulus Ægineta
Pauſanias
Philemon
Philoſtratus
Philoxenus
Phocion grāmaticus
Pindarus
Plutarchus
Poſidonius
Pſellus
Ptolomæus
Pytheas
Quadrigarius
Satyrus
Seneca
Serapio
Sex. Pompejus Feſtus
Solinus
Sophocles
Sotacus
Stephanus
Strabo
Sudines
Suetonius
Theognides
Theophraſtus
Theomenes
Theopompus
Timæus
Valerius Maximus
Verrius
Virgilius
Vitruvius
Xenocrates
Xenophon
Zenothemis
Zoroaſtres

FINIS.

ILLUSTRI

Agricola's 1546 bibliography alphabetized sensibly by first (and usually only) names because nearly all his sources represent classical authors identified by their initial names.

The common message of these stories—that our traditions may arise for small and sensible reasons but may then outlive their utility by persisting as oddities and impediments in an altered world—may strike most scientists as intriguing enough but irrelevant to their own professional practices. We all recognize that organisms, in their evolution, may be hampered by such historical baggage, not easily shed from the genetic and developmental systems of complex creatures now adapted to different environments. Whales cannot jettison their air-breathing lungs, while humans suffer herniations and lower-back pains because upright posture imposes such weight and stress upon weak muscles that, in our four-footed ancestors, never needed to bear such a burden. But we do not sufficiently honor the analogous principle that scientific ideas, in their history of growth and development, may also be stymied by superannuated beliefs and traditions—intellectual luggage that should be checked at the gate (and truly lost, on purpose for once, by an obliging airline).

We do eventually amend the worst injustices of our foundational documents, thus avoiding the tragedies of Lincoln's more drastic alternative of armed insurrection: women can now vote, and an African American can no longer count as only three-fifths of a person in our national census. But we do not always shed the burdens of less pernicious irrationalities. We have never, for example, amended the constitutional provision that presidents must have been born within the geographic confines of literal American territory. If your impeccably patriotic parents of Mayflower descent happened to be traveling in France when you decided to make your worldly entry, you had best find another way to place your mark upon history.

All the traditions that I have discussed in this essay—from airline listings to bibliographies and constitutional definitions—represent taxonomies, or classifications of related objects into an order that either helps us to retrieve information (the basic utilitarian reason for erecting taxonomies) or purports to explain the basis of variation (the scientist's more general rationale for devising systems of classification). My examples from the Constitution of the United States, discussed just above, illustrate this principle well. All the claims that now seem either cruel or merely senseless pose arbitrary answers to questions about categories and classifications: Who shall vote? How shall we count people for the census? How shall "true blue" be defined as a criterion for ultimate leadership?

I emphasize this principle because false taxonomies—based on sensible criteria at first but then persisting as traditions that can only be deemed arbitrary (at best) or harmful (at worst)—form a potent category of mental biases that becloud our view of empirical nature and our moral compass as well. My fel-

B.	C.
Barnaudus.	Cæſalpinus.
Barlæus.	de Caſtagnia.
Bartholinus.	Certaldus.
Beguinus.	de la Chambre.
Bernhardus.	Chriſtina Regina.
Boodt.	Chryſoſtomus.
Borellus.	Claveus.
Boyle.	Clazomerius.

Becher's bibliography of 1671, now sensibly ordered by last names.

low scientists seem particularly subject to this species of blindness because we have been trained to think that we see the world objectively. We therefore become specially subject to delusion by taxonomic schemes implanted in our minds by cultural traditions of learning but falsely regarded as expressing an objective natural reality.

To present an example of tradition's power to reach forward from arbitrary beginnings by imposing false taxonomies upon human thought, let me return to the Bauhins—this time, to the brother hidden by Caspar among his Johns. Strange as this fact may sound to modern ears, the first half-century of modern paleontology—from Agricola's first printed treatise of 1546 up to 1600—included virtually no illustrations of fossils in published sources, although botanical traditions for illustration had already led to the production of several elaborate, lavishly illustrated herbals. Agricola's long and elegantly printed treatise included no pictures—and neither did the great source from antiquity, Pliny's *Natural History.* I can think of only four or five sixteenth-century sources that printed any illustrations of fossils at all, and only two of these works feature series of drawings that could be called either extensive or systematic—*De Rerum Fossilium,* the 1565 treatise by the great Swiss polymath Conrad Gesner, and a 1598 monograph on the medicinal waters and surrounding natural environments of the German fountains at Boll by Jean Bauhin (who as a young man had studied and then collaborated with Gesner).

Gesner's work presents a general exemplification, not a report about a specific collection from a particular place. Gesner therefore provided simple woodcut illustrations for one or two specimens from each major group of "fossils" then known in Europe (a heterogeneous assemblage by modern standards, including paleolithic hand axes then interpreted as stones that fell from the sky in thunderstorms, and sea urchins then interpreted by some scholars as serpents' eggs).

Bauhin's treatise, on the other hand, represents a true beginning for an important tradition in science: the depiction not only of characteristic forms or representative specimens, but an attempt to present the full range of variety found in a particular fauna—in other words, to "draw 'em all just as one sees 'em," without any confusing selection or interpretation. In fact, in his very few paragraphs of introductory text, Bauhin tells us that he will simply show what he sees and not enter the brewing debate about the meaning of fossils. For that fascinating but (for his purposes) diversionary activity, readers will have to consult (he pointedly tells us) the aforementioned scholarly works of Agricola and Gesner. He, Jean Bauhin, nature's humble servant, will simply draw the fossils he has found and let readers draw their own conclusions.

Bauhin's fifty-three pages and 211 drawings therefore mark the first printed presentation of a complete set of fossil specimens from a particular place. In consulting his treatise, we are truly "present at the creation" of an important tradition in the depiction and classification of nature's bountiful variety. But however much we appreciate the privilege, and however appropriately we admire Bauhin's originality, we should also bear the theme of this essay in mind and ask some crucial questions about cultural practices: What conventions did Bauhin invent in creating this genre? Did his rules and customs make sense in his day? Did they then become arbitrary impediments to increasing knowledge, masquerading as an "obvious" way to present "objective" facts of nature?

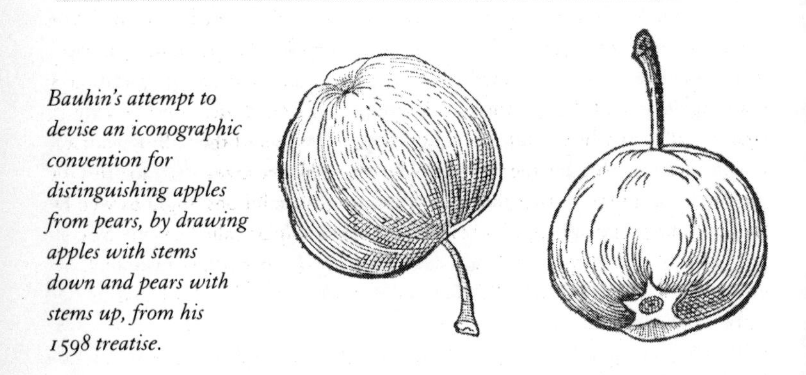

Bauhin's attempt to devise an iconographic convention for distinguishing apples from pears, by drawing apples with stems down and pears with stems up, from his 1598 treatise.

As a striking proof that our iconographic traditions may originate as arbitrary inventions of idiosyncratic beginners, we need only consider the chapter following the opening section on fossils in Bauhin's 1598 treatise—his discussion of variation in local pears and apples. The average apple couldn't be confused with the average pear, but so many forms of both fruits had been developed in this region of Germany that extensive overlap could lead to uncertainties for dumpy and elongated apples or for compressed and top-heavy pears. Bauhin therefore invented the practice of drawing all the apples stem down and all the pears stem up!

Bauhin's convention did not take hold, so we tend to view his illustrations as rather quaint—and the purely arbitrary character of his decision stands out clearly to us today. But suppose that his practice had endured. Wouldn't we be wondering today why Delicious goes down and Bartlett goes up? Or, more interesting to contemplate, would we be pondering this issue at all? Perhaps we would simply be accepting a printed orientation that we had seen throughout life, never bothering to question the evident discrepancy with nature's obedience to gravity, where both fruits hang *down* from stems. (Or perhaps, as city folks dwelling in concrete jungles, we would never even realize that nature works differently from art. Honest, growing up as a New York City street kid, I really didn't know for a long time that milk came out of cows' teats—yuck!—rather than from bottles *ab initio.*)

This example may strike readers as silly. But we follow similarly arbitrary conventions and mistake them for natural reality all the time. Anglophone publications, for example, always draw snails with the apex (the pointy end) on top and the aperture at the bottom. This orientation seems so obviously natural to me—apex up, aperture down. Of course—how else could a snail be? But French publications—and I do not know how or why the differences in practice began—usually draw snails in the opposite orientation, with the apex pointing down and the aperture up. So millions of Frenchmen must be wrong, *n'est-ce pas?* But when you learn about the difference and then allow yourself to consider the issue for the first time, you suddenly realize—and the insight can be quite salutary—that the French and English solutions might as well be Bauhin's pears and apples. Neither mode can possibly be called correct by correspondence to nature. Most snails crawl horizontally along the substrate. Both ends of the shell lie basically parallel to the sea floor. Neither can be labeled as intrinsically up or down.

In another example that caused me some personal embarrassment but also taught me something important about convention versus nature, I once wrote that the North Pole pointed up and that our planet rotated counterclockwise around this axis (viewed, as by God or an astronaut, from above). An Australian reader wrote me a letter, gently pointing out the absence of absolute "up" or "down" in the cosmos and reminding me that our cartographic convention only reflects where most map-making Europeans live. From his patriotic vantage (and accepting another dubious convention that equates "up" with "good"), the Antarctic Pole points up, and Earth rotates clockwise around this southern standard.

The situation becomes even more complicated, and even more evidently ruled by convention, when we consider the history of cartography. In many medieval maps, drawn under the Ptolemaic notion of a central and nonrotating Earth, east—the direction of the rising Sun—occupies the top of the map. The word *orient*— meaning "east," but with an etymology of "rising" in reference to the sun—gained the additional and more symbolic definition of "locating one's position," because east once occupied this favored top spot on our standard maps. (The Chinese used to be called Orientals for the same reason, before the term lapsed from political correctness, while Europeans became Occidentals, or westerners, in literal reference to the "falling down," or setting, of the Sun.)

When we survey Bauhin's more than two hundred fossil drawings, the largest single cache of sixteenth-century paleontological illustrations, we note the origin of several conventions that, although superseded today (and therefore unknown to most modern scientists), seriously impeded, for nearly two centuries, a proper understanding of the nature of fossils and the history of life. Consider just three classes of examples, all based on the taxonomic conventions of sixteenth-century paleontology. The recognition of Bauhin's illustrations as conventional rather than natural, and their replacement, by the end of the eighteenth century, with "modern" figures that clearly depict fossils as ancient organisms, virtually defines the primary shift in understanding that led to our greatest gain in knowledge during the early history of paleontology.

1. *Conflation of categories.* In Bauhin's day, the word *fossil*— derived from the past participle of the Latin verb *fodere,* meaning "to dig up"—referred to any object of distinctive form found within the earth, thus placing the remains of ancient organisms in the same general category as crystals, stalactites, and a wide range of other inorganic objects. Until organic remains could be recognized as distinctive, placed in a category of their own, and properly interpreted as the products of history, modern geology, with its central concept of continuous change through deep time, could not replace the reigning paradigm of an Earth only a few thousand years old and created pretty much as we find it today, with the possible exception of changes wrought by Noah's universal flood.

If fossils originated within rocks as products of the mineral kingdom, just as crystals grow in mines and stalactites form in caves, then a petrified "shell" may just denote one kind of inorganic object manufactured in its proper place within the mineral kingdom. Thus, when Bauhin places his drawing of a fossil snail shell right next to a conical mound of crystals because both share a roughly similar shape and supposed mode of inorganic origin, this taxonomic convention does not merely record a neutral "fact" of pure observation, as Bauhin claimed. Rather, his juxtaposition of two objects now viewed as fundamentally different in origin and meaning expresses a theory about the structure of nature and the pattern of history—a worldview, moreover, that stood firmly against one of the great revolutions in the history of scientific understanding: the depth of time and the extent of change.

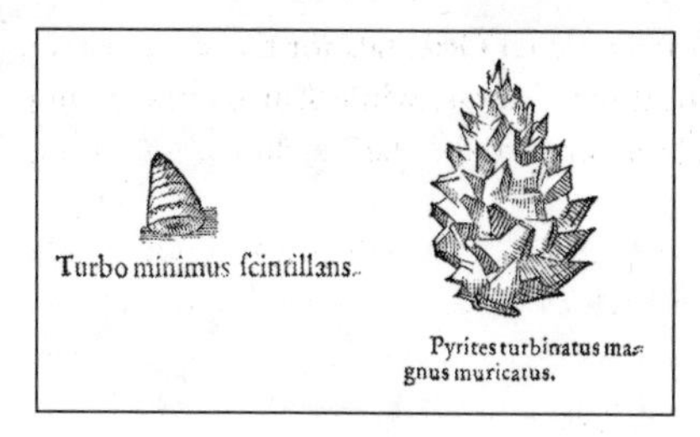

Bauhin's taxonomy places a fossil snail shell next to an inorganic mass of crystals because both have roughly the same shape.

2. *Failure to distinguish accidental resemblance from genuine embodiment.* Contrary to a common impression, Jean Bauhin and his contemporaries did not claim that all fossils must be inorganic in origin and that none could represent the petrified remains of former plants and animals. Rather, they failed to make a sharp taxonomic distinction between specimens that they did regard as organic remains and other rocks that struck them as curious or meaningful for their resemblance to organisms or human artifacts, even though they had presumably formed as inorganic products of the mineral kingdom. Thus, Bauhin includes an entire page of six rocks that resemble male genitalia, while another page features an aggregate of crystals with striking similarity in form to the helmet and head covering in a suit of armor. He does not regard the rocks as actual fossilized penises and testicles, and he certainly doesn't interpret the "helmet" as a shrunken trophy from the Battle of Agincourt. But his taxonomic juxtaposition of recognized mineral accidents with suspected organic remains does lump apples and oranges together (or, should I say, apples and pears, without a stem-up-or-down convention to permit a fruitful separation), thereby strongly impeding our ability to identify distinct causes and modes of origin for genuine fossil plants and animals.

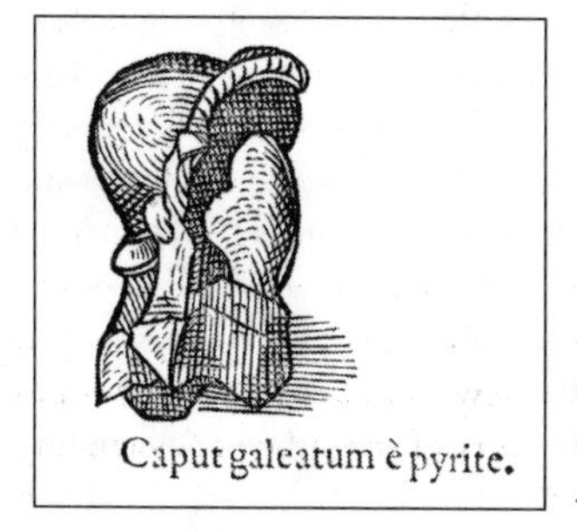

Bauhin's caption reads: "a helmeted head made of pyrite"—an accidental resemblance that he regarded as meaningful, at least in a symbolic sense.

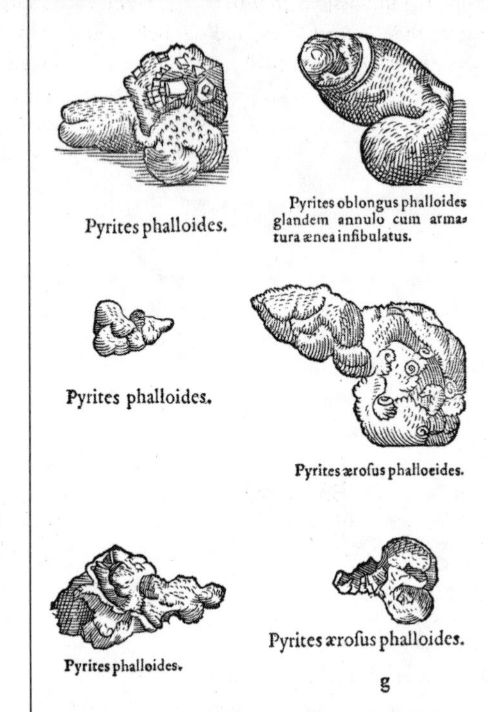

Bauhin's 1598 illustration of six rocks that resemble penises. He did not regard them as actual fossils of human parts, but he did interpret, as causally meaningful, a likeness that we now regard as accidental.

3. *Drawing organic fossils with errors that preclude insight into their origins.* Bauhin claimed that he drew only what he saw with his eyes, unencumbered by theories about the nature of objects. We may applaud this ideal, but we must also recognize the practical impossibility of full realization. What can be more intricate and complex than a variegated rock filled with fossils (most in fragmentary condition), mineral grains, and sedimentary features of bedding followed by later cracking and fracturing? Accurate drawing requires that an artist embrace some kind of theory about the nature of these objects, if only to organize such a jumble of observations into something coherent enough to draw.

Since Bauhin did not properly interpret many of his objects as shells of ancient organisms that had grown bigger during their lifetimes, he drew several specimens with errors that, if accepted as literal representations, would have precluded their organic status. For example, he tried to represent the growth lines of a fossil clam, but he drew them as a series of concentric circles—

implying impossible growth from a point on the surface of one of the valves—rather than as a set of expanding shell margins radiating from a starting point at the edge of the shell where the two valves hinge together.

Bauhin also drew many ammonite shells fairly accurately, but these extinct relatives of the chambered nautilus grew with continually enlarging whorls (as the animal inside increased in size). Bauhin presents several of his ammonites, however, with a final whorl distinctly smaller than preceding volutions from a younger stage of growth. Reading this error literally, an observer would conclude that these shells could not have belonged to living and growing organisms. Finally, in the most telling example of all, Bauhin drew three belemnites (cylindrical internal shells of squidlike animals) in vertical orientation, covered with a layer of inorganic crystals on top—clearly implying that these objects grew inorganically, like stalactites hanging from the roof of a cave.

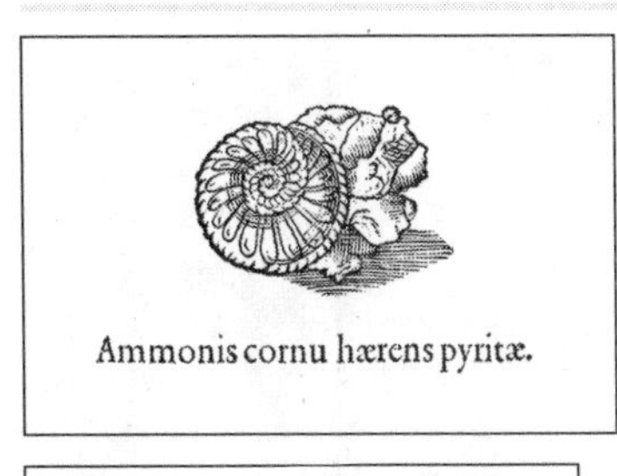

The final whorl of this fossil ammonite becomes smaller, showing that Bauhin did not recognize the object as a fossil organism whose shell would have increased in width during growth.

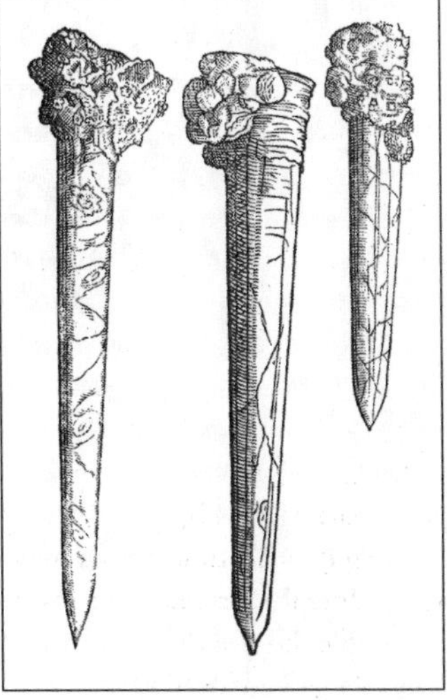

Bauhin's most telling example of iconographic conventions that he established and that show his misunderstanding of the organic nature of fossils. He draws these three belemnites (internal shells of extinct squidlike animals) as if they were inorganic stalactites hanging down from the ceiling of a cave.

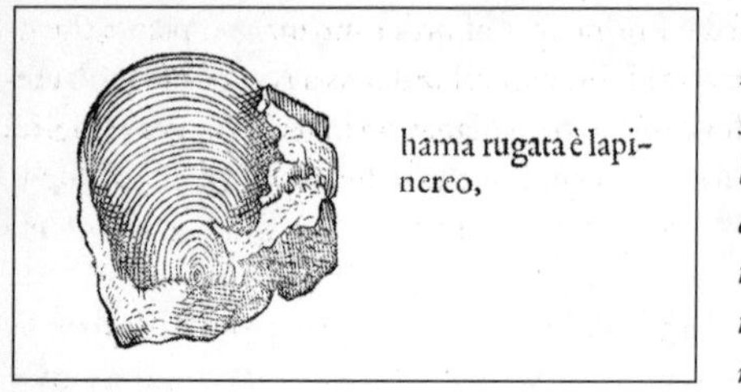

Bauhin draws this fossil clam shell incorrectly, with growth lines as concentric circles. He did not recognize the object as the remains of an organism that could not grow in such a manner.

So long as later scientists followed these three iconographic conventions that Bauhin developed in 1598, paleontology could not establish the key principle for a scientific understanding of life's history: a clear taxonomic separation of genuine organic remains from all the confusing inorganic objects that had once been lumped together with them into the heterogeneous category of "figured stones"—an overextended set of specimens far too diffuse in form, and far too disparate in origin, to yield any useful common explanations. These early conventions of drawing and classification persisted until the late eighteenth century, thus impeding our understanding about the age of the Earth and the history of life's changes. Even the word *fossil* did not achieve its modern restriction to organic remains until the early years of the nineteenth century.

I do not exhume this forgotten story to blame Jean Bauhin for establishing a tradition of drawing that made sense when naturalists did not understand the meaning of fossils and had not yet separated organic remains from mineral productions—a tradition that soon ceased to provide an adequate framework and then acted as an impediment to more-productive taxonomies. The dead bear no responsibility for the failures of the living to correct their inevitable errors.

I would rather praise the Bauhin brothers for their greatest accomplishment in the subject of their primary joint expertise—botanical taxonomy. Brother Caspar of the bezoar stone published his greatest work, *Pinax,* in 1623—a taxonomic system for the names of some six thousand plants, representing forty years of his concentrated labor. Brother Jean of the fossils of Boll had been dead for thirty-seven years before his greatest work, *Historia plantarum universalis,* achieved posthumous publication in 1650, with even more elaborate descriptions and synonyms of 5,226 distinct kinds of plants.

Botanical taxonomy before the Bauhin brothers had generally followed capricious conventions of human convenience rather than attempting to determine any natural basis for resemblances among various forms of plants (several previous naturalists had simply listed the names of plants in alphabetical order). The Bauhin brothers dedicated themselves to the first truly systematic search for a "natural" taxonomy based on principles of order intrinsic to plants themselves. (They would have interpreted this natural order as a record of God's creative intentions; we would offer a different explanation in terms of genealogical affinity produced by evolutionary change. But the value of simply deciding to search for a "natural" classification precedes and transcends the virtue of any subsequent attempt to unravel the causes of order.)

Caspar Bauhin may have slightly impeded the progress of bibliography by retaining an outmoded system in his bezoar book. Jean Bauhin may have stymied the development of paleontology in a more serious manner by establishing iconographic conventions that soon ceased to make sense but that later scientists retained for lack of courage or imagination. But the Bauhin brothers vastly superseded these less successful efforts with their brilliant work on the fruitful basis of botanical taxonomy. Their system, in fact, featured a close approach to the practice of binomial nomenclature, as later codified by Linnaeus in mid-eighteenth-century works that still serve as the basis for modern taxonomy of both plants and animals.

Science honored the Bauhin brothers when an early Linnaean botanist established *Bauhinia* as the name for a genus of tropical trees. Linnaeus himself then provided an ultimate accolade when he named a species of this genus *Bauhinia bijuga,* meaning "Bauhins linked together," to honor the joint work of these two remarkable men. We might also recall Abraham Lincoln's famous words (from the same inaugural address that opened this essay) about filial linkage of a larger kind—between brethren now at war, who must somehow remember the "mystic chords of memory" and reestablish, someday, their former union on a higher plane of understanding.

The impediments of outmoded systems may sow frustration and discord, but if we force our minds to search for more fruitful arrangements and to challenge our propensity for passive acceptance of traditional thinking, then we may expand the realms of conceptual space by the most apparently humble, yet most markedly effective, intellectual device: the development of a new taxonomic scheme to break a mental logjam. "Rise up . . . and come away. For, lo, the winter is past, the rain is over and gone; the flowers appear on the earth" (Song of Solomon 2:10–12)—fruits of nature for the Bauhin brothers, and all their followers, to classify and relish.

11

Syphilis and the Shepherd of Atlantis

We usually manage to confine our appetite for mutual recrimination to merely petty or mildly amusing taunts. Among English speakers, unannounced departures (especially with bills left unpaid), or military absences without permission, go by the epithet of "taking French leave." But a Frenchman calls the same, presumably universal, human tendency *s'en filer à l'Anglaise,* or "taking English leave." I learned, during an undergraduate year in England, that the condoms I had bought (for no realized purpose, alas) were "French letters" to my fellow students. In France that summer, my fellow students of another nation called the same item a *chapeau Anglais,* or "English hat."

But this form of pettiness can escalate to danger. Names and symbols enflame us, and wars have been fought over flags or soccer matches. Thus, when syphilis first began to ravage Europe in the

1480s or 1490s (the distinction, as we shall see, becomes crucial), a debate erupted about naming rights for this novel plague—that is, the right to name the disease for your enemies. The first major outbreak had occurred in Naples in the mid 1490s, so the plague became, for some, the Italian or the Neapolitan disease. According to one popular theory, still under debate in fact, syphilis had arrived from the New World, brought back by Columbus's sailors who had pursued the usual activities in novel places—hence the Spanish disease. The plague had been sufficiently acute a bit northeast of Columbus's site of return—hence the German disease. In the most popular moniker of all, for this nation maintained an impressive supply of enemies, syphilis became the French disease (*morbus Gallicus* in contemporary medical publications, usually written in Latin), with blame cast upon the troops of the young French king, Charles VIII, who had conquered Naples, where the disease first reached epidemic proportions, in 1495. Supporters of this theory then blamed the spread through the rest of Europe upon the activities of Charles's large corps of mercenary soldiers who, upon demobilization, fanned out to their homes all over the continent.

I first encountered this debate in a succinct summary written by Ludovico Moscardo, who described potential herbal remedies in the catalog of his museum, published in 1682: *"Ne sapendo, a chi dar la colpa, li spagnuoli lo chiamorono mal Francese, li Francesi mal Napolitano, e li Tedeschi, mal Spagnuolo"* (not knowing whom to blame, the Spaniards call it the French disease, the French the Neapolitan disease, and the Germans the Spanish disease). Moscardo then adds that other people, citing no specific human agent, attribute the origins of syphilis to bad airs generated by a conjunction of the three most distant planets, Mars, Jupiter, and Saturn, in the nighttime sky.

How, then, did the new plague receive its universal modern appellation of syphilis; and what does syphilis mean, anyway? The peculiar and fascinating tale of the name's origin can help us to understand two key principles of intellectual life that may seem contradictory at first, but that must be amalgamated into a coherent picture if we hope to appreciate both the theories of our forebears and the power of science to overcome past error: first, that the apparently foolish concepts of early scientists made sense in their times and can therefore teach us to respect their struggles; and, second, that those older beliefs were truly erroneous, and that science both progresses, in any meaningful sense of the term, and holds immense promise for human benefit through correction of error and discovery of genuine natural truths.

Syphilis, the proper name of a fictional shepherd, entered our language in a long poem, 1,300 verses in elegant Latin hexameter, written in 1530 by the greatest physician of his generation (and my second-favorite character of the time, after Leonardo da Vinci)—a gentleman from Verona (also the home of Romeo and Juliet), Girolamo Fracastoro (1478–1553). Fracastoro dabbled in astronomy (becoming friendly with Copernicus when both studied medicine at Padua in 1501), made some crucial geological observations about the nature of fossils, wrote dense philosophical treatises and long classical poems, and held high status as the most celebrated physician of his time. (In his role as papal doctor, for example, he supervised the transfer of the Council of Trent to Bologna in 1547, both to honor His Holiness's political preferences and to avoid a threatened epidemic.) In short, a Renaissance man of the Renaissance itself.

I cannot imagine a starker contrast—my original inspiration for writing this essay, by the way—between Fracastoro's christening of 1530 and the sequencing of the genome of *Treponema pallidum,* the bacterial spirochete that truly causes syphilis, in 1998. Fracastoro could not resolve the origins of syphilis and didn't even recognize its venereal mode of transmission at first. So he wrote a poem and devised a myth, naming syphilis therein to honor a fictional shepherd of his own invention. In greatest contrast, the sober paper, published in *Science* magazine (July 17, 1998) in an article with thirty-three coauthors, resolves the 1,138,006 base pairs, arranged in a sequence of 1,041 genes, in the genome of *T. pallidum,* the undoubted biological cause of syphilis.

Fracastoro's shepherd may have ended an acrimonious debate by donating his neutral name, but Fracastoro himself, as a Veronese patriot, made his own allegiances clear in the full title of his epic poem: *Syphilis sive Morbus Gallicus*—"Syphilis, or the French Disease." To epitomize some horrendous complexities of local politics: Verona had long been controlled by the more powerful neighboring city of Venice. Italy did not yet exist as a nation, and the separate kingdom of Naples maintained no formal tie to Venice. But commonalities of language and interest led the citizens of Verona to side with Naples against the invading French forces of Charles VIII, while general French designs on Italian territory prompted nearly a half-century of war, and strong Italian enmity, following Charles's temporary occupation of Naples.

Meanwhile, Maximilian I, the Habsburg Holy Roman Emperor (a largely German confederation in central Europe, despite the name), added Spain to his extensive holdings by marrying both his son and daughter to Spanish rulers. He also allied himself with the Pope, Venice, and Spain to drive Charles VIII out of Italy. Ten years later, given the shifting alliances of *Realpolitik,* Maximilian had made peace with France and even sought their aid to wage war on Venice. His successful campaign split Venetian holdings, and Maximilian occupied Fracastoro's city of Verona from 1509 until 1517, when control reverted to Venice by treaty.

Fracastoro had fled the territory to escape Maximilian's war with Venice. But he returned in 1509, and began to prosper both immediately and mightily—so I assume that his allegiances lay with Maximilian. But, to shorten the tale and come to the relevant point, Maximilian (at least most of the time) controlled Spain and regarded France as his major enemy. Fracastoro, as a Veronese patriot and supporter of Maximilian, also despised the French presence and pretensions. Fracastoro's interest therefore lay with absolving Spain for the European spread of syphilis by denying the popular theory that Columbus's men had inadvertently imported the "Spanish disease" with their other spoils from the New World. Hence, for Fracastoro, his newly christened syphilis would carry the subtitle of *Morbus Gallicus.*

I can't boast nearly enough Latin to appreciate Fracastoro's literary nuances, but experts then and now have heaped praise upon his Virgilian style. J. C. Scaliger, perhaps the greatest scholar of Fracastoro's generation, lauded the work as "a divine poem"; and Geoffrey Etough, the major translator of our time, writes that "even Fracastoro's rivals acclaimed him second only to Virgil." In this essay, I will use Nahum Tate's English version of 1686, the first complete translation ever made into any other language, and a highly influential work in its own right (despite the clunkiness of Tate's heroic couplets, in utterly unrelieved iambic pentameter). This version remained a standard source for English readers for more than two centuries. Tate, one of England's least celebrated poets laureate (or should we pluralize this title as poet laureates, or even poets laureates?), wrote the libretto for Henry Purcell's short operatic jewel *Dido and Aeneas.* A few devout choristers may also know his texts for "While shepherds watched" or "As pants the hart." We shall pass by his once popular "adaptation" of *King Lear,* with its happy ending in Cordelia's marriage to Edgar.

Syphilis sive Morbus Gallicus includes three parts, each with its own form and purpose. Part 1 discusses origins and causes, while parts 2 and 3 narrate myths in closely parallel structure, devised to illustrate the two most popular (though, in retrospect, not particularly effective) cures. Fracastoro begins by defending his choice of *morbus Gallicus:*

> . . . To Naples first it came
> From France, and justly took from France his name
> Companion from the war. . . .

He then considers the theory of New World transmission on Spanish ships and admits the tragic irony, if true:

If then by Traffick thence this plague was brought
How dearly dearly was that Traffick bought!

But Spanish shipping cannot be blamed, Fracastoro holds, because the disease appeared too quickly and in too many places (including areas that had never received products from the New World) to validate a single point of origin:

Some instances in divers lands are shown
To whom all Indian Traffick is unknown
Nor could th'infection from the Western Clime
Seize distant nations at the self same time.

Spain must therefore be absolved:

Nor can th'infection first be charged on Spain
That sought new worlds beyond the Western main.
Since from Pyrene's foot, to Italy
It shed its bane on France, While Spain was free . . .
From whence 'tis plain this Pest must be assigned
To some more pow'rful cause and hard to find.

The remainder of part 1 presents Fracastoro's general view of nature as complex and puzzling, but intelligible—thereby giving us fascinating insight into the attitudes of Renaissance humanism, an approach that tried to break through the strictures of scholastic logical analysis to recover the presumed wisdom of classical times (renaissance means "rebirth"), but had not developed the belief in primacy of empirical documentation that would characterize the rise of modern science more than a century later. Fracastoro tells us that we must not view syphilis as divine retribution for human malfeasance (a popular theory at the time), a plague that must be endured but cannot, as a departure from nature's usual course, be comprehended.

Rather, syphilis has a natural cause that can, in principle, be understood. But nature is far more complex, and unattuned to human sensibilities, than we had been willing to admit, and explanation will not come easily—for nature works in strange ways, and at scales far from our easy perception. For example, Fracastoro argues, syphilis probably had no single point of origin followed by later spread (thus absolving Spain once again). Its particles of contagion (whatever they may be) must be carried by air, but may remain latent for centuries before breaking out. Thus the plague of any moment may emerge from causes set long before. Moreover, certain potent causes—planetary conjunctions, for example, that may send poisonous emanations to earth—remain far from our potential observation or understanding. In any case, and on a note of hope, Fracastoro depicts plagues as comprehensible phenomena of complex nature. And just as they ravage us with sudden and unanticipated fury, the fostering conditions will change in time, and our distress shall lift:

> Since nature's then so liable to change
> Why should we think this late contagion strange? . . .
> The offices of nature to define
> And to each cause a true effect assign
> Must be a task both hard and doubtful too . . .
> [But] nature always to herself is true.

Part 2 continues the central theme of natural causation and potential alleviation, but in a very different manner. Fracastoro, following traditions of Latin epic poetry, now constructs a myth to illustrate both the dangers of human hubris and the power of salvation through knowledge. He begins by giving the usual sage advice about alleviation via good living: lots of vigorous exercise, healthy and frugal diet, and no sex. (This regimen, addressed to males alone, proscribes sex only as a strain upon bodily energy, not as a source of infection—for Fracastoro did not yet understand the venereal transmission of syphilis.) But the cure also requires pharmacological aid. Fracastoro upheld the traditional Galenic theory of humors and regarded all disease, including syphilis, as a disharmony among essential components—an imbalance that must be restored by such measures as bleeding, sweating, and purging:

> At first approach of Spring, I would advise,
> Or ev'n in Autumn months if strength suffice,
> to bleed your patient in the regal vein,
> And by degrees th'infected current drain.

Part 2 then extols the virtues of mercury as a cure in this context. Mercury can, in fact, retard the spread of the syphilis spirochete; but Fracastoro interpreted its benefits only in terms of humoral rebalancing, and the purging of poisons—for mercury plasters induced sweating, while ingestion encouraged copious spitting. The treatment, he admitted, can only be called unpleasant in the extreme, but ever so preferable to the dementia, paralysis, and death imposed by syphilis in the final stages of worst cases:

Nor let the foulness of the course displease.
Obscene indeed, but less than your disease . . .
The mass of humors now dissolved within,
To purge themselves by spittle shall begin,
Till you with wonder at your feet shall see,
A tide of filth, and bless the remedy.

Finally, Fracastoro spins his myth about human hubris, repentance, and the discovery of mercury. A hunter named Ilceus kills one of Diana's sacred deer. Apollo, Diana's brother, becomes royally infuriated and inflicts the pox of syphilis upon poor Ilceus. But the contrite hunter prays mightily and sincerely for relief, and the goddess Callirhoe, feeling pity, carries Ilceus underground far from the reach of the sun god's continuing wrath. There, in the realms of mineralogy, Ilceus discovers the curative power of mercury.

The title page of an early treatise on syphilis and the ineffective "remedies" then touted.

Fracastoro wrote these first two parts in the early 1510s and apparently intended to publish them alone. But by the 1520s, a new (and ultimately ineffective) "wonder cure" had emerged, and Fracastoro therefore added a third part to describe the new remedy in the same mythic form previously applied to mercury—the same basic plot, but this time with a shepherd named Syphilus in place of the hunter Ilceus. And thus, with thanks to readers for their patience, we finally come to Fracastoro's reason and motives for naming syphilis. (An excellent article by R. A. Anselment supplied these details of Fracastoro's composition: "Fracastoro's Syphilis: Nahum Tate and the realms of Apollo," *Bulletin of the John Rylands University Library of Manchester* 73 [1991]: 105–18.)

Fracastoro's derivation of his shepherd's name has never been fully resolved (although much debated), but most scholars regard syphilis (often spelled Syphilus) as a medieval form of Sipylus, a son of Niobe in Ovid's *Metamorphosis*—a classical source that would have appealed both to Fracastoro's Renaissance concern for ancient wisdom, and to his abiding interest in natural change. In part 3 of Fracastoro's epic, the sailors of a noble leader (unnamed, but presumably Columbus) find great riches in a new world, but incur the wrath of the sun god by killing his sacred parrots (just as Ilceus had angered the same personage by slaying Diana's deer). Apollo promises horrible retribution in the form of a foul disease—syphilis again. But just as the sailors fall to their knees to beg the sun god's forgiveness, a group of natives arrives—"a race with human shape, but black as jet" in Tate's translation. They also suffer from this disease, but they have come to the grove of birds to perform an annual rite that both recalls the origin of their misfortune and permits them to use the curative power of local botany.

These people, we learn, are the degraded descendants of the race that inhabited the lost isle of Atlantis. They had already suffered enough in losing their ancestral lands and flocks. But a horrendous heat wave then parched their new island and fell with special fury on the king's shepherd:

> A shepherd once (distrust not ancient fame)
> Possessed these downs, and Syphilus his name.
> A thousand heifers in these vales he fed,
> A thousand ewes to those fair rivers led . . .
> This drought our Syphilus beheld with pain,
> Nor could the sufferings of his flock sustain,
> But to the noonday sun with upcast eyes,
> In rage threw these reproaching blasphemies.

Syphilus cursed the sun, destroyed Apollo's altars, and then decided to start a new religion based on direct worship of his local king, Alcithous. The king, to say the least, approved this new arrangement:

> Th'aspiring prince with godlike rites o'erjoyed,
> Commands all altars else to be destroyed,
> Proclaims himself in earth's low sphere to be
> The only and sufficient deity.

Apollo becomes even angrier than before (for Ilceus alone had inspired his wrath in part 2)—and he now inflicts the disease upon everyone, but first upon Syphilus who gains eternal notoriety as namebearer thereby:

> Th'all-seeing sun no longer could sustain
> These practices, but with engaged disdain
> Darts forth such pestilent malignant beams,
> As shed infection on air, earth and streams;
> From whence this malady its birth received,
> And first th'offending Syphilus was grieved . . .
> He first wore buboes dreadful to the sight,
> First felt strange pains and sleepless passed the night;
> From him the malady received its name,
> The neighboring shepherds caught the spreading flame:
> At last in city and in court 'twas known,
> And seized th'ambitious monarch on his throne.

A shepherd or two could be spared, but the suffering of kings demands surcease. The high priest therefore suggests a human sacrifice to assuage the wrath of Apollo (now given his Greek name of Phoebus)—and guess whom they choose? But fortunately the goddess Juno decides to spare the unfortunate shepherd, and to make a substitution in obvious parallel to the biblical tale of Abraham and Isaac:

> On Syphilus the dreadful lot did fall,
> Who now was placed before the altar bound
> His head with sacrificial garlands crowned,
> His throat laid open to the lifted knife,
> But interceding Juno spared his life,

Commands them in his stead a heifer slay,
For Phoebus's rage was now removed away.

Ever since then, these natives, the former inhabitants of Atlantis, perform an annual rite of sacrifice to memorialize the hubris of Syphilus and the salvation of the people by repentance. The natives still suffer from syphilis, but their annual rites of sacrifice please Juno, who, in return, allows a wondrous local cure, the guaiacum tree, to grow on their isle alone. The Spanish sailors, now also infected by the disease, learn about the new cure—ever so much more tolerable than mercury—and bring guaiacum back to Europe.

Thus the imprecation heaped upon Spain by calling syphilis "the Spanish disease" becomes doubly unfair. Not only should the Spaniards be absolved for importation (because the disease struck Europe all at once, and from a latent contagion that originated well before any ships reached the New World); but the same Spanish sailors, encountering a longer history of infection and treatment in the New World, had discovered a truly beneficent remedy.

Most people know about the former use of mercury in treating syphilis, for the substance had some benefit, and the remedy endured for centuries. But the guaiacum cure has faded to a historical footnote because, in a word, this magical New World potion flopped completely. (Paracelsus himself had branded guaiacum as useless by 1530, the year of Fracastoro's publication.) But Fracastoro devised his myth of Syphilus in the short period of euphoria about the power of the new nostrum. The treatment failed, but the name stuck.

We should not be surprised to learn that Fracastoro's attraction to guaiacum owed as much to politics as to scientific hope. The powerful Fugger family, the great bankers of German lands, had lent vast sums to Maximilian's grandson Charles V in his successful bid to swing election as Holy Roman Emperor over his (and Fracastoro's) arch-enemy, Francis I of France. Among the many repayments necessitated by Charles's debt, the Fuggers won a royal monopoly for importing guaiacum to Europe. (The Habsburg Charles V also controlled Spain and, consequently, all shipping to Hispaniola, where the guaiacum tree grew.) In fact, the Fuggers built a chain of hospitals for the treatment of syphilis with guaiacum. Fracastoro's allegiances, for reasons previously discussed, lay with Charles V and the Spanish connection—so his tale of the shepherd Syphilus and the discovery of guaiacum suited his larger concerns as well. (Guaiacum, also known as *lignum vitae* or *lignum sanctum* [wood of life, or holy wood] has some medicinal worth, although not for treating syphilis. As an extremely hard wood, of the quality of ebony, guaiacum also has value in building and decoration.)

Fracastoro did proceed beyond his politically motivated poetry to learn more about syphilis. In the later work that secured his enduring fame (but largely for the wrong reason)—his *De contagione et contagiosis morbis et curatione* (on contagion and contagious diseases and their cure) of 1546—Fracastoro finally recognized the venereal nature of syphilis, writing that infection occurs *"verum non ex omni contactu, neque prompte, sed tum solum, quum duo corpora contactu mutuo plurimum incalvissent, quod praecipue in coitu eveniebat"* (truly not from all contact, nor easily, but only when two bodies join in most intense mutual contact, as primarily occurs in coitus). Fracastoro also recognized that infected mothers could pass the disease to their children, either at birth or through suckling.

Treating himself diplomatically and in the third person, Fracastoro admitted and excused the follies of his previous poem, written *quum iuniores essemus* (when we were younger). In this later prose work of 1546, Fracastoro accurately describes both the modes of transmission and the three temporal stages of symptoms—the small, untroublesome (and often overlooked) genital sore in the primary stage; the secondary stage of lesions and aches, occurring several months later; and the dreaded tertiary stage, developing months to years later, and leading to death by destruction of the heart or brain (called *paresis,* or paralysis accompanied by dementia) in the worst cases.

In the hagiographical tradition still all too common in textbook accounts of the history of science, Fracastoro has been called the "father" of the germ theory of disease for his sensitive and accurate characterization, in this work, of three styles of contagion: by direct contact (as for syphilis), by transmission from contaminated objects, and at a distance through transport by air. Fracastoro discusses particles of contagion or *semina* (seeds), but this term, taken from ancient Greek medicine, carries no connotation of an organic nature or origin. Fracastoro does offer many speculations about the nature of contagious *semina*—but he never mentions microorganisms, a hypothesis that could scarcely be imagined more than a century before the invention of the microscope.

In fact, Fracastoro continues to argue that the infecting *semina* of syphilis may arise from poisonous emanations sparked by planetary conjunctions. He even invokes a linguistic parallel between transmission of syphilis by sexual contact *(coitus),* and the production of bad seeds by planetary overlap in the sky, for he describes the astronomical phenomenon with the same word as *"coitum et conventum syderum"* (the coitus and conjunction of stars), particularly, for syphilis, *"nostra trium superiorum, Saturni, Iovis et Martis"* (our three most distant bodies, Saturn, Jupiter, and Mars).

Nonetheless, we seem to need heroes, defined as courageous iconoclasts who discerned germs of modern truth (literal "germs" in this case) through strictures of ancient superstition—and Fracastoro therefore wins false accolades under our cultural myth of clairvoyance "ahead of his time," followed by rejection and later rediscovery, long after death and well beyond hope of earthly reward. For example, the *Encyclopaedia Britannica* entry on Fracastoro ends by proclaiming:

> Fracastoro's was the first scientific statement of the true nature of contagion, infection, disease germs, and modes of disease transmission. Fracastoro's theory was widely praised during his time, but its influence was soon obscured by the mystical doctrines of the Renaissance physician Paracelsus, and it fell into general disrepute until it was proved by Koch and Pasteur.

But Fracastoro deserves our warmest praise for his brilliance and compassion *within* the beliefs of his own time. We can only appreciate his genius when we understand the features of his work that strike us as most odd by current reckonings—particularly his choice of Latin epic poetry to describe syphilis, and his christening of the disease for a mythical shepherd whose suffering also reflected Fracastoro's political needs and beliefs. In his article on Fracastoro for the *Dictionary of Scientific Biography,* Bruno Zanobio gives a far more accurate description, properly rooted in sixteenth-century knowledge, for Fracastoro's concept of contagious seeds:

> They are distinct imperceptible particles, composed of various elements. Spontaneously generated in the course of certain types of putrefaction, they present particular characteristics and faculties, such as increasing themselves, having their own motion, propagating quickly, enduring for a long time, even far from their focus of origin, exerting specific contagious activity, and dying.

A good description, to be sure, but not buttressed by any hint that these *semina* might be living microorganisms. "Undoubtedly," Zanobio continues, "the *seminaria* derive from Democritean atomism via the *semina* of Lucretius and the gnostic and Neoplatonic speculations renewed by Saint Augustine and

Saint Bonaventura." Fracastoro, in short, remained true to his Renaissance conviction that answers must be sought in the wisdom of classical antiquity.

Fracastoro surely probed the limits of his time, but medicine, in general, made very little progress in controlling syphilis until the twentieth century. Guaiacum failed and mercury remained both minimally effective and maximally miserable. (We need only recall Erasmus's sardonic quip that, in exchange for a night with Venus, one must spend a month with Mercury.) Moreover, since more than 50 percent of people infected with the spirochete never develop symptoms of the dreaded third stage, the disease, if left untreated, effectively "cures" itself in a majority of cases (although spirochetes remain in the body). Thus one can argue that traditional medicine usually did far more harm than good—a common situation, recalling Benjamin Franklin's quip that, although Dr. Mesmer was surely a fraud, his ministrations should be regarded as benevolent because people who followed his "cures" by inducing "animal magnetism" didn't visit "real" physicians, thereby sparing themselves such useless and harmful remedies as bleeding and purging.

No truly effective treatment for syphilis existed until 1909, when Paul Ehrlich introduced Salvarsan 606. Genuine (and gratifyingly easy) cures only became available in 1943, with the discovery and development of penicillin. Identification in the first stage, followed by one course of penicillin, can control syphilis; but infections that proceed to later stages may still be intractable.

I make no apologies for science's long record of failure in treating syphilis—a history that includes both persistent, straightforward error (the poisoning and suffering of millions with ineffective remedies based upon false theories) and, on occasion, morally indefensible practices as well (most notoriously, in American history, the Tuskegee study that purposely left a group of black males untreated as "controls" for testing the efficacy of treatments upon another group. In a moving ceremony, President Clinton apologized for this national disgrace to the few remaining survivors of the untreated group). But syphilis can now be controlled, and may even be a good candidate for total elimination (as we have done for smallpox), at least in the United States, if not in the entire world. And we owe this blessing, after so much pain, to knowledge won by science. There is no other way.

And so, while science must own its shame (along with every other institution managed by that infuriating and mercurial creature known as *Homo sapiens*), science can also discover the only genuine mitigation for human miseries caused by external agents that must remain beyond our control until their factual nature and modes of operation become known. The sequential character of this duality—failures as necessary preludes to success, given the stepwise nature of progress in scientific knowledge—led me to contrast Fracastoro's Latin hexameter with the stodgy prose of the 1998 article on the genome of *Treponema pallidum,* the syphilis spirochete.

The recent work boasts none of Fracastoro's grace or charm (even in Tate's heroic couplets)—no lovely tales about mythical shepherds who displease sun gods, and no intricate pattern of dactyls and spondees. In fact, I can't imagine a duller prose ending than the last sentence of the 1998 article, with its impersonal subject and its entirely conventional plea for forging onward to further knowledge: "A more complete understanding of the biochemistry of this organism derived from genome analysis may provide a foundation for the development of a culture medium for *T. pallidum,* which opens up the possibility of future genetic studies." Any decent English teacher would run a big blue pencil through these words.

But consider the principal, and ever so much more important, difference between Fracastoro's efforts and our own. In an article written to accompany the genomic presentation, M. E. St. Louis and J. N. Wasserheit of the Centers for Disease Control and Prevention in Atlanta write:

> Syphilis meets all of the basic requirements for a disease susceptible to elimination. There is no animal reservoir; humans are the only host. The incubation period is usually several weeks, allowing for interruption of transmission with rapid prophylactic treatment of contacts, whereas infectiousness is limited to less than twelve months even if untreated. [Tertiary syphilis may be both dreadful and deadly, but the disease is not passed to others at this stage.] It can be diagnosed with inexpensive and widely available blood tests. In its infectious stage, it is treatable with a single dose of antibiotics. Antimicrobial resistance has not yet emerged.

Interestingly, Fracastoro knew that syphilis infected only humans, but he regarded this observation as a puzzle under his theory of poisonous airborne particles that might, in principle, harm all life. He discusses this anomaly at length in part 1 of *Syphilis sive Morbus Gallicus:*

> Sometimes th'infected air hurts trees alone,
> To grass and tender flowers pernicious known . . .
> When earth yields store, yet oft some strange disease

Shall fall and only on poor cattle seize . . .
Since then by dear [in the British sense of "costly"] experiment we find
Diseases various in their rise and kind
Of this contagion let us take a view
More terrible for being strange and new

Thus the very property that so puzzled Fracastoro, and that he couldn't fit into his concept of disease, becomes a great blessing under the microbial theory.

Similarly, the deciphering of a genome guarantees no automatic or rapid panacea, but what better source of information could we desire for a reservoir of factual hope? Already, several features of this study indicate potentially fruitful directions of research. To cite just three items that caught my attention as I read the technical article on the decipherment:

1. A group of genes that promote motility—and may help us to understand why these spirochetes become so invasive into so many tissues—have been identified and found to be virtually identical to known genes in *B. burgdorferi,* the spirochete that causes Lyme disease.
2. The *T. pallidum* genome includes only a few genes coding for integral membrane proteins. This fact may help us to explain why the syphilis spirochete can be so successful in evading the human immune response. For if our antibodies can't detect *T. pallidum* because the invader, so to speak, presents too "smooth" an outer surface, then our natural defenses can become crippled. But if these membrane proteins, even though few, can be identified and characterized, then we may be able to develop specific remedies, or potentiators for our own immunity.
3. *T. pallidum*'s genome includes a large family of duplicated genes for membrane proteins that act as porins and adhesins—in other words, as good attachers and invaders. Again, genes that can be identified and characterized thereby "come out of hiding" into the realm of potential demobilization.

Science may have needed nearly five hundred years, but we should look on the bright side of differences between then and now. Fracastoro wrote verse and invented shepherds because he knew effectively nothing about the causes of a frightening plague whose effects could be specified and described in moving detail well suited for poetic treatment. The thirty-three modern authors, in maximal contrast, have obtained the goods for doing good. We may judge their prose as uninspired, but the greatest "poetry" ever composed about syphilis lies not in Fracastoro's hexameter of 1530, but in the intricate and healing details of a schematic map of 1,041 genes made of 1,138,006 base pairs, forming the genome of *Treponema pallidum* and published with the 1998 article—the adamantine beauty of genuine and gloriously complex factuality, full of life-saving potential. Fracastoro did his best for his time; may he be forever honored in the annals of human achievement. But the modern map embodies far more beauty, both for its factuality and utility, and as Fracastoro's finest legacy in the history of increasing knowledge that we must not shy from labeling by its right and noble name of progress.

V

Casting the Die: Six Evolutionary Epitomes

Defending Evolution

12

Darwin and the Munchkins of Kansas

IN 1999 THE KANSAS BOARD OF EDUCATION VOTED 6 TO 4 to remove evolution, and the big bang theory as well, from the state's science curriculum. In so doing, the board transported its jurisdiction to a never-never land where a Dorothy of a new millennium might exclaim, "They still call it Kansas, but I don't think we're in the real world anymore." The new standards do not forbid the teaching of evolution, but the subject will no longer be included in statewide tests for evaluating students—a virtual guarantee, given the realities of education, that this central concept of biology will be diluted or eliminated, thus reducing biology courses to something like chemistry without the periodic table, or American history without Lincoln.

The Kansas skirmish marks the latest episode of a long struggle by religious fundamentalists and their allies to restrict or eliminate the teaching of evolution in public schools—a misguided effort that our courts have quashed at each stage, and that saddens both scientists and the vast majority of theologians as well. No scientific theory, including evolution, can pose any threat to religion—for these two great tools of human understanding operate in complementary (not contrary) fashion in their totally separate realms: science as an inquiry about the factual state of the natural world, religion as a search for spiritual meaning and ethical values.

In the early 1920s, several states simply forbade the teaching of evolution outright, opening an epoch that inspired the infamous 1925 Scopes trial (leading to the conviction of a Tennessee high school teacher), and that ended only in 1968, when the Supreme Court declared such laws unconstitutional on First Amendment grounds. In a second round in the late 1970s, Arkansas and Louisiana required that if evolution be taught, equal time must be given to Genesis literalism, masquerading as oxymoronic "creation science." The Supreme Court likewise rejected those laws in 1987.

The Kansas decision represents creationism's first—and surely temporary*—success with a third strategy for subverting a constitutional imperative: that by simply deleting, but not formally banning, evolution, and by not demanding instruction in a biblical literalist "alternative," their narrowly partisan religious motivations might not derail their goals by legal defeat.

Given this protracted struggle, Americans of goodwill might be excused for supposing that some genuine scientific or philosophical dispute motivates this issue: Is evolution speculative and ill-founded? Does evolution threaten our ethical values or our sense of life's meaning? As a paleontologist by training, and with abiding respect for religious traditions, I would raise three points to alleviate these worries:

First, no other Western nation has endured any similar movement, with any political clout, against evolution—a subject taught as fundamental, and without dispute, in all other countries that share our major sociocultural traditions.

Second, evolution is as well documented as any phenomenon in science, as firmly supported as the earth's revolution around the sun rather than vice versa. In this sense, we can call evolution a "fact." (Science does not deal in certainty, so "fact" can only mean a proposition affirmed to such a high degree that it would be perverse to withhold one's provisional assent.)

---

*This "viewpoint" appeared in *Time* magazine on August 23, 1999. As noted elsewhere (see page 242 n.), supporters of good scientific education defeated the creationists in the next school board election in 2000. This newly elected board immediately restored evolution to the biology curriculum.

The major argument advanced by the school board—that large-scale evolution must be dubious because the process has not been directly observed—smacks of absurdity and only reveals ignorance about the nature of science. Good science integrates observation with inference. No process that unfolds over such long stretches of time (mostly, in this case, before humans evolved), or beneath our powers of direct visualization (subatomic particles, for example), can be seen directly. If justification required eyewitness testimony, we would have no sciences of deep time—no geology, no ancient human history, either. (Should I believe Julius Caesar ever existed? The hard, bony evidence for human evolution surely exceeds our reliable documentation of Caesar's life.)

Third, no factual discovery of science (statements about how nature "is") can, in principle, lead us to ethical conclusions (how we "ought" to behave), or to convictions about intrinsic meaning (the "purpose" of our lives). These last two questions—and what more important inquiries could we make?—lie firmly in the domains of religion, philosophy, and humanistic study. Science and religion should operate as equal, mutually respecting partners, each the master of its own domain, and with each domain vital to human life in a different way.

Why get excited over this latest episode in the long, sad history of American anti-intellectualism? Let me suggest that, as patriotic Americans, we should cringe in embarrassment that, at the dawn of a new, technological millennium, a jurisdiction in our heartland has opted to suppress one of the greatest triumphs of human discovery. Evolution cannot be dismissed as a peripheral subject, for Darwin's concept operates as the central organizing principle of all biological science. No one who has not read the Bible or the Bard can be considered educated in Western traditions; similarly, no one ignorant of evolution can understand science.

Dorothy followed her yellow brick road as the path spiraled outward toward redemption and homecoming (to the true Kansas of our dreams and possibilities). The road of the newly adopted Kansas curriculum can only spiral inward toward restriction and ignorance.

13

Darwin's More Stately Mansion*

A famous Victorian story reports the reaction of an aristocratic lady to the primary heresy of her time: "Let us hope that what Mr. Darwin says is not true; but, if it is true, let us hope that it will not become generally known." Teachers continue to relate this tale as both a hilarious putdown of class delusions (as if the upper crust could protect public morality by permanently sequestering a basic fact of nature) and an absurdist homily about the predictable fate of ignorance versus enlightenment. And yet I think we should rehabilitate this lady as an acute social analyst and

*I wrote this short piece as an editorial to introduce a special issue of *Science* magazine on evolution. This editorial, like the preceding piece for *Time* magazine, represents my immediate reaction to the Kansas School Board's rejection of evolution in the state curriculum. I include both pieces, and present them sequentially, because I thought that readers might be interested in my sense of how the same subject can be presented to general and popular audiences *(Time)* and to professional colleagues (*Science*—America's leading technical journal, published by the American Association for the Advancement of Science, the profession's primary "umbrella" organization).

at least a minor prophet. For what Mr. Darwin said is, indeed, true. It has also not become generally known, at least in our nation.

What strange set of historical circumstances, what odd disconnect between science and society, can explain the paradox that organic evolution—the central operating concept of an entire discipline and one of the firmest facts ever validated by science—remains such a focus of controversy, even of widespread disbelief, in contemporary America?

In a wise statement that will endure beyond the fading basis of his general celebrity, Sigmund Freud argued that all great scientific revolutions feature two components: an intellectual reformulation of physical reality and a visceral demotion of *Homo sapiens* from arrogant domination atop a presumed pinnacle to a particular and contingent result, however interesting and unusual, of natural processes. Freud designated two such revolutions as paramount: the Copernican banishment of Earth from center to periphery and the Darwinian "relegation" (Freud's word) of our species from God's incarnated image to "descent from an animal world." Western culture adjusted to the first transformation with relative grace (despite Galileo's travails), but Darwin's challenge cuts so much closer (and literally) to the bone. The geometry of an external substrate, a question of real estate after all, carries much less emotional weight than the nature of an internal essence. The biblical Psalmist evoked our deepest fear by comparing our bodily insignificance with cosmic immensity and then crying out: "What is man, that thou art mindful of him?" (Psalm 8). But he then vanquished this spatial anxiety with a constitutional balm: "Thou hast made him a little lower than the angels . . . thou madest him to have dominion . . . thou hast put all things under his feet." Darwin removed this keystone of false comfort more than a century ago, but many people still believe that they cannot navigate our earthly vale of tears without such a crutch.

Denigration and disrespect will never win the minds (not to mention the hearts) of these people. But the right combination of education and humility might extend a hand of fellowship and eventually end the embarrassing paradox of a technological nation entering a new millennium with nearly half its people actively denying the greatest biological discovery ever made. Three principles might guide our pastoral efforts: First, evolution is true—and the truth can only make us free. Second, evolution liberates the human spirit. Factual nature cannot, in principle, answer the deep questions about ethics and meaning that all people of substance and valor must resolve for themselves. When we stop demanding more than nature can logically provide (thereby freeing ourselves for genuine dialogue with the outside world, rather than clothing nature with false projections of our needs), we liberate ourselves to look within. Science can then forge true partnerships with philosophy, religion, and the arts and humanities, for each must supply a patch in that ultimate coat of many colors, the garment called wisdom. Third, for sheer excitement, evolution, as an empirical reality, beats any myth of human origins by light-years. A genealogical nexus stretching back nearly 4 billion years and now ranging from bacteria in rocks several miles under Earth's surface to the tip of the highest redwood tree, to human footprints on the moon. Can any tale of Zeus or Wotan top this? When truth value and visceral thrill thus combine, then indeed, as Darwin stated in closing his great book, "there is grandeur in this view of life." Let us praise this evolutionary nexus—a far more stately mansion for the human soul than any pretty or parochial comfort ever conjured by our swollen neurology to obscure the source of our physical being, or to deny the natural substrate for our separate and complementary spiritual quest.

14

A Darwin for All Reasons

As a paleontologist by trade and (dare I say it?) a card-carrying liberal in politics, I have been amused, but also a bit chagrined, by the current fad in conservative intellectual circles for invoking the primary icon of my professional world—Charles Darwin—as either a scourge or an ally in support of cherished doctrines.

Since Darwin cannot logically fulfill both roles at the same time, and since the fact of evolution in general (and the theory of natural selection in particular) cannot legitimately buttress any particular moral or social philosophy in any case, I'm confident that this greatest of all biologists will remain silent no matter how loudly conservatives may summon him.

At one extreme, the scourging of Darwin—the idea that if we drive him away, then we can awaken—has animated a religious faction that views an old-style Christian revival as central to a stable and well-ordered polity. In *Slouching Towards Gomorrah,* for example, Robert Bork writes, "The major obstacle to a religious renewal is the intellectual classes," who "believe that science has left atheism as the only respectable intellectual stance. Freud, Marx, and Darwin, according to the conventional account, routed the believers. Freud and Marx are no longer taken as irrefutable by intellectuals, and now it appears to be Darwin's turn to undergo a devaluation."

Then, exhibiting as much knowledge of paleontology as I possess of constitutional law—effectively zero—Bork cites as supposed evidence for Darwin's forthcoming fall the old and absurd canard that "the fossil record is proving a major embarrassment to evolutionary theory." If Bork will give me a glimpse of that famous pillar of salt on the outskirts of Gomorrah, I shall be happy, in return, to show him the abundant evidence we possess of intermediary fossils in major evolutionary transitions—mammals from reptiles, whales from terrestrial forebears, humans from apelike ancestors.

Meanwhile, and at an opposite extreme, the celebration of Darwin—the claim that if we embrace him, he will validate the foundations of our views—motivates the efforts of some secular believers determined to enshrine conservative political dogmas as the dictates of nature. In the *National Review,* for example, John O. McGinnis recently argued that "the new biological learning holds the potential for providing stronger support for conservatism than any other new body of knowledge has done.

"We may fairly conclude," wrote McGinnis, "that a Darwinian politics is a largely conservative politics." McGinnis then listed the biological bases—including self-interest, sexual differences, and "natural inequality"—as examples of right-wing ideology resting on the foundations of evolutionary theory.

Moreover, according to McGinnis, Darwinism seems tailor-made not only to support conservative politics in general, but also to validate the particular brand favored by McGinnis himself. For example, he uses specious evolutionary arguments to excoriate "pure libertarianism." Thus he invokes Darwin to assert that the state maintains legitimate authority to compel people to save for their declining years or to rein in their sexual proclivities.

"The younger self is so weakly connected to the imagination of the older self (primarily because most individuals did not live to old age in hunter-gatherer societies) that most people cannot be expected to save sufficiently for old age," McGinnis writes. "Therefore there may be justification for state intervention to force individuals to save for their own retirements." In addition, "society may need to create institutions to channel and restrain sexual activity."

Misuse of Darwin has not been confined to the political right. Liberals have also played both contradictory ends of the same game—either denying Darwin when they found the implications of his theory displeasing, or invoking him to interpret their political principles as sanctioned by nature.

Some liberals bash Darwin because they misconceive his theory as a statement about overt battle and killing in a perpetual "struggle for existence." In fact, Darwin explicitly identified this "struggle" for existence as metaphorical—best pursued by cooperation in some circumstances and by competition in others. Using the opposite strategy of embracing Darwin, many early-twentieth-century liberals lauded reproduction among the gifted, while discouraging procreation among the supposedly unfit.

The Darwin bashers and boosters can both be refuted with simple and venerable arguments. To the bashers, I can assert only that Darwinian evolution continues to grow in vibrancy and cogency as the centerpiece of the biological sciences—and, more generally, that no scientific truth can pose any threat to religion rightly conceived as a search for moral order and spiritual meaning.

To those who would rest their religious case on facts of nature, I suggest that they take to heart the wise words of Reverend Thomas Burnet, the seventeenth-century scientist: "'Tis a dangerous thing to engage the authority of Scripture in disputes about the Natural world . . . lest Time, which brings all things to light, should discover that to be evidently false which we had made scripture to assert." So the Roman Catholic Church learned in the seventeenth century after accusing Galileo of heresy—and so should modern fundamentalists note and understand today when they deny the central conclusion of biology.

Those who recruit Darwin to support a particular moral or political line should remember that, at best, evolutionary biology may give us some insight into the anthropology of morals—why some (or most) peoples practice certain values, perhaps for their Darwinian advantage. But science can never decide the morality of morals. Suppose we discovered that aggression, xenophobia, selective infanticide, and the subjugation of women offered Darwinian advantages to our hunter-gatherer ancestors a million years ago on the African savannahs. Such a conclusion would not validate the moral worth of these or any other behaviors, either then or now.

Perhaps I should be flattered that my own field of evolutionary biology has usurped the position held by cosmology in former centuries—and by Freudianism earlier in our own times—as the science deemed most immediately relevant to deep questions about the meaning of our lives. But we must respect the limits of science if we wish to profit from its genuine insights. G. K. Chesterton's famous epigram—"art is limitation; the essence of every picture is the frame"—applies equally well to science.

Darwin himself understood this principle in suspecting that the human brain, evolved for other reasons over so many million years, might be ill equipped for solving the deepest and most abstract questions about life's ultimate meaning. As he wrote to the American botanist Asa Gray in 1860: "I feel most deeply that the whole subject is too profound for the human intellect. A dog might as well speculate on the mind of Newton."

Those who would misuse Darwin to advance their own agendas should remember the biblical injunction that provided the title to a great play about the attempted suppression of evolutionary theory in American classrooms: "He that seeketh mischief, it shall come unto him. . . . He that troubleth his own house shall inherit the wind."

Evolution and Human Nature

15

When Less Is Truly More

ON MONDAY, FEBRUARY 12, 2001, TWO GROUPS OF researchers released the formal report of data for the human genome. They timed their announcement well—and purposely—for February 12 is the birthday of Charles Darwin, who jump-started our biological understanding of life's nature and evolution when he published the *Origin of Species* in 1859. For only the second time in thirty-five years of teaching, I dropped my intended schedule—to discuss the importance of this work with my undergraduate course on the history of life. (The only other case, in a distant age of the late sixties, fell a half-hour after radical students had seized University Hall and physically ejected the deans; this time, at least, I told my students, the reason for the change lay squarely within the subject matter of the course.)

I am no lover, or master, of sound bites or epitomes, but I began by telling my students that we were sharing a great day in the history of science and of human understanding in general. (My personal joy in a scientific event had only been matched once before in my lifetime—at the lunar landing in 1969.)

The fruit fly *Drosophila,* the staple of laboratory genetics, possesses between 13,000 and 14,000 genes. The roundworm *C. elegans,* the staple of laboratory studies in development, contains only 959 cells, looks like a tiny, formless squib with virtually no complex anatomy beyond its genitalia, and possesses just over 19,000 genes.

The general estimate for *Homo sapiens* — sufficiently large to account for the vastly greater complexity of humans under conventional views—had stood at well over 100,000, with a more precise figure of 142,634 widely advertised and considered well within the range of reasonable expectation. But *Homo sapiens,* we now learn, possesses between 30,000 and 40,000 genes, with the final tally probably lying nearer the lower figure. In other words, our bodies develop under the directing influence of only half again as many genes as the tiny roundworm needs to manufacture its utter, if elegant, outward simplicity.

Human complexity cannot be generated by thirty thousand genes under the old view of life embodied in what geneticists literally called (admittedly with a sense of whimsy) their "central dogma": DNA makes RNA makes protein—in other words, one direction of causal flow from code to message to assembly of substance, with one item of code (a gene) ultimately making one item of substance (a protein), and the congeries of proteins making a body. Those 142,000 messages no doubt exist, as they must to build the complexity of our bodies. Our previous error may now be identified as the assumption that each message came from a distinct gene.

We may envision several kinds of solutions for generating many times more messages than genes, and future research will target this issue. In the most reasonable and widely discussed mechanism, a single gene can make several messages because genes of multicellular organisms are not discrete and inseparable sequences of instructions. Rather, genes are composed of coding segments (exons) separated by noncoding regions (introns). The resulting signal that eventually assembles the protein consists only of exons spliced together after elimination of introns. If some exons are omitted, or if the order of splicing changes, then several distinct messages can be generated by each gene.

The implications of this finding cascade across several realms. The commercial effects will be obvious, as so much biotechnology, including the rush to patent genes, has assumed the old view that "fixing" an aberrant gene would cure a specific human ailment. The social meaning may finally liberate us from the simplistic and harmful idea, false for many other reasons as well, that each aspect of our being, either physical or behavioral, may be ascribed to the action of a particular gene "for" the trait in question.

But the deepest ramifications will be scientific or philosophical in the largest sense. From its late-seventeenth-century inception in modern form, science has strongly privileged the reductionist mode of thought that breaks overt complexity into constituent parts and then tries to explain the totality by the properties of those parts and from simple interactions fully predictable from the parts. (*Analysis* literally means "to dissolve into basic parts.") The reductionist method works triumphantly for simple systems—predicting eclipses or the motion of planets (but not the histories of their complex surfaces), for example. But once again—and when will we ever learn?—we fell victim to hubris, as we imagined that, in discovering how to unlock some systems, we had found the key for the conquest of all natural phenomena. Will Parsifal ever learn that only humility (and a plurality of strategies for explanation) can locate the Holy Grail?

The collapse of the doctrine of one gene for one protein, and one direction of causal flow from basic codes to elaborate totality, marks the failure of reductionism for the complex system that we call biology—and for two major reasons.

First, the key ingredient for evolving greater complexity is not more genes, but more combinations and interactions generated by fewer units of code—and many of these interactions (as emergent properties, to use the technical jargon) must be explained at the level of their appearance, for they cannot be predicted from the separate underlying parts alone. So organisms must be explained as organisms, and not as a summation of genes.

Second, the unique contingencies of history, not the laws of physics, set many properties of complex biological systems. Our thirty thousand genes make up only one percent or so of our total genome. The rest—including bacterial immigrants and other pieces that can replicate and move—originated more as accidents of history than as predictable necessities of physical laws. Moreover, these noncoding regions, disrespectfully called "junk DNA," also build a pool of potential for future use that, more than any other factor, may establish any lineage's capacity for further evolutionary increase in complexity.

The deflation of hubris is blessedly positive, not cynically disabling. The failure of reductionism doesn't mark the failure of science, but only the replacement of an ultimately unworkable set of assumptions by more-appropriate styles of explanation that study complexity at its own level and respect the influences of unique histories. Yes, the task will be much harder than reductionistic science imagined. But our thirty thousand genes—in all the glorious ramifications of their irreducible interactions—have made us sufficiently complex and at least potentially adequate for the task ahead.

We may best succeed in this effort if we can heed some memorable words spoken by that other great historical figure born on February 12—on the very same day as Darwin, in 1809. Abraham Lincoln, in his first inaugural address, urged us to heal division and seek unity by marshaling the "better angels of our nature"—yet another irreducible and emergent property of our historically unique mentality, but inherent and invokable all the same, even though not resident within, say, gene 26 on chromosome number 12.

16

Darwin's Cultural Degree

We can embrace poetical reminders of our connection to the natural world, whether expressed as romantic effusions about oneness, or in the classical meter of Alexander Pope's heroic couplet:

> All are but parts of one stupendous whole
> Whose body Nature is, and God the soul.

Yet, when we look into the eyes of an ape, our perception of undeniable affinity evokes an eerie fascination that we usually express as laughter or as fear. Our discomfort then increases when we confront the loss of former confidence in our separate and exalted creation "a little lower than the angels . . . crowned . . . with glory and honor" (Psalm 8), and must own the evolutionary alternative, with a key implication stated by Darwin himself (in *The*

Descent of Man): "The difference in mind between man and the higher animals, great as it is, certainly is one of degree and not of kind."

We have generally tried to unite our intellectual duty to accept the established fact of evolutionary continuity with our continuing psychological need to see ourselves as separate and superior, by invoking one of our worst and oldest mental habits: dichotomization, or division into two opposite categories, usually with attributions of value expressed as good and bad, or higher and lower. We therefore try to define a "golden barrier," a firm criterion to mark an unbridgeable gap between the mentality and behavior of humans and all other creatures. We may have evolved from them, but at some point in our advance, we crossed a Rubicon that brooks no passage by any other species.

Thus, throughout the history of anthropology, we have proposed many varied criteria—and rejected them, one by one. We tried behavior—the use of tools, and, after the failure of this broad standard, the use of tools explicitly fashioned for particular tasks. (Chimps broke this barrier when we discovered their ability to strip leaves off twigs, and then use the naked sticks for extracting termites from nests.) And we considered distinctive mental attributes—the existence of a moral sense, or the ability to form abstractions. All proposed criteria have failed as absolutes of human uniqueness (while a complex debate continues to surround the meaning and spread of language and its potential rudiments).

The development of "culture"—defined as distinct and complex behavior originating in local populations and clearly passed by learning, rather than by genetic predisposition—has persisted as a favored candidate for a "golden barrier" to separate humans from animals, but must now be rejected as well. A study published in a recent issue of the journal *Nature* proves the existence of complex cultures in chimpanzees. This research demonstrates that chimpanzees learn behaviors through observation and imitation and then teach these traits to other chimpanzees. The study represents a cooperative effort of all major research groups engaged in the long-term study of particular groups of chimpanzees in the wild (with Jane Goodall's nearly forty-year study of the Gombe chimps as the flagship of these efforts).

Isolated examples of cultural transmission have long been recognized—with local "dialects" of songbirds and the potato-washing of macaques on a small Japanese island as classical cases. But such rudimentary examples scarcely qualified as arguments against a meaningful barrier between humans and animals. However, the chimpanzee study, summarizing 151 years of observation at seven field sites, found culturally determined, and often quite complex, differences among the sites for thirty-nine behavioral patterns that must have originated in local groups and then spread by learning.

To cite just one example, contrasting the two most intensively studied sites (Goodall's at Gombe and Toshisada Nishida's in the Mahale Mountains 170 kilometers away, with no recorded contact between the groups), the Mahale chimps clap two hands together over their heads as part of the grooming ritual, while no Gombe chimp has ever so behaved (at least while under human observation. Grooming itself may be genetically enjoined, but such capricious variations in explicit style must be culturally invented and transmitted). In a commentary accompanying the *Nature* article, Frans de Waal of the Yerkes Regional Primate Research Center in Atlanta summarizes the entire study by writing, "The evidence is overwhelming that chimpanzees have a remarkable ability to invent new customs and technologies, and that they pass these on socially rather than genetically."

The conventional commentary on such a conclusion would end here, leaving a far more important issue unaddressed. Why are we so surprised by such a finding? The new documentation may be rich and decisive—but why would anyone have doubted the existence of culture in chimps, given well-documented examples in other animals and our expanding knowledge of the far more sophisticated mental lives of chimpanzees?

Our surprise may teach us as much about ourselves as the new findings reveal about chimpanzees. For starters, the basic formulation of them versus us, and the resulting search for a "golden barrier," represents a deep fallacy of human thought. We need not fear Darwin's correct conclusion that we differ from other animals only in degree. A sufficient difference in quantity translates to what we call difference in quality *ipso facto*. A frozen pond is not the same object as a boiling pool—and New York City does not represent a mere extension of the tree nests at Gombe.

In addition, evolution does provide a legitimate criterion of genuine and principled separation between *Homo sapiens* and any other species. But the true basis of distinction lies in topology and genealogy, not in any functional attribute marking our superiority. We are linked to chimpanzees (and more distantly to any other species) by complete chains of intermediate forms that proceed backward from our current state into the fossil record until the two lineages meet in a common ancestor. But all these intermediate forms have become extinct, and the evolutionary gap between modern humans and chimps therefore stands as absolute and inviolate. In this crucial genealogical sense, all humans share equal fellowship as members of *Homo sapiens*. In biological terms, with species defined by historical and genealogical connection, the most mentally deficient person on earth remains as fully human as Einstein.

If we grasped this fundamental truth of evolution, we might finally make our peace with Alexander Pope's location of human nature on an "isthmus of a middle state"—that is, between bestiality and mental transcendence.

We might also become comfortable with his incisive characterization of our peculiar status as "the glory, jest, and riddle of the world."

17

The Without and Within of Smart Mice

Every age must develop its own version of the unobtainable and chimerical quick fix: the right abracadabra to select the winning lottery number, the proper prayer to initiate the blessed millennium, the correct formula to construct the philosopher's stone. In a technological age, we seek the transforming gene to elicit immediate salvation from within.

An excellent and provocative study of Joe Tsien and his colleagues will, one may safely predict, be widely misread in the false light of this age-old hope—combined with some equally age-old fallacies of human reasoning.

These scientists bred strains of mice with extra copies of a gene coded for a protein that can facilitate communication between neurons. Since one popular theory of memory relates this primary mental capacity to an organism's ability to make associations—say, between the buzz of a bee and the pain of its bite—this enhanced communication might promote a recording of associations within the brain, thus creating memories.

Pundits in our age of rapid misinformation will surely transmit the story as a claim that *the* gene for intelligence has been cloned and that a human smart pill for routine production of kiddie geniuses lies just around the millennial corner.* None of this punditry, however, will bear any relationship to current realities or to reasonable prospects for the short-term future. Even so, the mice studied by Tsien and his colleagues could help us to correct two common errors in our thinking about genetics and intelligence:

1. *The labeling fallacy.* Complex organisms cannot be construed as the sum of their genes, nor do genes alone build particular items of anatomy or behavior by themselves (see essay 15). Most genes influence several aspects of anatomy and behavior—as they operate through complex interactions with other genes and their products, and with environmental factors both within and outside the developing organism. We fall into a deep error, not just a harmless oversimplification, when we speak of genes "for" particular items of anatomy or behavior.

No single gene determines even the most concrete aspect of my physical being, say the length of my right thumb. The very notion of a gene "for" something as complex as "intelligence" lapses into absurdity. We use the word *intelligence* to describe an array of largely independent and socially defined mental attributes, not a quantity of a single something, secreted by one gene, measurable as one number, and capable of arranging human diversity into one line ordered by relative mental worth.

To cite an example of this fallacy, in 1996 scientists reported the discovery of a gene for novelty-seeking behavior—generally regarded as a good thing. In 1997 another study detected a linkage between the same gene and a propensity for heroin addiction. Did the "good" gene for enhanced exploration become the "bad" gene for addictive tendencies? The biochemistry may be constant, but context and background matter.

2. *The compositional fallacy*. Just as each gene doesn't make a separate piece of an organism, the entire organism cannot be regarded as a simple summation of relevant building codes and their action (a skeleton cannot be generated by a head gene added to a neck gene added to a rib gene, etc.). The fact that com-

---

*This "viewpoint" appeared in *Time* magazine, 13 September 1999.

plex systems like human mentality or anatomy can be disrupted easily by deficiencies in single factors does not validate the opposite claim that enhancement of the same factors will boost the system in a harmonious and beneficial manner. The potential "fixing" of specific abnormalities—the realistic hope of certain gene therapies for the near future—does not imply that we will be able to bioengineer superathletes or superscholars. The remedy for a specific deficiency does not become an elixir for general superiority. I can save a drowning man's mind if I hold his head above water, but I can't make him a genius by continually adding more oxygen to his ordinary surroundings.

Ironically, Tsien's mice disprove these two fallacies of genetic determinism from within. By identifying their gene and charting the biochemical basis of its action, Tsien has demonstrated the value and necessity of environmental enrichment for yielding a beneficial effect. This gene doesn't make a mouse "smart" all by its biochemical self. Rather, the gene's action allows adult mice to retain a neural openness for learning that young mice naturally possess but then lose in aging.

Even if Tsien's gene exists, and maintains the same basic function in humans (a realistic possibility), we will need an extensive regimen of learning to potentiate any benefit from its enhanced action. In fact, we try very hard—often without success, in part because false beliefs in genetic determinism discourage our efforts—to institute just such a regimen during a human lifetime. We call this regimen "education." Perhaps Jesus expressed a good biological insight when he stated (Matthew 18:3), "Except ye be converted, and become as little children, ye shall not enter into the kingdom of heaven."

VI

The Meaning and Drawing of Evolution

Defining and Beginning

18

What Does the Dreaded "E" Word Mean Anyway?

EVOLUTION POSED NO TERRORS IN THE LIBERAL constituency of New York City when I studied biology at Jamaica High School in 1956. But our textbooks didn't utter the word either—a legacy of the statutes that had brought William Jennings Bryan and Clarence Darrow to legal blows at Tennessee's trial of John Scopes in 1925. The subject remained doubly hidden within my textbook—covered only in chapter 63 of sixty-six, and described in euphemism as "the hypothesis of racial development."

The anti-evolution laws of the Scopes era, passed during the early 1920s in several Southern states, remained on the books until 1968, when the Supreme Court declared them unconstitutional. The laws were never strictly enforced, but their existence cast a pall over American education, as textbook publishers capitulated to produce "least common denominator" versions acceptable in all states—so schoolkids in New York got short shrift because the statutes of some distant states had labeled evolution as dangerous and unteachable.

Ironically, at the very end of the millennium (I wrote this essay in late November 1999), demotions, warnings, and anathemas have again come into vogue in several regions of our nation. The Kansas school board has reduced evolution, the central and unifying concept of the life sciences, to an optional subject within the state's biology curriculum (an educational ruling akin to stating that English will still be taught, but grammar may henceforth be regarded as a peripheral frill, permitted but not mandated as a classroom subject).* Two states now require that warning labels be pasted (literally) into all biology textbooks, alerting students that they might wish to consider alternatives to evolution (although no other well-documented scientific concept evokes similar caution). Finally, at least two states have retained their Darwinian material in official pamphlets and curricula, but have replaced the dreaded "E" word with a circumlocution, thus reviving the old strategy of my high school text.

As our fight for good (and politically untrammeled) public education in science must include the forceful defense of a key word—for inquisitors have always understood that an idea can be extinguished most effectively by suppressing all memory of a defining word or an inspirational person—we might consider an interesting historical irony that, properly elucidated, might even aid our battle. We must not compromise *our* showcasing of the "E" word, for we give up the game before we start if we grant our opponents control over defining language. But we should also note that Darwin himself never used the word "evolution" in his epochal book of 1859, the *Origin of Species,* where he calls this fundamental biological process "descent with modification." Darwin, needless to say, did not shun "evolution" for motives of fear, conciliation, or political savvy—but rather for an opposite and principled reason that can help us to appreciate the depth of the intellectual revolution that he inspired, and some of the reasons (understandable if indefensible) for persisting public unease.

Pre-Darwinian concepts of evolution—a widely discussed, if unorthodox, view of life in early-nineteenth-century biology—generally went by such names as "transformation," "transmutation," or "the development hypothesis." In choosing a label for his very different account of genealogical change, Darwin would never have considered "evolution" as a descriptor because this

*As a happy footnote, I can now report (while editing this essay for republication in book form at the outset of the next millennium) that some good old-fashioned political activism turned the fundamentalist rascals out in the subsequent school board election of 2000—and that evolution has now been restored to the Kansas curriculum.

vernacular English word implied a set of consequences contrary to the most distinctive features of his own revolutionary mechanism of change—the hypothesis of natural selection.

"Evolution," from the Latin *evolvere,* literally means "to unroll"—and clearly implies an unfolding in time of a predictable or prepackaged sequence in an inherently progressive, or at least directional, manner. (The "fiddlehead" of a fern unrolls and expands to bring forth the adult plant—a true "evolution" of preformed parts.) The *Oxford English Dictionary* traces the word to seventeenth-century English poetry, where the key meaning of sequential exposure of prepackaged potential inspired the first recorded usages in our language. For example, Henry More (1614–1687), the British poet and philosopher responsible for most seventeenth-century citations in the *OED,* stated in 1664: "I have not yet evolved all the intangling superstitions that may be wrapt up."

The few pre-Darwinian English citations of genealogical change as "evolution" all employ the word as a synonym for predictable progress. For example, in describing Lamarck's theory for British readers (in the second volume of his *Principles of Geology* in 1832), Charles Lyell generally uses the neutral term "transmutation"—except in one passage, when he wishes to highlight a claim for progress: "The testacea [shelled invertebrates] of the ocean existed first, until some of them by gradual evolution were improved into those inhabiting the land."

Although the word *evolution* does not appear in the first edition of the *Origin of Species,* Darwin does use the verbal form "evolved"—clearly in the vernacular sense and in an especially prominent spot: as the very last word of the book! Most students have failed to appreciate the incisive and intended "gotcha" of these closing lines, which have generally been read as a poetic reverie, a harmless linguistic flourish essentially devoid of content, however rich in imagery. In fact, the canny Darwin used this maximally effective location to make a telling point about the absolute glory and comparative importance of natural history as a calling.

We usually regard planetary physics as the paragon of rigorous science, while dismissing natural history as a lightweight exercise in dull, descriptive cataloging that any person with sufficient patience might accomplish. But Darwin, in his closing passage, identified the primary phenomenon of planetary physics as a dull and simple cycling to nowhere, in sharp contrast with life's history, depicted as a dynamic and upwardly growing tree. The earth revolves in uninteresting sameness, but life *evolves* by unfolding its potential for ever-expanding diversity along admittedly unpredictable, but wonderfully various, branchings. Thus Darwin ends his great book:

> Whilst this planet has gone cycling on according to the fixed law of gravity, from so simple a beginning endless forms most beautiful and most wonderful have been, and are being, evolved.

But Darwin could not have described the *process* regulated by his mechanism of natural selection as "evolution" in the vernacular meaning then conveyed by this word. For the mechanism of natural selection only yields increasing adaptation to changing local environments, not predictable "progress" in the usual sense of cosmic or general "betterment" expressed in favored Western themes of growing complexity or augmented mentality. In Darwin's causal world, an anatomically degenerate parasite, reduced to a formless clump of feeding and reproductive cells within the body of a host, may be just as well adapted to its surroundings, and just as well endowed with prospects for evolutionary persistence, as the most intricate creature, exquisitely adapted in all parts to a complex and dangerous external environment. Moreover, since natural selection can only adapt organisms to *local* circumstances, and since local circumstances change in an effectively random manner through geological time, the pathways of adaptive evolution cannot be predicted.

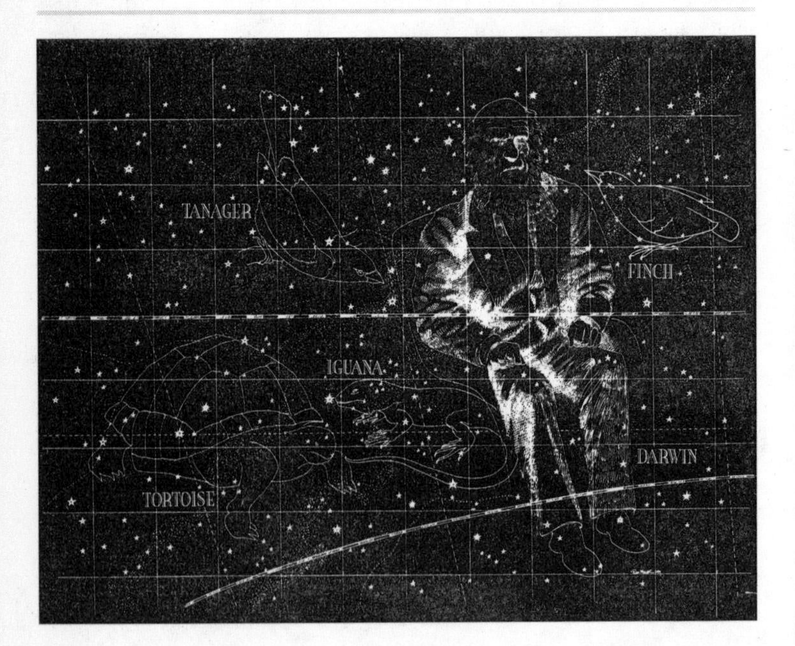

Thus, on these two fundamental grounds—absence of inherent directionality and lack of predictability—the process regulated by natural selection could scarcely have suggested, to Darwin, the label of "evolution," an ordinary English word for sequences of predictable and directional unfolding. We must then, and obviously, ask how "evolution" achieved its coup in becoming the name for Darwin's process—a takeover so complete that the word has now almost (but not quite, as we shall soon see) lost its original English meaning of unfolding, and has transmuted (or should we say "evolved") into an effective synonym for biological change through time?

This interesting shift, despite Darwin's own reticence, occurred primarily because a great majority of his contemporaries, while granting the overwhelming evidence for evolution's factuality, could not accept Darwin's radical views about the causes and patterns of biological change. Most important, they could not bear to surrender the comforting and traditional view that human consciousness must represent a predictable (if not a divinely intended) summit of biological existence. If scientific discoveries enjoined an evolutionary reading of human superiority, then one must bow to the evidence. But Darwin's contemporaries (and many people today as well) would not surrender their traditional view of human domination, and therefore could only conceptualize genealogical transmutation as a process defined by predictable progress toward a human acme—in short, as a process well described by the term "evolution" in its vernacular meaning of unfolding an inherent potential.

Herbert Spencer's progressivist view of natural change probably exerted most influence in establishing "evolution" as the general name for Darwin's process—for Spencer held a dominating status as Victorian pundit and grand panjandrum of nearly everything conceptual. In any case, Darwin had too many other fish to fry, and didn't choose to fight a battle about words rather than things. He felt confident that his views would eventually prevail, even over the contrary etymology of word imposed upon his process by popular will. (He knew, after all, that meanings of words can transmute within new climates of immediate utility, just as species transform under new local environments of life and ecology!) Darwin never used the "E" word extensively in his writings, but he did capitulate to a developing consensus by referring to his process as "evolution" for the first time in *The Descent of Man,* published in 1871. (Still, Darwin never cited "evolution" in the title of any book—and he chose, in labeling his major work on our species, to emphasize our genealogical "descent," not our "ascent" to higher levels of consciousness.)

When I was a young boy, growing up on the streets of New York City, the Museum of Natural History became my second home and inspiration. I loved two exhibits most of all—the *Tyrannosaurus* skeleton on the fourth floor and the star show at the adjacent Hayden Planetarium. I juggled these two passions for many years, and eventually became a paleontologist. (Carl Sagan, my near contemporary from the neighboring neverland of Brooklyn—I grew up in Queens—weighed the same two interests in the same building, but opted for astronomy as a calling. I have always suspected a basic biological determinism behind our opposite choices. Carl was tall and looked up toward the heavens; I am shorter than average and tend to look down at the ground.)

My essays generally follow the strategy of selecting odd little tidbits as illustrations of general themes. I wrote this piece to celebrate the reopening of the Hayden Planetarium in 2000, and I followed my passion for apparent trivia with large tentacles of implication by marking this great occasion with a disquisition on something so arcane and seemingly irrelevant as the odyssey of the term "evolution" in my two old loves of biology and astronomy. In fact, I chose to write about "evolution" in the biological domain that I know in order to explicate a strikingly different meaning of "evolution" in the profession that I put aside, but still love avocationally. I believe that such a discussion of the contrast between biological and cosmological "evolution" might highlight an important general point about alternative worldviews, and also serve as a reminder that many supposed debates in science arise from confusion engendered by differing uses of words, and not from deep conceptual muddles about the nature of things.

Interdisciplinary unification represents a grand and worthy goal of intellectual life, but greater understanding can often be better won by principled separation and mutual respect, based on clear definitions and distinctions among truly disparate processes, than by false unions forged with superficial similarities and papered over by a common terminology. In our understandable desire to unify the sciences of temporal change, we have too often followed the Procrustean strategy of enforcing a common set of causes and explanations upon the history of a species and the life of a star—partly, at least, for the very bad reason that both professions use the term "evolution" to designate change through time. In this case, the fundamental differences embody far more interest and insight than the superficial similarities—and true unity will only be achieved when we acknowledge the disparate substrates that, taken together, probe the range of possibilities for theories of historical order.

The Darwinian principle of natural selection yields temporal change—"evolution" in the biological definition—by a twofold process of generating copious and undirected variation within a population, and then passing only a biased (selected) portion of this variation to the next generation. In this manner, the variation *within a population* at any moment can be converted into differences in mean values (such as average size or average braininess) *among successive populations* through time. For this fundamental reason, we call such theories of change "variational" as opposed to more conventional, and more direct, models of "transformational" change imposed by natural laws that mandate a particular trajectory based on inherent, and therefore predicable, properties of substances and environments. (A ball rolling down an inclined plane does not reach the bottom because selection has favored the differential propagation of moving versus stable elements of its totality, but because gravity dictates this temporal sequence and result when round balls roll down smooth planes.)

To illustrate the peculiar properties of variational theories like Darwin's in an obviously caricatured, but not inaccurate, description: Suppose that a population of elephants inhabits Siberia during a warm interval before the advance of an ice sheet. The elephants vary, at random and in all directions, in their amount of body hair. As the ice advances and local conditions become colder, elephants with more hair will tend to cope better, by the sheer good fortune of their superior adaptation to changing climates—and they will leave more surviving offspring on average. (This differential reproductive success must be conceived as broadly statistical, and not guaranteed in every case. In any generation, the hairiest elephant of all may, in the flower of youthful strength but before any reproductive action, fall into a crevasse and die.) Because offspring inherit their parental degree of hairiness, the next generation will include a higher proportion of more densely clad elephants (who will continue to be favored by natural selection as the climate becomes still colder). This process of increasing average hairiness may continue for many generations, leading to the evolution of woolly mammoths.

This little fable can help us to understand how peculiar, and how contrary to all traditions of Western thought and explanation, the Darwinian theory of evolution, and variational theories of historical change in general, must sound to the common ear. All the odd and fascinating properties of Darwinian evolution flow from the variational basis of natural selection—including the sensible and explainable, but quite unpredictable, nature of the outcome (dependent upon complex and contingent changes in local environments), and the nonprogressive character of the alteration (adaptive only to these unpredictable local circumstances and not building a "better" elephant in any cosmic or general sense).

Transformational theories work in a much simpler and more direct manner. If I want to go from *a* to *b,* I will have so much less conceptual (and actual) trouble if I can postulate a mechanism that will push me there directly, than if I must rely upon selection of "a few good men" from a random cloud of variation about point *a,* then constitute a new generation around an average point one step closer to *b,* then generate a new cloud of random variation about this new point, then select "a few good men" once again from this new array—and then repeat this process over and over until I finally reach *b.* When one adds the oddity of variational theories in general to our strong cultural and psychological resistance against their application to our own evolutionary origin—leading to the reconceptualization of *Homo sapiens* as an unpredictable and not necessarily progressive little twig on life's luxuriant tree—then we can better understand why Darwin's revolution surpassed all other scientific discoveries in reformatory power, and why so many people still fail to understand, or even actively to resist, its truly liberating content. (I must leave the issue of liberation for another time, but once we recognize that the specification of morals and the search for a meaning in our lives cannot be resolved by scientific data in any case, then Darwin's variational mechanism will no longer seem threatening, and may even become liberating as a rationale for abandoning a chimerical search for the purpose of our lives, and the source of our ethical values, in the external workings of nature.)

These difficulties in grasping Darwin's great insight became exacerbated when our Victorian forebears made their unfortunate choice of a defining word—"evolution," with its vernacular meaning of directed unfolding. We would not face this additional problem today if "evolution" had undergone a complete transformation to become a strict and exclusive definition of biological change—with earlier, and etymologically more appropriate, usages then abandoned and forgotten. But important words rarely undergo such a clean switch of meaning, and "evolution" still maintains its original definition of predictable unfolding in several nonbiological disciplines—including astronomy.

When astronomers talk about the evolution of a star, they clearly do not invoke a variational theory like Darwin's. Stars do not change through time because mama and papa stars generate broods of varying daughter stars, followed by the differential survival of daughters best adapted to their particular region of the cosmos. Rather, theories of stellar "evolution" could not be more relentlessly transformational in positing a definite and predictable sequence of changes unfolding as simple consequences of physical laws. (No biological process operates in exactly the same manner, but the life cycle of an organism certainly works better than the evolution of a species as a source of analogy.)

Ironically, astronomy undeniably trumps biology in faithfulness to the etymology and vernacular definition of "evolution"—even though the term now holds far wider scientific currency in the radically altered definition of the biological sciences. In fact, astronomers have been so true to the original definition that they confine "evolution" to historical sequences of predictable unfolding, and resolutely shun the word when describing historical cosmic changes that *do* exhibit the key features of biological "evolution"—unpredictability and lack of inherent directionality.

As an illustration of this astronomical usage, consider the most standard and conventional of all sources—the *Encyclopaedia Britannica* article (fifteenth edition, 1990) on "stars and star clusters." The section on "star formation and evolution" begins by analogizing stellar "evolution" to a preprogrammed life cycle, with degree of "evolution" defined as position along the predictable trajectory:

> Throughout the Milky Way Galaxy . . . astronomers have discovered stars that are well evolved or even approaching extinction, or both, as well as occasional stars that must be very young or still in the process of formation. Evolutionary effects on these stars are not negligible.

The fully predictable and linear sequence of stages in a stellar "lifetime" ("evolution" to astronomers) records the consequences of a defining physical process in the construction and history of stars: the conversion of mass to energy, with a depletion of hydrogen and transformation to helium:

> The spread of luminosities and colors of stars within the main sequence can be understood as a consequence of evolution. . . . As the stars evolve, they adjust to the increase in the helium-to-hydrogen ratio in their cores. . . . When the core fuel is exhausted, the internal structure of the star changes rapidly; it quickly leaves the main sequence and moves towards the region of giants and supergiants.

The same basic sequence unfolds through stellar lives, but the rate of change ("evolution" to astronomers) varies as a predictable consequence of differences in mass:

> Like the rate of formation of a star, the subsequent rate of evolution on the main sequence is proportional to the mass of the star; the greater the mass, the more rapid the evolution.

More-complex factors may determine variation in some stages of the life cycle, but the basic directionality ("evolution" to astronomers) does not alter, and predictability from natural law remains precise and complete:

> The great spread in luminosities and colors of giant, supergiant, and subgiant stars is also understood to result from evolutionary events. When a star leaves the main sequence, its future evolution is precisely determined by its mass, rate of rotation (or angular momentum), chemical composition, and whether or not it is a member of a close binary system.

In the most revealing verbal clue of all, the discourse of this particular scientific "culture" seems to shun the word "evolution" when historical sequences become too meandering, too nondirectional, or too complex to explain as simple consequences of controlling laws—even though the end result may be markedly different from the beginning state, thus illustrating significant change through time. For example, the same *Britannica* article on stellar evolution notes that one can often reach conclusions about the origin of a star or a planet from the relative abundance of chemical elements in its present composition. But the earth's geological history has so altered its original state that we cannot make such inferences for our own planet.

In other words, the earth has undergone a set of profound and broadly directional changes—alterations so extensive that we can no longer utilize the present state to make inferences about our planet's original composition. However, since this current configuration developed through complex contingencies, and could not have been predicted from simple laws, this style of change apparently does not rank as "evolution"—but only as being "affected"—in astronomical parlance:

> The relative abundances of the chemical elements provide significant clues regarding their origin. The Earth's crust has been affected severely by erosion, fractionation, and other geologic events, so that its present varied composition offers few clues as to its early stages.

I don't mention these differences to lament, to complain, or to criticize astronomical usage. After all, their concept of "evolution" remains more faithful to etymology and the original English definition; whereas our Darwinian reconstruction has virtually reversed the original meaning. In this case, since neither side will or should give up its understanding of "evolution"—astronomers because they have retained an original and etymologically correct meaning, evolutionists because their redefinition expresses the very heart of their central and revolutionary concept of life's history—our best solution lies simply in exposing and understanding the legitimate differences, and in explicating the good reasons behind the disparity of use.

In this way, at least, we may avoid confusion and the special frustration generated when prolonged wrangles arise from misunderstandings about words, rather than genuine disputes about things and causes in nature. Evolutionary biologists must remain especially sensitive to this issue, because we still face considerable opposition, based on conventional hopes and fears, to our emphasis on an unpredictable history of life evolving in no inherently determined direction. Since astronomical "evolution" upholds both contrary positions—predictability and directionality—evolutionary biologists need to emphasize their own distinctive meaning, especially since the general public feels much more comfortable with the astronomical sense—and will therefore impose this more congenial definition upon the history of life if we do not clearly explain the logic, the evidence, and the sheer fascination of our challenging conclusion.

Two recent studies led me to this topic because each discovery confirms the biological, variational, and Darwinian "take" on evolution, while also, and quite explicitly, refuting a previous transformational interpretation—rooted in our culturally established prejudices for the more comforting astronomical view—that had blocked our understanding and skewed our thought about an important episode in life's history.

1. *Vertebrates all the way down.* In one of the most crucial and enigmatic episodes in the history of life—and a challenge to the old and congenial idea that life has progressed in a basically stately and linear manner through the ages—nearly all animal phyla make their first appearance in the fossil record at essentially the same time, an interval of some five million years (about 525 to 530 million years ago) called the Cambrian Explosion. (Geological firecrackers have long fuses when measured by the inappropriate scale of human time.) Only one major phylum with prominent and fossilizable hard parts does not appear either in this incident, or during the Cambrian period at all—the Bryozoa, a group of colonial marine organisms unknown to most nonspecialists today (although still relatively common), but prominent in the early fossil record of animal life.

One other group, until a discovery published in 1999, had also yielded no record within the Cambrian explosion, although late-Cambrian representatives (well after the explosion itself) have been known for some time. But, whereas popular texts have virtually ignored the Bryozoa, the absence of this other group had been prominently showcased and proclaimed highly significant. No vertebrates had ever been recovered from deposits of the Cambrian explosion, although close relatives within our phylum (the Chordata), but not technically vertebrates, had been collected. (The Chordata includes three major subgroups: the tunicates, *Amphioxus* and its relatives, and the vertebrates proper.)

This absence of vertebrates from strata bearing nearly all other fossilizable animal phyla provided a ray of hope for people who wished to view our own group as "higher" or more evolved in a predictable direction. If evolution implies linear progression, then later is better—and uniquely later (or almost uniquely, given those pesky bryozoans) can only enhance the distinction. But the November 4, 1999, issue of *Nature* includes a persuasive article by D.-G. Shu, H.-L. Luo, S. Conway Morris, X.-L. Zhang, S.-X. Hu, L. Chen, J. Han, M. Zhu, Y. Li, and L.-Z. Chen ("Lower Cambrian vertebrates from South China," volume 402, pages 42–46), reporting the discovery of two vertebrate genera within the Lower Cambrian Chengjiang Formation of South China, right within the temporal heart of the Cambrian explosion. (The Burgess Shale of Western Canada, the celebrated site for most previous knowledge of early Cambrian animals, postdates the actual explosion by several million years. The recently discovered Chengjiang fauna, with equally exquisite preservation of soft anatomy, has been yielding comparable or even greater treasures for more than a decade.)

These two creatures—each only an inch or so in length, and lacking both jaws and a backbone, in fact possessing no bony skeleton at all—might not strike a casual student as worthy of inclusion within our exalted lineage. But jaws and backbones, however much they may command our present focus, arose later in the history of vertebrates, and do not enter the central and inclusive taxonomic definition of our group. The vertebrate jaw, for example, evolved from hard parts that originally fortified the gill openings just behind, and then moved forward to surround the mouth. All early fishes lacked jaws—as do the two modern survivors of this initial radiation, the lampreys and hagfishes.

The two Chengjiang genera possess all defining features of vertebrates—the stiff dorsal supporting rod or notochord (subsequently lost in adults after the vertebral column evolved), the arrangement of flank musculature in a series of zigzag elements from front to back, the set of paired openings piercing the pharynx (operating primarily as respiratory gills in later fishes, but used mostly for filter-feeding in ancestral vertebrates). In fact, the best reconstruction of branching order on the vertebrate tree places the origin of these two new genera after the inferred ancestors of modern hagfishes, but before the presumed forebears of lampreys. If this inference holds, then vertebrates already existed in substantial diversity within the Cambrian explosion. In any case, we now know two distinct and concrete examples of vertebrates all the way down. We vertebrates do not stand higher and later than our invertebrate cousins, for all "advanced" animal phyla made their debut in the fossil record at essentially the same time. The vaunted complexity of vertebrates did not require a special delay to accommodate a slow series of progressive steps, predictable from the general principles of evolution.

2. *An ultimate parasite, or "how are the mighty fallen."* The phyla of complex multicellular animals enjoy a collective designation as Metazoa (literally, higher animals). Mobile, single-celled creatures bear the name Protozoa (or "first animals"—actually a misnomer, since most of these creatures stand as close to multicellular plants and fungi as to multicellular animals on the genealogical tree of life). In a verbal in-between stand the Mesozoa (or "middle animals"). Many taxonomic and evolutionary schemes for the organization of life rank the Mesozoa exactly as their name implies—as a persistently primitive group, intermediate between the single-celled and multicellular animals, and illustrating a necessary transitional step in a progressivist reading of life's history.

But the Mesozoa have always been viewed as enigmatic—primarily because they live as parasites within truly multicellular animals, and parasites often adapt to their protected surroundings by evolving an extremely simplified anatomy, sometimes little more than a glob of absorptive and reproductive tissue cocooned within the body of a host. Thus the extreme simplicity of parasitic anatomy could represent the evolutionary degeneration of a complex, free-living ancestor rather than the maintenance of a primitive state.

The major group of mesozoans, the Dicyemida, live as microscopic parasites in the renal organs of squid and octopuses. Their adult anatomy could hardly be simpler—a single axial cell (which generates the reproductive cells) in the center, enveloped by a single layer of ciliated outer cells, some ten to forty in number, and arranged in a spiral around the axial cell, except at the front end, where two circlets of cells (called the *calotte*) form a rough "mouth" that attaches to the tissues of the host.

The zoological status of the dicyemids has always been controversial. Some scientists, including Libbie Hyman, who wrote the definitive multivolume text on invertebrate anatomy for her generation, regarded their simplicity as primitive, and their evolutionary status as intermediate in the rising complexity of evolution. She wrote in 1940: "their characters are in the main primitive and not the result of parasitic degeneration." But even those researchers who viewed the dicyemids as parasitic descendants of more-complex, free-living ancestors never dared to derive these ultimately simple multicellular creatures from a *very* complex metazoan. For example, Horace W. Stunkard, the leading student of dicyemids in the generation of my teachers, thought that mesozoans had descended from the simplest of all Metazoa above the grade of sponges and corals—the platyhelminth flatworms.

Unfortunately, the anatomy of dicyemids has become so regressed and specialized that no evidence remains to link those creatures firmly with other animal groups, so the controversy of persistently primitive versus degeneratively parasitic could never be settled until now. But newer methods of gene sequencing can solve this dilemma, because even though visible anatomy may fade or transform beyond genealogical recognition, evolution can hardly erase all traces of complex gene sequences. If genes known only from advanced Metazoa—and known to operate only in the context of organs and functions unique to Metazoa—also exist in dicyemids, then these creatures should be interpreted as degenerated metazoans. But if, after extensive search, no sign of distinctive metazoan genomes can be detected in dicyemids, then the Mesozoa may well rank as intermediates between single and multicelled life after all.

In the October 21, 1999, issue of *Nature,* M. Kobayashi, H. Furuya, and P. W. H. Holland present an elegant solution to this old problem ("Dicyemids Are Higher Animals"). These researchers located a *Hox* gene—a member of a distinctive subset known only from metazoans and operating in the differentiation of body structures along the antero-posterior (front to back) axis—in *Dicyema orientale.* These particular *Hox* genes occur only in triploblastic, or "higher," metazoans with body cavities and three cell layers, and not in any of the groups (such as the Porifera, or sponges, and the Cnidaria, or corals and their relatives) traditionally placed "below" triploblasts. Thus the dicyemids are descended from "higher," triploblastic animals and have become maximally simplified in anatomy by adaptation to their parasitic lifestyle. They do not represent primitive vestiges of an early stage in the linear progress of life.

In short, if the traditionally "highest" of all triploblasts—the vertebrate line, including our exalted selves—appears in the fossil record at the same time as all other triploblast phyla in the Cambrian explosion; and if the most anatomically simplified of all parasites can evolve, as an adaptation to local ecology, from a free-living lineage within the "higher" triploblast phyla; then the biological, variational, and Darwinian meaning of "evolution" as unpredictable and nondirectional gains powerful support from two cases that, in a former and now disproven interpretation, once bolstered an opposite set of transformational prejudices.

As a final thought to contrast the predictable unfolding of stellar "evolution" with the contingent nondirectionality of biological "evolution," I should note that Darwin's closing line about "this planet . . . cycling on according to the fixed law of gravity," while adequate for now, cannot hold for all time. Stellar "evolution" will, one day, enjoin a predictable end, at least to life on Earth. Quoting one more time from the *Britannica* article on stellar evolution:

> The Sun is destined to perish as a white dwarf. But before that happens, it will evolve into a red giant, engulfing Mercury and Venus in the process. At the same time, it will blow away the earth's atmosphere and boil its oceans, making the planet uninhabitable.

The same predictability also allows us to specify the timing of this catastrophe—about 5 billion years from now! A tolerably distant future to be sure, but consider the issue in comparison with the very different style of change known as biological evolution. Our planet originated about 4.6 billion years ago. Thus, half of the earth's potential history unfolded before contingent biological evolution produced even a single species with consciousness sufficient to muse over such matters. Moreover, this single lineage arose within a marginal group of mammals (among two hundred species of primates amid only four thousand or so species of mammals overall; by contrast, the world holds at least half a million species of beetles alone among insects). If a meandering process consumed half of all available time to build such an adaptation even once, then mentality at a human level certainly doesn't seem to rank among the "sure bets," or even mild probabilities, of history.

We must therefore contrast the good fortune of our own "evolution" with the inexorable "evolution" of our nurturing sun toward a spectacular climax that might make our further evolution impossible. True, the time may be too distant to inspire any practical concern, but we humans do like to ponder and to wonder. The contingency of our "evolution" offers no guarantees against the certainties of the sun's "evolution." We shall probably be long gone by then, perhaps taking a good deal of life with us, and perhaps leaving those previously indestructible bacteria as the highest mute witnesses to a stellar expansion leading to unicellular Armageddon as well. Or perhaps we, or our successors, will have colonized the universe by then, and will only shed a brief tear for the destruction of a little cosmic exhibit titled "the museum of our geographic origins." Somehow, I prefer the excitement of wondering and cogitation—not to mention the power inherent in action upon things that *can* be changed—to the certainty of distant dissolution.

19

The First Day of the Rest of Our Life

THE COMPARISON OF THE HUMAN BODY AND THE UNIverse—the microcosm with the macrocosm—has served as a standard device for explicating both the factuality and the meaning of nature throughout most of Western history. When Leonardo da Vinci, for example, likened our bodily heat, breath, blood, and bones to the lavas of volcanic eruptions, the effusions of interior air in earthquakes, the emergence of streams from underground springs, and the rocks that build the earth's framework—and then interpreted these sequences as particular expressions of the four Greek elements of fire, air, water, and earth—he did not view his argument as an excursion into poetry or metaphorical suggestion, but as his best understanding of nature's actual construction.

We now take a more cynical, or at least a more bemused, view of such analogistic reveries—for we recognize that the cosmos, in all its grandness, does not exist for us, or as a mirror of our centrality in the scheme of universal things. That is, we would now freely admit that most attempts to understand such geological or astronomical scales of size and time in terms of comfortable regularities noted in our short life spans or puny dimensions can only represent, in the most flattering interpretation, an honorable "best try" within our own mental and perceptual limits or, at worst, yet another manifestation of the ancient sin of pride.

As a striking example, however unrecognized by most people who could scarcely avoid both walking the walk and talking the talk, the recent fuss over our millennial transition cannot be entirely ascribed to modern commercial hype because the taproot of concern draws upon one of the oldest surviving arguments about deep and meaningful coincidence between the human microcosm and the surrounding macrocosm of universal time and space—in this case, an explicit comparison of human secular calendars to the full sweep of the creation and subsequent history of the earth and life. By this reckoning, January 1, 2000, should have marked the termination of the old order, and the inception of something new and at least potentially glorious. This momentous turning of calendrical dials should therefore have inspired our attention for reasons almost immeasurably deeper than the simple visual attraction of changing all four markers from 1999 to 2000—the "odometer rationale," if you will. (Of course, the vast majority of people, in our secular and technological age, have forgotten this old, and factually discarded, Christian argument for the significance of millennial turnings. But vestiges of these historical claims still affect both our calendars and our discourse. Moreover, and with potentially tragic results, the vestiges of a majority persist as literal portents for a few "true believers," leading, in the most extreme case, to the suicide of thirty-nine members of the Heaven's Gate cult in 1997.)

The traditional Christian linkage of human calendrical microcosms to universal historical macrocosms proceeds as an argument in five stages.

1. The original millennium, as expressed in the famous biblical prophecy of Revelation, chapter 20, referred to a *future* one-thousand-year period of bliss following the return of Jesus and the binding of Satan, not to a secular passage of one thousand years in recorded human history. How, then, did the primary meaning of "millennium" change from the duration of a future epoch to the ticking of current calendars?

2. The earliest Christians expected an imminent inception of the millennium, as Jesus had apparently stated in foreseeing his quick return after bodily death: "Verily I say unto you, There shall be some standing here, which shall

not taste of death, till they see the Son of Man coming in his kingdom" (Matthew 16:28). The failure of this expectation unleashed an extended discussion among early Christians on the meaning of the millennium and the true timing of the second coming of Christ.

3. Opinions varied widely, but the most popular claim rested upon several biblical passages suggesting an equation of God's days with a thousand human years, as in the admonition of 2 Peter 3:8—"But, beloved, be not ignorant of this one thing, that one day is with the Lord as a thousand years, and a thousand years as one day."

4. The link between human calendars and the inception of the true millennium then rested upon an analogistic argument that we, by modern standards, would tend to regard as fuzzy, indefinite, and metaphorical, but that seemed quite satisfactory to many of our forebears (who used their equally powerful brains in different conceptual contexts): If God created the earth in six days and rested on a seventh, and if each of God's days equals one thousand human years, then the earth's full history must mirror God's complete span of creation by lasting six thousand years, while God's seventh day of rest must correspond to the forthcoming blissful millennium of one thousand additional years. If, therefore, we can count the earth's history in millennia (periods of one thousand years representing God's days), we will know, with precision, the end of the current order and the time of inception for the millennium—for this transition will occur exactly six thousand years after the earth's beginning.

5. This argument inspired a burst of scholarship (culminating in the seventeenth century) that tried to use the Bible and other ancient records to construct a true and exact chronology for universal history. In the most popular scheme, Christ was born exactly four thousand years after creation, and the current order may therefore persist for an additional two thousand years. Finally, if the birth of Jesus occurred at the B.C.–A.D. transition of our calendar, then the end of this secular millennium should terminate our current order and initiate the millennium (in the original meaning of a forthcoming period of bliss) of Jesus' second coming. Clearly, then, we should care about microcosmal human calendars because they mark the epochs of macrocosmal universal history and prepare us for the fears of apocalypse followed by a better world to come.

I have presented this influential argument of Christian history as a prologue to a segue devoted to reminding readers about the most boring of all general topics—one that we would all rather forget, but remember so well, from our primary school years: the inevitable "what I did on my . . ." assigned upon every return to school after an extended absence (with "my summer vacation" and "my Christmas break" as the most common particulars). I shall now dare to regale you with an essay in precisely this dreaded form: "What I did on the millennial day of January 1, 2000." I can only hope and pray that my prologue, combined with a forthcoming explication, may build an apparatus to overcome the inherent limitations of this general topic.

The purely factual resolution requires but a phrase: I sang in a performance of Joseph Haydn's great oratorio *The Creation,* presented by the Boston Cecilia at Jordan Hall. For my larger topic, let me try to explain (an effort, alas, that will take a bit more space than the factual assertion stated just above) why the conjunction of this particular piece with the millennial day strikes me as so optimally appropriate in a general sense; why the privilege of participation meant so much to me personally (an otherwise private matter, but vouchsafed to essayists ever since Montaigne defined this genre as personal commentary upon generalities more than four centuries ago); and why a topic so off the left-field wall (to combine two common metaphors for the bizarre)—namely a musical composition on a text drawn from the same creation narrative, Genesis 1, now urged by our antiscientific opponents as an alternative to teaching evolution in America's public schools—might find a truly fitting place in a book of essays about natural history.

Now, if I may try your patience for just one more round of annoyingly necessary (and prefatory) footnotes, let me dismiss three little niggling issues about dates before we come to Haydn's magisterial creation of light in C major.

1. With apologies for shining the factual torch of modern science on the best-laid intellectual schemes of ancestral mice and men, the earth is really about 4.7 billion years old, and life's known fossil record extends back to about 3.6 billion years—so days and millennia scarcely qualify as terms for a serious discussion of factual matters related to life's origin and history.

2. Even within the system that exalted millennial transitions as God's days, and the end of the sixth transition as the termination of our current universe, the year 2000 really doesn't qualify for much consideration. Unfortunately, poor Dionysius Exiguus (Dennis the Short), the sixth-century monk who devised the B.C.–A.D. calendrical system, made a little error in setting Christ's birth. We have no direct testimony about the historical Jesus, and no eyewitness account can set his time of birth. But we do know that Herod died in 4 B.C. (Kings tend to leave better written evidence of their lives than poor kids born in stables.) Now, if Herod and Jesus overlapped—and some of the most rousing biblical stories must be discounted if they did not (the Slaughter of the Innocents, the return of the Magi to their own country, rather than to Herod)—

then Jesus, despite the oxymoronic nature of the claim, must have been born in 4 B.C. or earlier. Thus, by the millennial chronology, the current order should actually have ended a few years ago—and it didn't.

3. Even if we didn't know about this inconvenient issue of Jesus' birth, or just wish to maintain a polite fiction about his appearance right at the B.C.–A.D. junction, we have still erred in concentrating our millennial fears on the 1999–2000 transition. Again, we must recognize Dionysius Exiguus as the culprit, although we cannot cast much blame this time. No zero existed in Western mathematics when Dionysius performed his calendrical duties, so he began A.D. time on January 1, year one—and our calendar never experienced a year zero. Now, if you believe that the blessed millennium of Jesus' second coming will begin exactly two thousand years after his birth, then you still have another year to wait. For the completion of two thousand years since Jesus' birth occurs at the 2000–2001 transition, not on the day of fear that has just passed.

Of course, as most folks know by now, this same issue underlies the great, unresolvable, and basically silly debate about whether the new millennium starts at the beginning of 2000 or 2001. I won't rehearse this particular well-beaten and very dead horse—although you may all consult my now remaindered book, *Questioning the Millennium,* if the subject still holds any interest for you. I will only observe—and then promise never to raise the subject again—that this debate expresses nothing new, but has erupted at the end of every century (admittedly with greater intensity this time because this particular turning encompasses a millennium as well, and also occurs in our age of media overkill for everything). I merely append an illustration from a French pamphlet, published in 1699, and titled "Dissertation on the beginning of the next century and the solution to the problem: to know which one of the two years, 1700 or 1701, is the first of the century." As our Gallic cousins like to say: *plus ça change, plus c'est la même chose* (the more things change, the more they stay the same).

Haydn's text faithfully follows the six-day sequence of creation in Genesis 1—the basis (by the traditional argument outlined above) for regarding the day of our singing as the end of history and the inception of a new order. (Haydn wrote *The Creation* in German, but based on a translated English text taken mostly from Genesis and from some paraphrases of Milton's *Paradise Lost.* Haydn published the text in both languages and apparently intended his piece for bilingual performance.)

One can easily formulate the obvious and legitimate excuses: "such great music . . ." and "you can't blame Haydn in 1798 for not anticipating what Darwin would publish in 1859." But shouldn't a paleontologist and evolution-

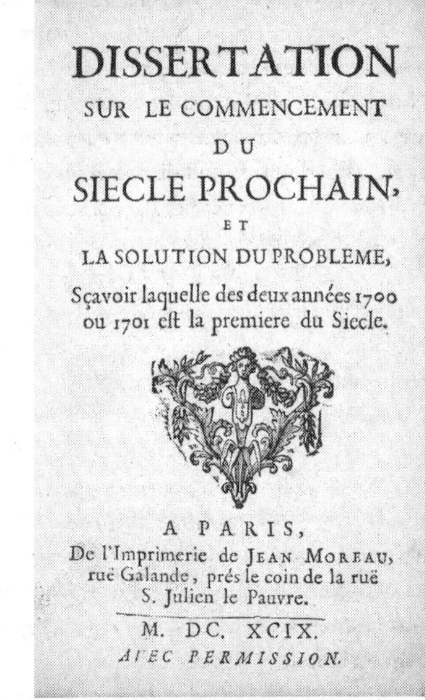
DISSERTATION
SUR LE COMMENCEMENT
DU
SIECLE PROCHAIN,
ET
LA SOLUTION DU PROBLEME,
Sçavoir laquelle des deux années 1700 ou 1701 est la premiere du Siecle.

A PARIS,
De l'Imprimerie de JEAN MOREAU, ruë Galande, prés le coin de la ruë S. Julien le Pauvre.
M. DC. XCIX.
AVEC PERMISSION.

The cover page for a French pamphlet, published in 1699, and proving the antiquity of the argument about whether centuries (and millennia) begin in the '00 or the '01 year.

ary biologist, sitting on stage in the chorus, become at least a bit uncomfortable when the angel Raphael, recounting the origin of land animals on the sixth day, explicitly proclaims their sudden creation "in perfect forms, and fully grown"?

I don't deny that participation in some great music raises difficult issues and considerable emotional distress—particularly the strongly anti-Semitic choral passages, representing the Jewish crowd taunting Jesus or demanding his death, in J. S. Bach's sublime St. Matthew and St. John Passions, perhaps the greatest choral works ever written. (The power and quality of the music, of course, only enhances the discomfort.) I find the "blood guilt" passage from the Matthew Passion particularly disturbing because I know that these very words served, for centuries, as a primary rationale—often with explicitly deadly consequences for my people—for labeling Jews as the killers of Christ. (For my own personal resolution, I decided long ago that whenever we sang this work, I would at least mention the historical context during our first rehearsal of this text—based on the statement of the Jewish crowd, after Pilate finds no guilt in Jesus and literally washes his hands of the affair: *sein Blut komme über uns and unsre Kinder*—let his blood then be upon us and upon our children).

I do, by the way, accept the different historical context of Bach's time. I feel no enmity toward this great man, who may never have known a Jew, and who probably never considered the issue as he simply set the literal text of Matthew. Nor would I ever consider changing the words for any modern performance, lest an understandable deed for a particular purpose establish a precedent and open a floodgate for wholesale revision of any great work to suit the whims of fashion. But I do think that the issue should never be avoided, and should always be explicitly discussed—in pre-concert lectures, program notes, etc.

But I feel not the slightest tinge of discomfort—and, quite to the contrary, experience nothing but joy—in singing the text of Haydn's *Creation.* In explaining these different reactions, I must begin by saying that I don't use factual accuracy as a major criterion for judging a musical libretto any more than I would look for aesthetic beauty (to my personal sensitivities) or moral rectitude in assessing the validity of a scientific conclusion. (Much of nature's factuality strikes us as both messy and unpleasant—but no less true or fascinating thereby.) I recoil from the anti-Semitic Passion texts because they express the worst aspects of our common nature, and because these words have wreaked actual death and havoc. Similarly, I embrace Haydn's *Creation* text for its moral and aesthetic qualities, while regarding its factual inaccuracies as quite irrelevant and beside the point.

After all, we read the Bible as a source of moral debate and instruction, not as a treatise in natural history. Moreover, even if Haydn had decided to express the science of his day, he would not have written a libretto about evolution. As for creation in six days, Haydn, as a devout Catholic, surely never conceived the text as a set of statements about twenty-four-hour periods—for no literalist tradition existed within the doctrines of his church, and such interpretations never gained currency after Saint Augustine's entirely persuasive denials more than one thousand years before. Our currently active scourge of fundamentalism, or biblical literalism, arose later and from different traditions. The basic analogy of God's days to human millennia, while still ungenerous by the standards of geological time, surely illustrates a Catholic consensus for reading "days" of creation as sequential intervals, not as equal and predetermined tickings of God's stopwatch.

All cultures generate creation myths, and such stories play a part in the drama of human life far different from the role that we grant to the fascination and utility of factual discoveries made by science. With this perspective, I can summarize my case for Haydn's text in a paragraph: The Book of Genesis presents two entirely different creation myths, told in chapters 1 and 2. I find two aspects of the second myth morally troubling, whereas (with one exception) I rejoice in the meanings and implications of the first story. Interestingly, Haydn's text uses only the first story, and explicitly deletes the one theme (human hegemony over the rest of God's creation) that disturbs me (and has troubled so much of human history as well). I do not think that these textual decisions were accidental, and I therefore regard Haydn's *Creation* as an affirmation of all the themes that a wise and maximally useful creation myth should stress—joy, generosity, optimism—while not forgetting the dark side and our resulting capacity to make a horrid mess out of such promise.

The second creation myth of Genesis 2—the text that Haydn did *not* set—emphasizes two themes that I find less than inspiring: God's order (by fiat and not by explanation) that we not seek certain kinds of knowledge, and an anatomical rationale for the subjugation of women. We tend to forget the profound differences between the two stories of Genesis, and we usually amalgamate parts of this second tale with out primary memory of the first story (see essay 7 for a very different context, but longer analysis, of these two distinct creation myths). In Genesis 2, God creates Adam first, and then builds the Garden of Eden. To assuage Adam's loneliness, he then creates the animals and permits

Creation of Adam and Eve*, Flemish school, seventeenth century.*

Adam to assign their names. But Adam is still lonely, so God creates Eve from his rib. (Genesis 1, Haydn's text, says nothing about forbidden fruits, and describes the simultaneous creation of man and woman: "So God created man in his own image . . . male and female created he them.")

The theme of forbidden access to knowledge occurs only in Genesis 2. (I recognize, of course, that some exegetes can and have suggested a benign meaning for these passages in terms of moral restraint upon our darker capacities. But most people, throughout Western history, have read these words as a divine indiction against questioning certain forms of authority and seeking certain forms of knowledge—injunctions that cannot be congenial to any scientist.) "And the Lord God commanded the man, saying, Of every tree of the garden thou mayest freely eat: But of the tree of the knowledge of good and evil, thou shalt not eat of it: for in the day that thou eatest thereof thou shalt surely die" (Genesis 2:16–17).

Similarly, no statement in Genesis 1 speaks about inequality between the sexes, but Adam uses Eve's status as both subsequent and partial to hint at such a claim in Genesis 2: "And Adam said, This is now bone of my bones, and flesh of my flesh: she shall be called Woman, because she was taken out of Man" (Genesis 2:23).

Haydn's text divides the creation myth of Genesis 1 into three sensible and dramatic units. We usually view the six-day sequence as a story of successive additions, but I think that such a reading seriously mistakes the form of this particular myth (see essay 20 for a detailed development of this argument). Creation myths, based on limits of our mental powers, and also on the structural possibilities of material objects, can only "go" in a few basic ways—and Genesis 1 invokes the primary theme of successive differentiation from initial chaos, not sequential addition. The universe begins in undefined confusion ("without form and void"). God then constructs a series of separations and divisions to mark the first four days. On day one, he divides light from darkness. Haydn's amazing overture violates many contemporary musical traditions of tonality and structure in order to depict this initial chaos. He then, at the end of the first chorus, describes the creation of light with a device both amazingly simple and (to this day) startlingly evocative: a series of crashing chords in bright and utterly unsophisticated C major. (A virtual cliché among statements in the history of classical music designates this passage—but so truly—as the most stunningly effective C-major chords ever written.)

On day two, God divides the waters of earth and sky; on day three, he separates the earth into water and land (and also allows the land to bring forth plants). On day four, he returns to the heavens to concentrate the diffuse light

Creation of the Stars and Planets, *from the Sistine ceiling, by Michelangelo, sixteenth century.*

into two great sources, the sun and the moon ("he made the stars also" as an afterthought). Soloists describe the work of each day, and each sequence finishes with a wonderful chorus. Part 1 therefore ends with Haydn's most famous ensemble: "The heavens are telling the glory of God"—the heavens, that is, because no animals have yet been formed!

Part 2 describes the work of days five and six, the creation of animals: creatures of the water and air on day five, and of the earth, including humans, on day six. Soloists and chorus alternate as in Part 1. Haydn's music exudes beauty, power, exaltation, etc., but he can also be whimsical, earthy, and ordinary—a combination that captures the essence of the humanistic (or should I say naturalistic) spirit by acknowledging that glory and fascination lie as much in the little foibles as in the grand overarchings. Haydn shows this bumptious and quotidian side of totality in describing the creation of animals. First, the soprano soloist, in a charming and idyllic aria, describes the birds—the noble eagle, the cooing dove, the merry lark, and the nightingale, who had not yet learned (but, alas, soon will) to sing an unhappy note: "No grief affected yet her breast, Nor to a mournful tale were tuned her soft enchanting lays." The bass soloist, alternating between the bucolic and the simply funny, then describes the tawny lion, the flexible tiger, the nimble stag, and, finally, "in long dimension, creeps with sinuous trace the worm" (usually ending, if the bass soloist can, and ours could, on a low D—actually set an octave higher by Haydn, but taken down by soloists as a traditional bass license corresponding to those annoying high C's that tenors forever interpolate).

The shorter Part 3 then uses Milton's style (if not exactly his words) for two long and rapturous duets between Adam and Eve, interlaced with choral praises and culminating in a paean of thanks with a final musical device that always thrills me as a singer (and, I hope, pleases the audience as well). The final expostulation of joy for the glorious diversity of the earth and its life—"praise the Lord, utter thanks, Amen"—runs twice, first as an alternation of passages for a quartet of soloists and the full chorus, and then, even louder, for the full chorus alone. This acceleration or promotion—more an emotional device than a compositional beauty per se (but mastery of such devices also marks a composer's skill)—always leaves me feeling that we should mount even higher, thus allowing the performance to reverberate beyond its formal ending (which can only be deemed quite grand enough already!).

Haydn's text stands revealed as a great document of optimism and humanism as much for its omissions as for its inclusions. Interestingly, although nearly the entire text of Genesis 1 enters the narrative, one long passage has been conspicuously (and, I assume, consciously) omitted—the set of "objectionable" (to me at least) statements about divinely ordained human domination over nature: "And God said, Let us make man in our image, after our likeness; and let them have dominion over the fish of the sea, and over the fowl of the air, and over the cattle, and over all the earth. . . . And God blessed them, and God said unto them, Be fruitful, and multiply, and replenish the earth, and subdue it; and have dominion . . . over every living thing that moveth upon the earth" (Genesis 1:26 and 28).

Instead, following the creation of land animals, Haydn's text, in a nonbiblical interpolation, suggests an entirely different reason for the creation of men and women. In Genesis 1, God fashions us to have dominion over everything else. But in Haydn's text, God creates humans simply because the living world remains unfulfilled, even a bit sad, after such effort in making everything up to the tawny lion and the sinuous worm. In nearly six full days of hard work, God has stuffed the earth with a glorious series of diverse and wonderful objects. But he then realizes that one omission precludes the fulfillment of this greatest architectural task ever set to music (far exceeding the accomplishments of Fasolt and Fafnir in building Valhalla). Not a single item in his creation has enough mental power to appreciate the beauty and glory of these optimal surroundings. God has to make men and women—so that some creature can know and praise the grandeur of existence. And so Raphael, just following his low D for the sinuous worm, exclaims: "But all the work was not complete; there wanted yet that wondrous being; that, grateful, could God's power admire, with heart and voice his goodness praise."

I don't want to make either this recitation, or Haydn's text, sound too saccharine or devoid of complexity. The humanistic tradition does not deny the dark side, but chooses to use these themes as warnings for potential correction, rather than statements about innate depravity. Thus, Haydn does not entirely neglect the common biblical subject (so prominent in Genesis 2) of dangers inherent in knowing too much. But he certainly reduces the point to a barest possible minimum. Just before the final chorus, the tenor soloist sings a quick passage in the least impassioned narrational style of "dry recitative" (with only keyboard and continuo as accompaniment): "O happy pair and happy still might be if not misled by false conceit, ye strive at more than is granted and desire to know more than you should know." Modern listeners might also be discomfited by Eve's promises of obedience to Adam in their second duet (from Milton, not from Genesis 1)—even though her inspiration follows Adam's promise to "pour new delights" and "show wonders everywhere" with every step they take together upon this newly created world. We can't, after all, impose the sensibilities of 2000 upon 1798. And who would want to defend 2000 before any truly just court of universal righteousness?

But while we identify Haydn's text as a creation myth in the most expansive and optimistic spirit of love and wonder for all works of earth and life, we must also confront a historical puzzle. Haydn began his work in 1794, and the first performance took place in 1798 (with Haydn conducting and none other than Antonio Salieri, the unfairly maligned villain of *Amadeus,* at the harpsichord). Such an expansively optimistic text seems entirely out of keeping with the conservative gloom that spread throughout Europe after the excesses of the French Revolution, culminating in the guillotining of the guillotiner Robespierre in 1794. Moreover, the spread of romanticism in music and art, for all its virtues, scarcely sanctioned such old-fashioned joy in the objective material world.

The apparent solution to this problem rests upon an interesting twist. Haydn wrote *The Creation* as a result of inspiration received during trips to London, particularly in 1791, when he heard (and felt overwhelmed by) the power of Handel's oratorios. This source has always been recognized, and the pleasure of singing Haydn's *Creation* lies at least partly in the wonderful Handelian anachronisms included amid the lush classical and near-romantic orchestration. But Handel's posthumous influence may have run far deeper. The source of Haydn's text has always presented a mystery. Who wrote it, and how did Haydn obtain the goods and the rights? (We know that Haydn's friend, Baron Gottfried van Swieten, translated the text into German from an English original—but whence the original?) Latest scholarship indicates that the text may have been written for Handel more than forty years earlier

(Handel died in 1759), but never set by the greatest master of the oratorio, and therefore still available for Haydn two generations later.

Such an earlier source would solve the problems of content. For if Haydn's libretto really dates from the 1740s or 1750s, then the incongruities disappear. The text becomes a document composed during the heart of the Enlightenment, an intellectual and artistic movement that embodied all the optimism, all the pleasure in nature's beauty, all the faith that a combination of human reason and moral potential might ensure both goodness and justice. The text of *The Creation* reflects this hopeful world, when Linnaeus worked in Uppsala, classifying all plants and animals for the glory of God and the knowledge of men, while Ben Franklin promoted the virtues of fire departments, public libraries, and universities in Philadelphia. The Enlightenment may have veered toward naïveté in its optimism about human and worldly possibilities, but the goals still seem attainable—and we will never get there if we lose the hope and spirit. Ya gotta believe.

The difficulty of this task (so well epitomized in some great words of another famous Enlightenment thinker as the actual provision for all people of our unalienable rights of "life, liberty and the pursuit of happiness") requires that all facets of human achievement be mobilized in the great work. We will surely need the benefits of science—if only to feed, and keep healthy, all the people that science has permitted us to rear to adulthood. We will also need—and just as much—the moral guidance and ennobling capacities of religion, the humanities, and the arts, for otherwise the dark side of our capacities will win, and humanity may perish in war and recrimination on a blighted planet.

Art and science provide different and equally legitimate perspectives on the same set of saving subjects, and we need both approaches. Thus, for me as a scientist who has devoted a full career to the study of evolution (but who also fancies himself a serious and competent choral singer, and not just an occasional duffer at a Saturday-night piano bar), I see no contradiction, but only harmony, in integrating the final line from a great work of science, a statement that Darwin chose to make in personal terms of poetic awe, with Haydn's decision to write an inspirational choral work based on an Enlightenment version of a creation myth that seems to use (in its different way) the same subject of Darwin's scientific studies. The factual truth of evolution cannot conflict with the search for meaning embodied in a good creation myth. "There is grandeur in this view of life," Darwin wrote. And, as Haydn said, "the heavens are telling the glory of God."

The task before us remains so daunting that we need to find tools beyond the integration of science, morality, and all the other separate patches on what

I like to designate as our coat of many colors called wisdom. We also need symbols to intensify and epitomize this grand effort that, ultimately, must lead us all to hang together or hang separately (a great pun by the Enlightened Mr. Franklin). Given my propensities and proclivities, I do not know how, in this symbolic sense, I could have spent the inception of the millennium in a more meaningful way. And so, Mrs. Ponti, my truly beloved fifth grade teacher, I dedicate this version of "what I did on . . ." to your memory and to the inspiration that you so freely provided with your love and skill.

I spent the millennial day that, in long tradition and the persisting fears of some folks, would mark the end of the world, by sharing and blending my efforts with a group of colleagues who had worked long and hard to prepare a performance of the greatest musical work ever written about the joyful, glorious, and optimistic *beginnings* of an order that can end (on our time scale) only if we fail to join the spirit of Darwin and Haydn, thereby potentiating all the saving graces of our nature. We express this union in many ways. The last words of Genesis 1 do not represent my personal choice, but who can doubt the nobility of the sentiments, and what person of goodwill can fail to be horrified by the prospect (and therefore be inspired to devote some personal effort toward prevention) that one species might eviscerate something so wonderful that we did not create, and that was not fashioned for us? "And God saw every thing that he had made, and, behold, it was very good."

20

The Narthex of San Marco and the Pangenetic Paradigm

I DO REALIZE THAT THE BIGGEST OF ALL BOSSES LABORED with maximal sweat and diligence during those first six days. So "perhaps it would be wise not to carp or criticize," as Gilbert and Sullivan's Chief of Police once remarked in a different context (in *The Pirates of Penzance*). Still, I must confess that I've always been puzzled by the relative paltriness of accomplishment on the second day, for the import of this episode seems almost derisory compared with the scope of all others—light created on the first day (in the grandest of all opening acts, and especially necessary for noticing any subsequent event); earthly land, waters, and even plants in a foretaste of life on the third day; sun and moon, with stars as a mere afterthought, to populate the heavens on the fourth day; animal denizens of the sea and air on the fifth day; inhabitants of the land, culminating in our exalted selves, on the sixth day.

But on this second day of relative nodding, God limited his efforts to installing a plane of division (called the "firmament" in the King James Bible, the "sky" in many modern translations, but closer to a thin metal plate in the original Hebrew), simply to distinguish the water above the plate from the water below. Big deal. Compared with all the makings and shakings of the other days, this second effort created nothing new, but only constructed an artificial division within an existing homogeneity. Did God need a breather right after his initial effort in the creation business? Did he have to pause after the first step in order to recoup his courage, or to gain a second wind for pushing through to the end?

The light dawned on my pugnacious ignorance (thus validating the product of the first day) several years ago when I studied the thirteenth-century mosaics of this creation story (Genesis 1, of course) in the south dome of the narthex (the covered western porch just in front of the main entrance) of the great cathedral of San Marco in Venice—thus also explaining the first part of my cryptic title (I shall soon shed light on the second and even more cryptic part, but all in good time).

In this wonderfully animistic set of scenes, radiating downward in three circles from the apex of the dome, each episode of creation features the youthful, beardless God of Greek and Byzantine traditions doing his appointed tasks for the day, while an appropriate number of angels either help out or look on (Otto Demus's authoritative four-volume work on the San Marco mosaics argues convincingly that the scenes of the creation dome derive from an illuminated fifth-century Greek manuscript of the Book of Genesis).

In the scene for the second day—and begging an authorial pass for my irreverent anachronism in explication—God goes bowling to divide the waters by rolling a globe horizontally through the middle of the homogeneous mass, thus carving a barrier to define "above" and "below." The two angels of the day stand awkwardly at the right of the scene, more in the way (for the waters remain undivided to their right) than as auxiliaries in this case. (Demus explains that the mosaic has been heavily restored, and that the separation may originally have extended all the way to the right edge.)

But I only understood my error when I backed up a scene to contemplate the work of the first day. God, again at the left edge, creates light by making a vertical division this time—with a dark globe representing night at the right edge, and the light globe of day closer to God on the left. The single angel of the first day, in a lovely touch, bears one luminescent wing in the realm of light

Mosaic from the narthex of San Marco in Venice showing the second day of creation—as God rolls a ball horizontally to divide the waters above the firmament from waters below the firmament.

above his right arm (the angel faces us, so his right arm lies in the left domain of light), and a dark wing in the nighttime of his left side. Six rays, representing the work of creation's six days, emanate from each globe.

And now I grasped the maximal depth of my error. I had demoted the second day because I had failed to appreciate the controlling theme of the entire story! I had always regarded the narrative of Genesis 1 as a creation myth based upon the theme of sequential addition—as I think most people do, given our current cultural preferences for viewing history, at least for technological achievements, as a tale of accreting progress, and as my childhood instructors, both secular and religious, had certainly taught: light on day one, something about water on day two, land and plants on day three, sun and moon on day four, birds and fishes on day five, humans and other mammals on day six.

But the mosaics in the San Marco narthex clearly expressed a quite different organizing theme in their charmingly naïve iconography. At least for the first three days, these mosaics tell a story about the successive separation and

The first day of creation. God (at left edge) divides darkness (to the right) from light (to the left). The angel in the middle has one wing in the realm of light, the other in the realm of darkness. In the scene to the left, a dove flies above the initial chaos and void before God's first act of division.

precipitation of concrete items from an initial inchoate mass that must have contained, right from the beginning, all the seeds or prototypes for later realizations—in other words, a tale of *progressive differentiation* from unformed potential, rather than *successive addition,* piece by novel piece, in a sequence of increasing excellence.

The striking device of drawing these divisions as alternating vertical and horizontal planes illustrates the case in a wonderfully direct way—thus emplacing, in a million mosaic tesserae, the organizing theme that, I feel confident, the author of the text intended, and that artistic translators understood and expressed for at least a millennium or two before changes in Western culture blurred this context and subtly led us to read the story in a very different manner.

The tale of creation in San Marco's narthex begins in the undifferentiated chaos of full potential as, in the first scene, a dove (representing the spirit of God) flies above homogeneous waves of inchoate stuff. Then the divisions begin: a vertical plane on the first day to separate light from darkness; a horizontal plane (the bowling ball through the waters) to separate rain from rivers on the second day; and, in confirming the theme of separation rather than addi-

Further divisions on the third day of creation. A vertical strip differentiates the earth below the firmament into land and water.

tion, both a horizontal and a vertical strip of land on the third day, as God differentiates the earth below the firmament into its primary components of land and sea.

With this different theme of separation and coagulation in mind, I could finally understand two aspects of the story that had long puzzled me when I erroneously regarded Genesis 1 as a myth about creation by progressive addition imbued with increasing excellence. First, and to resolve my opening conundrum, the doings on day two no longer seem anomalous, for the separation of waters becomes, under the model of differentiation, an eminently worthy and weighty episode in a coherent tale. Most prescientific conceptions of basic elements regarded water (along with earth, fire and air, at least in the classical Greek formulation) as one of the few fundamental principles of cosmic construction. The ancestors of our cultures did not understand that ocean surfaces evaporated to form clouds that return water as rain. Thus, for them, the location of life-giving water in two maximally separated realms—at their feet in rivers and seas, and from the upper reaches of the sky as rain—must have generated a major puzzle demanding resolution at the fundamental level of creation itself. The work of the second day achieves full centrality and importance once we recognize the story of Genesis 1 as a tale about differentiation, and once we acknowledge water as both a primary element and a source of all sustenance. The separation of life-giving waters into two great reservoirs, located at opposite poles in the geography of existence, certainly merits a full day of God's creative effort.

Second, the substitution of differentiation for addition also suggests that the author of Genesis 1 probably conceived the nature of plants in a manner radically different from our understanding today. We make a primary distinction between organic and inorganic, with plants unambiguously in the first category, although usually relegated to the bottom of a rising sequence of plant, animal, and human. (Sometimes, vernacular language even restricts "animal" to mammals alone—as in a kiddie card game called "Bird, Fish, Animal," which I, as a professional biologist, simply couldn't fathom until I thumbed through the deck and realized that "animal" only meant lions and tigers and bears, oh my!—and then the name of the game just drove me nuts. I shall die content if I ever persuade the manufacturers to rechristen this otherwise harmless and even instructive recreation as "Bird, Fish, Mammal.") Our conventional taxonomy of organic and inorganic reinforces the false view that Genesis 1 should be read as a tale of addition—for primitive plants on day three should originate before any more advanced animal on day five. But why, then, does the intervening day four neglect organisms altogether?

However, when we reinterpret Genesis 1 as a differentiation myth, an alternate taxonomy best explains the discontinuity between an appearance of plants on day three and the creation of a sequence of animals that begins on day five and then continues, with no further break, right to the end of the story. Day three, with both vertical and horizontal divisions in San Marco's narthex, marks the episode of differentiation for the earth's physical potential—the separation of the primal chaos below the firmament into its major components of water and land. The subsequent origin of plants on the same day (and in the next verse) suggests to me that the author of Genesis 1 viewed plants as the culminating aspect of the land's differentiation, and not as a later addition from a separate organic realm, merely rooted within the substrate of another category of material stuff. In short, I would bet that the taxonomy of Genesis 1 intends to rank plants with earth, and not with animals.

In advocating the importance of ancient creation myths for our modern understanding of natural history, I make no argument about truth value. Rather, I think that our primal myths teach us something important about the limits and capacities of the human mind for organizing complex material into sensible stories. All cultures must generate creation myths; how else can we infuse order into the buzzing and blooming confusion of nature's surrounding diversity and complexity? Anthropologists, ethnographers, and folklorists have long noted the striking similarities among creation stories devised by people living in distant lands and without any known contact.

Two explanations for these similarities have generally been offered: (1) perhaps the story really arose only once, and then passed from one culture to another by more efficient routes of sharing than anthropologists have recognized; or (2) perhaps the stories arose in true independence, but with striking similarity enforced by inherited mental preferences and images—the archetypes of Jungian psychology—deeply and innately embedded in the evolutionary construction of the human brain. But I think that we should add a third possibility, invoking logical limits to the structure of stories rather than explicitly similar content, derived either by direct transfer or evoked from common residence in all human brains. (Of course, such logical limits also lie within our mentality, but this third principle calls upon a different and much more general aspect of consciousness than the specific images of Jungian archetypes.) After all, comprehensible stories can only "go" in a certain number of ways—and the full list of creation myths, independently devised by our varied and separate cultures all over the globe, includes so many entries that, inevitably, several specific tales must fall under each of the few general rubrics.

These themes, although far fewer than the stories that they must organize, still span a fairly ample range. For example, creation stories may be primarily *eliminative,* as when a promiscuous creator begins by populating the cosmos with all conceivable forms, and then lets nature take its course by weeding out the malformed and the nonoperational. Or creation stories may be *cyclical,* when sets of fully adequate orders successively succumb as new generations arise to enjoy their own transient ascendancy—so the Gods of Jupiter's generation succeed their Saturnian ancestors, while Wotan and his cohort perish in Wagner's *Götterdämmerung,* as a new day dawns at the end of four very long operas.

But only two basic alternatives exist when a culture chooses to organize a creation myth as a successive and progressive series—surely a common, if not *the* preferred, theme among most human groups (for whatever set of complex reasons involving both mental preferences and natural appearances). Such tales of sequential improvement may invoke either *successive addition* (first make this, then add this at a higher level, etc.) or *refining differentiation* (start with a big soup containing all eventual products as unformed potential, and then separate-coagulate-harden, separate-coagulate-harden, etc.). Perhaps other alternatives exist, and creation myths certainly can (and usually do) include aspects of both primal tales. But addition and differentiation define the primary mental territory of creation myths constructed under the theme of sequential progression.

I now reach the point of necessary confession for my cryptic and self-indulgent title. You have probably excused me for the narthex part, already explained above. But "pangenetic," and the resulting totality in particular, require an abject plea for your indulgence. In using "pangenetic," first of all, I honor my hero, and the inspiration for all ten volumes in this series of essays, Charles Darwin. In his longest book of 1868, the two-volume *Variation of Animals and Plants under Domestication,* Darwin proposed, as a "provisional hypothesis" in his own words and judgment, a theory of heredity that he called *pangenesis.* History has forgotten this incorrect theory, both rightly and entirely—although, curiously and by a complex route, our most salient modern word *gene* explicitly honors Darwin's failed effort.

According to pangenesis, each organ of the body casts off tiny particles called *gemmules.* These gemmules circulate throughout the body, and each sex cell eventually accumulates a full set. Thus the fertilized egg contains a complete array of determinants for growing all parts of the adult body—and embryology becomes a process of making these gemmules manifest and letting them grow. In other words, pangenesis expresses a pure theory of differentiation rather than addition for the explanation of organic development. The initial fertilized cell (like the primal chaos of Genesis 1) includes all components of the complex adult, but in unexpressed and inchoate form. Embryology then unfolds as the realization of an initially unformed but completely self-contained potential. Thus the pangenetic paradigm, honoring Darwin's version of a larger theme, encompasses a class of models based on differentiation rather than addition for the generation of progressive complexity in a temporal series. And the creation story, as depicted in San Marco's narthex, clearly lies within this pangenetic class.

Now for the self-indulgent part: the most widely cited technical paper that I ever wrote (with the exception of my first article on punctuated equilibrium, coauthored with Niles Eldredge) bore the title "The spandrels of San Marco and the Panglossian paradigm" (written with my colleague Dick Lewontin, the smartest man I have ever known). For a variety of reasons happily irrelevant to the subject of this essay, the spandrels paper has been unmercifully attacked by a substantial group of biologists committed to the strictly adaptationist account of evolution that this paper questions. (Our spandrels paper has also been appreciated, I trust, by an even more substantial group of colleagues, both in numbers and perspicacity!) Among the many published attacks, several have parodied our original title—as in "the scandals of San Marco" and even "the spaniels of St. Marx." So I thought I'd indulge myself, after all this *tsuris,* by writing a paper with my own title parody, albeit on a quite different subject. Sorry, folks, but at least I have laid down all my cards—and I now bare my throat.

I have little doubt that the first three days of Genesis 1 should be read as a tale of differentiation rather than addition. I also tend to view the fourth day as a continuation in the same mode—that is, God differentiates the earth below into water and land (with plants) on the third day and then, on the fourth day, differentiates the light of the sky above into sun, moon, and stars. But I must confess mixed feelings and signals about the fifth and sixth days. Does God now switch modes to utilize the alternative theme of addition in populating a cosmos (prepared by differentiation) with living creatures? Or does the origin of animals continue the theme of differentiation—as the air and water precipitate their living counterparts on day five, while the land generates its own appropriate forms of life on day six?

The scene of Adam's creation in San Marco's narthex might be read as support for the continuing theme of differentiation. Adam arises, dark as the soil, from the substrate of his origin—brought forth from the earth (the literal meaning of *Adam* in Hebrew), rather than imposed upon the earth as a separate creation from astral realms. (A familiar Latin pun—*homo ex humo* (man from the earth)—expresses the same thought, as does the old injunction "for dust thou art, and unto dust shalt thou return.")

My own, utterly nonscholarly, intuitions lead me to view the first four days as pure differentiations: light divided from darkness on day one, the upper water of rain from the lower water of rivers and lakes on day two, land from sea on the earth of day three, and sun from moon (the coagulation of previously diffuse light) in the heavens of day four. I then view days five and six as, in part, emplacements (more additive than differentiative) into appropriate surroundings, but also as the final differentiations of each realm—as water, air, and land all bring forth their appropriate living expressions.

Other interpretations abound, of course. In one popular scheme, advocated in two books that I have read and in several letters received from readers of these essays, the six days of creation fall into two equal cycles: three days of preparation followed by three days of population (of the heavens by sun, moon, and stars on day four; of the sea and air by swimming and flying animals on day five; and of the earth by terrestrial animals and humans on day six). In this reading, one might view the first three days as differentiative (whereas I would so interpret the first four days under this theme), and the last three as primarily additive. Nonetheless, however one explicates the story, I don't see how our usual reading of progressive addition for all six days can possibly be supported. At least the first three days—probably the first four, and perhaps all six—must be reconceived as a creation myth based on the great alternative theme of differentiation from unformed potential, rather than addition piece by piece.

This contrast of differentiation and addition as the two primary modes of organizing stories about sequential and progressive development becomes relevant to students of natural history not only as a framework for analyzing our oldest classics of the discipline (the creation myths in our earliest historical documents), but also as a guide to understanding our current problems and conflicts. In particular, when we recognize that we do not derive our concepts of history only from the factual signals that scientific research has extracted from nature, but also from internal limits upon the logical and cognitive modes of human thought, then we can appreciate the complex interaction of mind and nature (or inside and outside) that all great theories must embody.

The ancient creation myths of our cultures become particularly interesting in this context because they originated when our ancestors possessed no direct data at all about the actual pathways of life's history as revealed in the fossil record. These myths therefore represent nearly pure experiments in the range of mental possibilities for explaining the natural world when no hard information constrained our field of speculation.

The "bottom line" or "take home" message—that mind and nature always interact to build our basic concepts of natural order—becomes especially relevant in our current scientific age, where prevailing beliefs about the sources of knowledge lead us to downplay the role of the mind's organizing potentials and limits, and therefore encourage us to regard our theories of nature as products of objective observation alone. In particular, the logical restriction of tales about sequential and progressive development to two basic modes—differentiation and addition in the terms of this essay—can help us to understand our current theories (and to probe their scientific weaknesses when our mental preferences have hidden a different factual reality).

Consider the two major processes in biology—embryology and evolution—that, when specifically applied to human history, will inevitably be depicted as tales of sequential development toward greater complexity. (I do not believe that either process, especially evolution, must yield stories in this mode, but I do not challenge this pattern for our own particular case.) Both the history and current array of views on human embryology and evolution may be regarded, without gross caricature or oversimplification, as one long exercise in the interplay of shifting preferences for stories about differentiation or addition.

The study of vertebrate embryology, from the invention of the microscope in the seventeenth century to our modern understanding of genetics, has featured a set of debates between differentiative and additive stories for a process of increasing size and complexity from a tiny homogeneous egg to a neonate with all the anatomical complexity of adulthood. During most of the seventeenth and eighteenth centuries, the debate between "epigenesis" and "preformationism" virtually defined the territory of study for embryology. The epigeneticists embraced an additive model, arguing that the initial egg should be interpreted as its literal appearance suggests—that is, merely as a mass of promiscuous potential, devoid of structure, and eventually shaped to the particular anatomy of the complex neonate only because formative principles then operate upon this initial homogeneity to build, step by step and in an unerring manner (so long as embryology follows its normal course) the complexity of the final product. (*Epigenesis* means, literally, "generated upon"—that is, one step after the other.)

By contrast, the preformationists rallied behind a story of differentiation that envisaged all the structural complexity of the neonate as already present within the initial cell, and only brought to visibility during embryology. In the caricatured version, preformationism has usually been ridiculed as the belief that a perfect homunculus lies within each sperm or egg cell. No serious scientific preformationist held such a view. Rather, they argued that all structures must be present in the initial cell, but in too tiny, too transparent, and too diffuse a state to be visible (like the chaos at the outset of Genesis 1, and not like a fully formed homunculus). Embryology then becomes a differentiative process of concentration, coagulation, solidification, and growth.

When evolutionary ideas pervaded embryology in the nineteenth century, the two leading interpretations continued to uphold contrasting stories of addition or differentiation. Haeckel's famous theory of recapitulation held, in a purely additive account, that sequential steps in embryonic complexification repeated the evolutionary accretion of successive adult stages to the ancestral lineage—so that a complex animal, in its embryology, literally climbed its own family tree. The primary alternative, von Baer's theory of differentiation, argued that the visual simplicity of an early stage does not represent an ancient ancestor that must then be augmented (as in Haeckel's additive theory), but rather a more general form of greater homogeneity and lesser differentiation, holding all potential for the definitive complexity that eventually develops in each lifetime. Thus, at an early stage of development, we know that the embryo will become a vertebrate, then (at a later stage) a mammal, then a primate, then a hominoid, and finally a human being—a process of increasingly finer specification, contrasted with Haeckel's additive model of ever-increasing complexity in accretion.

When we consider the other historical process that led to our human form—the much longer evolutionary construction of *Homo sapiens* in geological time, rather than the embryological generation of each individual *Homo sapiens* in nine months—we note that concepts of evolution may also be classified into additive and differentiative models. Darwinism embodies an additive view. Because the Darwinian style of explanation has prevailed within science, we tend to forget that several abandoned theories of evolution advocated differentiative models. These accounts imagined that the first vertebrate in the Cambrian explosion—a boneless and jawless creature just an inch or two in length—already contained all the parts and potentials that evolution would necessarily elaborate into human form in a distant future. The supposed mechanisms for such a "programmed" differentiation spanned the full gamut from God's direct actions (in a few overtly theological accounts) to principles embodied in unknown, but entirely physical, laws of nature (for some atheistic versions at an opposite speculative extreme).

If allegiance to an additive or differentiative model implied no consequences for skewing our views of life in disparate directions, then we could dismiss the entire subject as an effete intellectual game without meaning for scientific understanding. But our preferred theories often act as biases that strongly influence our basic conceptions of the natural world—and additive versus differentiative views of historical sequences do not hold the same intellectual weights, properties, and implications. We might summarize the differences, looking at lessons from the history of science, by saying that each basic model features a defining property and struggles with a major problem.

Stories of differentiation work primarily from the inside out. That is, the sequence begins with all eventual results already preformed, albeit unexpressed, within an initial homogeneity. How, then, can this potential become actualized? In explicit contrast, stories of addition operate primarily from the outside in. That is, the sequence begins with truly unformed stuff, promiscuous potential that could be drawn into any number of pathways by outside forces. How, then, can such a gloppy mass be carved by external agents into such an exquisitely complex final product (an even worse problem for embryology than for evolution, because the carving must follow the same basic path each time for normal embryos within a species, whereas each evolutionary result arises only once).

In the major weight of difference between the two models, stories of differentiation fit better for determined systems in a predictable world, whereas stories about addition hold the conceptual edge in a contingent world where each historical sequence may follow innumerable (and unpredictable) options, with the actual result conditioned by the particular set of external prods that a rolling ball of promiscuous potential happens to encounter in its trajectory through time. For this primary reason, our modern embryological models tend to be primarily differentiative, and our evolutionary models primarily additive.

After all, embryology does generally follow an internally prescribed route specified not by the preformed parts of preformationists, but by the programmed instructions of modern genetic understanding. (We should not accuse the eighteenth-century preformationists of stupidity for placing the right idea into the wrong substance. After all, their intellectual world did not include a concept of programmed information, except, perhaps, as embodied in the old trifle of music boxes or the newfangled invention of the Jacquard weaving loom—whereas no sentient person in our age of genes and computers could fail to assimilate such informational models as an intellectual centerpiece.)

By contrast, the evolution of any lineage wanders along contingent and unpredictable paths of a uniquely complex history. The few lineages, including our own, that do become more complex through time may add their increments of sophistication in a sequence that makes sense after the fact. But even an omniscient observer could never designate, for certain, the next step in an unpredictable future. Therefore, as a description of evolution, additive models that introduce sequential steps from the outside work better than differentiative models that must hypothesize an entire future as already implicit and enfolded within any current form.

Under this analysis, we should not be surprised that Genesis 1, despite our usual and unconsidered readings, tells a tale of differentiation rather than addition. After all, if God proceeded with the usual care and thought conventionally attributed to his might, he probably had a pretty accurate idea about the finished product even before he began the work. Biological evolution, on the other hand, at least as viewed under the limits of our eminently fallible mental machinery, seems to wander along a wondrously erratic set of specific pathways within its broad predictabilities.

Our preferred intellectual models do make a difference, and we must therefore be sensitive to the disparate implications of additive and differentiative models as we struggle to understand the history of life. Still, I think that any passionate and sentient person can feel the same emotional thrill that emanates from either intellectual interpretation. We live in one helluva fascinating universe, whatever its modalities of construction. Thus, if I may beg one last indulgence from my readers—this time for ending with the same image that I invoked in essay 19 for a different treatment of Genesis and evolution—I happily embrace the common sentiment behind two maximally different views of organic order: the differentiative model of Genesis 1, with its ending of sublime satisfaction: "And God saw every thing that he had made, and, behold, it was very good." And the additive model of natural selection, so lovingly described by Charles Darwin in the last paragraph of *The Origin of Species:* "There is grandeur in this view of life."

Parsing and Proceeding

21

Linnaeus's Luck?

Carolus Linnaeus (1707–1778), the founder of modern taxonomy and the focus of this essay, frequently cited an ancient motto to epitomize his view of life: *natura non facit saltum* (nature does not make leaps). Such unbroken continuity may reign in the material world, but our human passion for order and clear distinction leads us to designate certain moments or events as "official" beginnings for something discrete and new. Thus the signatures on a document define the birth of a nation on July 4, 1776, and the easily remembered eleventh hour of the eleventh day of the eleventh month (November 11, 1918) marks the armistice to a horrible war supposedly fought to end all contemplation of future wars. In a small irony of history, our apostle of natural continuity also became the author and guardian of a symbolic leap to novelty—for the modern taxonomy of animals officially began with the publication of the definitive tenth edition of Linnaeus's *Systema Naturae* in 1758.

The current classification of animals may boast such a formally recognized inauguration, but an agreement about beginnings does not guarantee a consensus about importance. In fact, the worth assigned to taxonomy by great scientists has spanned the full range of conceivable evaluations. When Lord Rutherford, the great British physicist (born in New Zealand), discovered that the dates of radioactive decay could establish the true age of the earth (billions rather than millions of years), he scorned the opposition of paleontologists by branding their taxonomic labors in classifying fossils as the lowest form of purely descriptive activity, a style of research barely meriting the name "science." Taxonomy, he fumed, could claim no more intellectual depth than "stamp collecting"—an old canard that makes me bristle from two sides of my being: as a present paleontologist and a former philatelist!

Rutherford's anathema dates to the first decade of the twentieth century. Interestingly, when Luis Alvarez, a physicist of similar distinction, became equally enraged by some paleontologists during the last decade of the twenti-

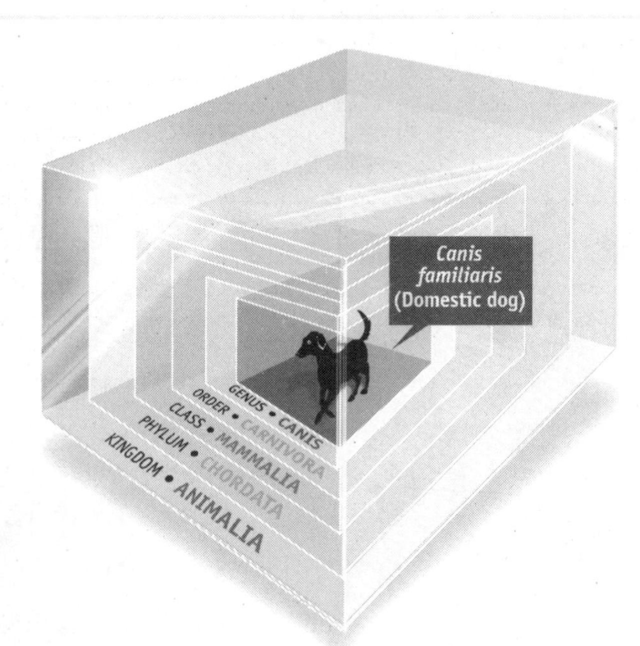

Linnaean taxonomy in one of its geometric portrayals as boxes within boxes for continually finer specifications.

eth century, he invoked the same image in denigration: "they're not very good scientists; they're just stamp collectors." I continue to reject both the metaphor and the damning of all for the stodginess of a majority—for Luis had exploded in righteous and legitimate frustration at the strong biases that initially led most paleontologists to reject, without fair consideration, his apparently correct conclusion that the impact of a large extraterrestrial body triggered the mass extinction of dinosaurs and about 50 percent of marine animal species 65 million years ago.

The phony assumption underlying this debasement of taxonomy to philately holds that the order among organisms stands forth as a simple fact plainly accessible to any half-decent observer. The task of taxonomy may then be equated with the dullest form of cataloging—the allocation of an admittedly large array of objects to their preassigned places: pasting stamps into the designated spaces of nature's album, putting hats on the right hooks of the world's objective hatrack, or shoving bundles into the proper pigeonholes in evolution's storehouse, to cite a standard set of dismissive metaphors.

In maximal contrast, the great Swiss zoologist Louis Agassiz exalted taxonomy as the highest possible calling of all, when he opened Harvard's Museum of Comparative Zoology in his adopted land in 1859. Each species, Agassiz argued, represents the material incarnation on earth of a single and discrete idea in the mind of God. The natural order among species—their taxonomy—therefore reflects the structure of divine thought. If we can accurately identify the system of interrelationships among species, Agassiz concluded, we will stand as close as rationality can bring us to the nature of God.

Notwithstanding their maximally disparate judgments of taxonomy, Rutherford and Agassiz rank as strange bedfellows in their shared premise that a single objective order exists "out there" in the "real world," and that a proper classification will allocate each organism to its designated spot in the one true system. (For Rutherford, this order represents a basically boring and easily ascertainable aspect of macroscopic nature—too far removed from the atomic world of fundamental laws and causes to generate much scientific interest or insight. For Agassiz, in greatest conceivable contrast, this order represents our best shot for grasping the otherwise arcane and inaccessible intellect of God himself.)

In framing a modern "Goldilocks" defense for the importance of taxonomy—far warmer than Rutherford's icy indifference, but not quite so hot as Agassiz's impassioned embrace—we must begin by refuting their shared assumption that one true order exists "out there," and that correct classifications may be equated with accurate maps. We can best defend the scientific vitality of taxonomy by asserting the opposite premise, that all systems of classification must express theories about the causes of order, and must therefore feature a complex mixture of concepts and percepts—that is, preferences in human thinking combined with observations of nature's often cryptic realities. Good taxonomies may be analogized with useful maps, but they reveal (as do all good maps) both our preferred mental schemes and the pieces of external reality that we have chosen to order and depict in our cartographic effort.

This acknowledgment that taxonomies can only express nature's objective realities in terms of theories devised by the human mind should not encourage any trendy postmodern pessimism about the relativity of knowledge. All taxonomies do not become equally valid because each must filter nature's facts through sieves of human thought and perception. Some popular attributions of former centuries may be dismissed as just plain wrong. (Corals, for example, are animals, not plants.) Other common schemes may be rejected as more confusing than helpful in nearly all situations. (We learn more about whales by classifying them genealogically with mammals than by amalgamating them with squids and sharks into an evolutionarily heterogeneous group of "things that swim fast in the ocean.")

Professional taxonomists have always recognized this inequality among systems of naming by proclaiming the search for a "natural" classification as the goal of their science. Although we may regard the word *natural* as a peculiar, or even arrogant, description for an optimal scheme of classification, the rationale for this verbal choice seems clear enough. If all taxonomies must express theories about nature's order, then we may define the most "natural" classification as the scheme that best respects, reveals, and reflects the causes that generated the diversity of organisms (thereby evoking our urge to classify in the first place)!

A zoo director might, for practical purposes, choose to classify organisms by size (as a convenience for selecting cages) or by climatic preferences (so that his polar bears won't asphyxiate in an exhibit on tropical rain forests). But we would label such taxonomic schemes as artificial because we know that evolution has generated the interrelationships among organisms by a process of genealogical descent through geological time. The most "natural" classification may therefore be defined as the scheme that best permits us to infer the genealogical connections among organisms—that is, the primary cause of their similarities and differences—from the names and forms of our taxonomies.

When we recognize all influential classifications as careful descriptions of organisms made in the light of fruitful theories about the causes of order, then we can finally appreciate the fascination of taxonomy as a source of insight about *both* mind and nature. In particular, the history of changing classifications becomes far more than a dull archive or chronicle of successive purchases from nature's post office (discoveries of new species), followed by careful sorting and proper pasting into preassigned spaces of a permanent album (taxonomic lists of objectively defined groups, with space always available for new occupants in a domicile that can always grow larger without changing its definitive style or structure). Rather, major taxonomic revisions often require that old mental designs be razed to their foundations, so that new conceptual structures may be raised to accommodate radically different groupings of occupants.

In the obvious example of this essay, Agassiz's lovely cathedral of taxonomic structure, conceived as a material incarnation of God's mentality, did not collapse because new observations disproved his central conviction about the close affinity of jellyfish and starfish (now recognized as members of two genealogically distant phyla, falsely united by Agassiz for their common property of radial symmetry). Instead, the greatest theoretical revolution in the history of biology—Darwin's triumphant case for evolution—revealed a fundamentally different causal basis for taxonomic order. Evolution fired the old firm and hired a new architect to rebuild the structure of classification, all the better to display the "grandeur" that Darwin had located in "this view of life." Ironically, Agassiz opened his museum in 1859, in the same year that Darwin published *The Origin of Species.* Thus, Agassiz's replica of God's eternal mind at two degrees of separation (from the structure of divine thought to the taxonomic arrangement of organisms to the ordered display of a museum) became an unintended pageant of history's genealogical flow and continuity.

But this argument, that the history of taxonomy wins its fascination, at least in large part, as a dynamic interplay of mind (changing theories about the causes of order) and matter (deeper and more accurate understanding of nature's factuality), now exposes a paradox that defines the second half of this essay, and leads us back to the official founder of taxonomy, Carolus Linnaeus. Darwinian evolution has set our modern theoretical context for understanding the causes of organic diversity. But if taxonomies always record theories about the causal order that underlies their construction, and if evolution generated the organic resemblances that our taxonomies attempt to express, then how can Linnaeus, a creationist who lived a full century before Darwin discovered the basis of biological order, be the official father of modern—that is, evolutionary—taxonomy? How, in short, can Linnaeus's system continue to work so well in Darwin's brave new world?

Perhaps we should resolve this paradox by demoting the role of theory in taxonomy. Should we embrace Rutherford's philatelic model after all, and regard organic interrelationships as simple, observable facts of nature, quite impervious to changing winds of theoretical fashion? Linnaeus, on this philatelic view, may have won success by simple virtue of his superior observational skills.

Or perhaps we should argue, in maximal contrast, that Linnaeus just lucked out in one of history's most felicitous casting of dice. Perhaps theories do specify the underlying order of any important taxonomic system, and Linnaeus's creationist account just happened to imply a structure that, by pure good fortune, could be translated without fuss or fracture into the evolutionary terms of Darwin's new biology.

I will advocate a position between these two extremes of exemplary observational skill in an objective world and pure good luck in a world structured by theoretical preferences. Linnaeus may, no doubt, be ranked as both the premier observer and one of the smartest scientists of his (or any) age. But, following my central claim that taxonomies should be judged for their intrinsic mixture of accurate observation linked to fruitful theory, I would argue that Linnaeus has endured because he combined the best observational skills of his time with a theoretical conception of organic relationships that happens to conform, but not by pure accident, with the topology of evolutionary systems—even though Linnaeus himself interpreted this organizing principle in creationist terms. (As for the fascinating and largely psychological question of whether Linnaeus devised a system compatible with evolution because he glimpsed "truth" through a glass darkly, or because his biological intuitions subtly, and unconsciously, tweaked his theoretical leanings in an especially fruitful direction—well, I suspect [as for all inquiries in this speculative domain of human motivations] that Linnaeus took this particular issue with his mortal remains to the grave.)

We refer to Linnaeus's system as "binomial nomenclature" because the formal name of each species includes two components: the generic designation, given first with an initial uppercase letter (*Homo* for us, *Canis* for dogs, etc.); and the so-called "trivial" name, presented last and in fully lowercase letters (*sapiens* to designate us within the genus *Homo,* and *familiaris* to distinguish dogs from other species within the genus *Canis*— for example the wolf, *Canis lupus*). Incidentally, and to correct a common error, the trivial name has no standing by itself, and does not define a species. The name of our species, using both parts of the binomial designation, is *Homo sapiens,* not *sapiens* all by itself. We regard the 1758 version of *Systema Naturae* as the founding document of modern animal taxonomy because, in this edition and for the first time, Linnaeus used the binomial system in complete consistency and without exception. (Previous editions had designated some species binomially and others by a genus name followed by several descriptive words.)

The binomial system includes several wise and innovative features that have ensured its continuing success. But, for the theme of this essay, the logical implications of this system for the nature of interrelationships among organisms stands out as the keystone of Linnaeus's uncanny relevance in Darwin's thoroughly altered evolutionary world. The very structure of a binomial name encodes the essential property that makes Linnaeus's system consistent with life's evolutionary topology.

Linnaeus's taxonomic scheme designates a rigorously nested hierarchy of groups (starting with species as the smallest unit) embedded within successively larger groups (species within genera within families within orders, etc.). Such a nested hierarchy implies the organizing geometry of a single branching tree, with a common trunk that then ramifies into ever-finer divisions of boughs, limbs, branches, and twigs. This treelike form also happens to express the hypothesis that interrelationships among organisms record a genealogical hierarchy built by evolutionary branching. Linnaeus's system thus embodies, quite apart from Linnaeus's own intentions or theoretical connections, the causality of Darwin's world.

This correspondence between the Linnaean hierarchy and life's evolutionary tree achieves its clearest expression in pictorial form. The accompanying drawings show a Linnaean ordering of box within box, as well as the alternate expression of this logical order in a branching diagram of sequential genealogical splitting—in this case, a successive carving of the kingdom of all animals into, first, chordates contrasted with all other animals; then vertebrates contrasted with all invertebrates; mammals contrasted with all other vertebrates; the order Carnivora contrasted with all other mammals; the family Canidae contrasted with all other carnivores; the genus *Canis* contrasted with all other doglike carnivores; and domestic dogs contrasted with all other members of the genus *Canis*. I stated that binomial nomenclature expresses the first step of this hierarchical ordering—and thus presents a microcosm of the entire scheme—because the two parts of a species's name record the first act of embedding smaller units within more inclusive groups of relatives. The name *Canis familiaris* states that the smallest unit, the dog species, ranks as one member of the next most inclusive group, the genus *Canis,* which unites all other species (e.g., the wolf *Canis lupus* and the coyote *Canis latrans*) that originated from a common ancestor shared by no other species in any other group.

Linnaeus thought that his chosen scheme of mapping biological relationships as smaller boxes within successively larger boxes, until all units nested within the

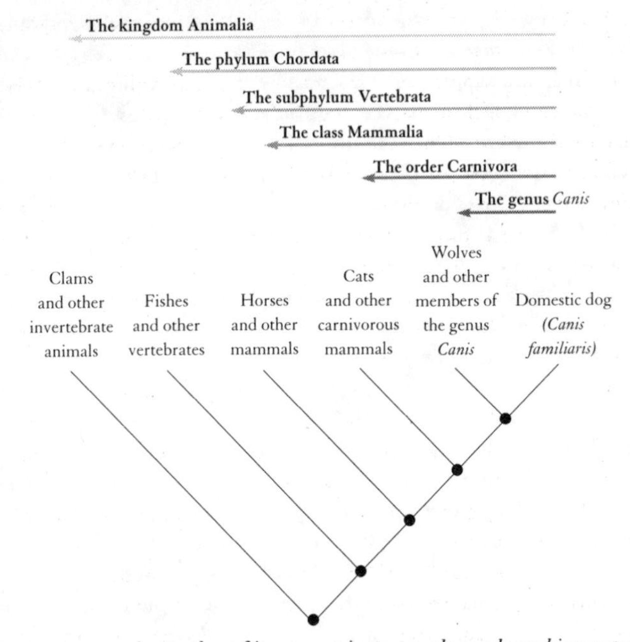

Linnaean taxonomy in another of its geometric portrayals as a branching system with all divisions beginning from a common trunk and no fusion of separated branches allowed.

most inclusive box of life itself, represented the best human device for expressing the eternal order that God had chosen when he populated the universe. I doubt that he ever explicitly said to himself (for I suspect that his mental mansion included no room for such a thought): "But if, quite to the contrary, life evolved by a process of ever-ramified branching from a single ancestor over a long period of time, then the hierarchical order of the binomial system will capture the topology of organic relationships just as well, because the logic of my system translates pictorially into a tree with a single trunk at the base, and subsequent division into branches that never coalesce thereafter. And such a topology might represent either God's permanent order, preconceived from the first, or the happenstances of historical change and development on an evolutionary tree growing from a single starting point under the constraint of unbroken continuity (although branches may die and fall from the tree as lineages become extinct), and continuous bifurcation without subsequent joining of lineages."

I emphasize this property of irrevocable branching without subsequent amalgamation because the Linnaean logic of placing small boxes into larger boxes—which just happens to conform to the historical reality of Darwin's system—establishes such a map of organic relationships as its primary and inevitable consequence. One can't, after all, cram big boxes into smaller boxes. Therefore, for example, two species in the same genus can't reside in different families, and two orders in the same class can't be placed in different phyla. If lions and tigers rank as two species in the same genus *(Panthera),* they cannot then be allocated to different families of higher rank (lions to the Felidae and tigers to the Canidae, for example)—for the two larger family boxes would then have to fit within the smaller box of the genus *Panthera,* and both the rules of Linnaean logic, and the requirements of Darwinian evolutionary history, would then be fractured. I can only be a monkey's uncle or a horse's ass in a metaphorical sense—for my species fits into the small box of the genus *Homo,* which must nest within the larger box of the family Hominidae; and one member of my species can't opt out of our box to join the Cercopithecidae or the Equidae, thus splitting a coherent lower group into two higher groups, and violating both Linnaean logic and Darwinian reality.

So perhaps Linnaeus enjoyed a little bit of luck in choosing the one logic for a creationist system that would also fit without fuss into a new universe of historical evolution by branching. At least Linnaeus demonstrated exemplary survival skills in passing the test of time as taxonomy's father. But I hesitate to ascribe his remarkable success to pure dumb fortune, and for a primary reason rooted in the key contention of this essay: that taxonomies transcend simple description and always embody particular theories about the causes of order, thus melding preferences of mind with perceptions of nature.

I think that Linnaeus succeeded because, however unconsciously or preconsciously, he made some excellent decisions about both the mental and perceptual aspects of taxonomic systems. On the perceptual side, he must have seen better than any of his colleagues that under the logic of hierarchy and branching, organisms could be arranged into a consistent order that might win general assent without constant bickering among practitioners. Other contemporaries had proposed very different logics for classification, but had never found a way to push them through to an unambiguous and consistent system. In the most telling example, Linnaeus's most famous contemporary and archrival, France's celebrated naturalist the Baron Georges Leclerc Buffon (1707–1788), had struggled through more than forty volumes of his *Histoire naturelle*—in my judgment, the greatest encyclopedia of natural science ever written—to develop, without conspicuous success, a nonhierarchical system that joined each species to some others by physiology, to a different group by anatomy, and to a still different set by ecology.

But I would, in addition, like to advance the unfamiliar argument that Linnaeus also succeeded because he made a very clever, and probably conscious, choice from the mental side of taxonomic requirements as well. In deciding to erect a hierarchical order based on continuous branching with no subsequent joining of branches, Linnaeus constructed his system according to the most familiar organizing device of Western logic since Aristotle (and, arguably, an expression of our innate and universal mental preferences as well): successive (and exceptionless) dichotomous branching as a system for making ever finer distinctions. In a logical tree of this form—often called a dichotomous key—one may move in either direction to place a particular basic object into ever larger groups by joining successive pairs, or to break down a large category into all component parts by successive twofold division.

One may, for example, interpret the diagram that I presented earlier as a dichotomous key. We can reach dogs by starting with the largest category of all animals, dividing this totality into vertebrates and invertebrates, then splitting the vertebrates into mammals and non-mammals, the mammals into carnivores and non-carnivores, the carnivores into canids and non-canids, and finally the canids into dogs versus others. (We can also work outward in the opposite direction to learn how dogs fit into the hierarchy of all animals.)

In fact, the idea for this essay came to me when I recently purchased an obscure late-sixteenth-century book on Aristotelian logic and noted that its numerous charts for working through the categories of reasoning, and the attributes of human form and behavior, had all been constructed as dichotomous maps bearing an uncanny resemblance to the taxonomic keying devices that I have seen and used in texts and guidebooks for naturalists throughout my career. Thus, Linnaeus gave himself quite a leg up by building his taxonomic system upon a familiar form of logic that scholars had applied to all subjects, scientific and otherwise, from the dawn of Western history—a style of reasoning, moreover, that may track the basic operation of our brains.

I took the accompanying chart from this 1586 treatise, published in Paris by the physician Nicolas Abraham, and entitled *Isogogethica ad rationis normam delineata* (an introduction to ethics as delineated by the rule of reason). Abraham first divides the domain of ethical decisions into the dichotomous pair of *mentis* (by the mind) and *moris* (by custom). He then splits the lower domain of custom into the two categories of *privatis* (above) and *publicis* (below). Interestingly, and I suspect consciously, authors of dichotomous keys also seem—at least in my limited study of such devices from pre-Darwinian

MENTIS, quæ facultatem animi rationalem illuminabt, vt quæ vera & falsa sint recte cognoscat. Est autem rationalis facultas duplex, scientifica & consultatrix, & ideo duæ virtutes sunt in rationalis animi parte,

SAPIENTIA, quæ rectus est facultatis sciendi habitus, quo res vniuersas, quæ secus euenire non possunt cognoscimus. Eius contrarium vitium est insipientia. Sapientiæ partes duæ sunt,

Intelligentia, qua principia rerum cognoscimus.

Scientia, qua conclusiones necessarias per suas causas & principia cognoscimus.

PRVDENTIA, quæ rectus est facultatis animæ consultatricis habitus quo singulas res agendas cognoscimus. Eius contrarium vitium est imprudentia. Prudentiæ partes duæ sunt,

Bona consultatio, qua bene ea quæ agenda sunt inquirimus & inuenimus.

Sagacitas, qua bene de rebus singulis iam inuentis iudicamus & discernimus.

A dichotomous key, as presented by Nicolas Abraham in 1586 to classify ethical decisions.

times—to order their pairings from the good and most valued on top to the least admirable on the bottom. In this case, reason beats custom, while, within custom, private decisions (presumably made for reasons of personal belief) trump public actions (that may be enforced by social pressure). I am particularly fond of the dichotomous key for birds of prey that the great English naturalist John Ray published in 1678—with a first division of day-flyers on top and night-flyers on the bottom; a second division of the preferred day-flyers into bigger species (above) and smaller (below); and a third division of the big species into "more generous" eagles above and "more cowardly and sluggish" vultures below.

But let us follow Abraham's key for the higher category of mental decisions, which then undergoes a further dichotomous split into *sapientia* (done by wisdom) above and *prudentia* (done for reasons of prudence) below. The third and final set of twofold divisions then separates judgments by *sapientia* into *intelligentia* (achieved by pure reason) above in preference to *scientia* (achieved by knowledge about material things) below. The lower judgments of *prudentia* then divide into a preferred category of *bona consultatio* above (derived from our seeking a good advice from others) and the less worthy dichotomous alternative of *sagacitas* below (determined only by our own judgment).

I do admire Linnaeus as an intellectually driven and brilliantly complicated, but arrogantly vainglorious, man. If I preferred the hagiographical mode of writing essays, I would stop here with a closing word of praise for Linnaeus's perspicacity in harnessing both the observational and theoretical sides of his mental skills to construct a flexible and enduring taxonomic system that could survive intact in sailing right through the greatest theoretical transformation in the history of biology.

But he who lives by the sword dies by the sword (as Jesus did not exactly say in a common misquotation that remains potent in truth and meaning despite a slight inaccuracy in citation). Linnaeus's consistency and wisdom in developing and defending the binomial system of hierarchical classification carried him through to intellectual victory. But, like so many originators of grand and innovative systems, he reached too far (whether by arrogance or overexcitement) and became too committed to his procedure as the one true way for classifying any collection of related objects. (I cannot help recalling my experience with a customs official on a small West Indian island who classified my land snails as turtles because his forms only permitted a distinction between warm-blooded and cold-blooded "animals"—and the word "animal," in his personal understanding, only designated vertebrates. Thus, snails became turtles because both are cold-blooded and move with legendary torpor.)

Once Linnaeus had fully developed the binomial system and its supporting logic of a consistently nested hierarchy, he supposed that he had discovered the proper way to classify any group of natural objects, and he therefore began to apply binomial nomenclature to several classes of inappropriate phenomena, including rocks and even human diseases. Clearly, he had become overenamored with his own device, and had lost sight of the key principle that hierarchical embedding by dichotomous branching only captures the causal order within *certain kinds* of systems—particularly those that develop historically by successive branching in unbroken genealogical continuity (with no later amalgamation of branches) from a common ancestor. Since Linnaeus tried to apply his binomial system to several groups of objects that, by their own rules of order or development, patently violate the required hierarchical logic, perhaps he never really did grasp the limitations (and therefore the essence) of his binomial system. So maybe Linnaeus did prevail partly by the luck of organic conformity to his general logic, rather than by his correct and conscious reasoning about distinctive causes of relationships among plants and animals.

For example, the accompanying page from the seventh (1748) edition of *Systema Naturae* designates binomial species of the genus *Quartzum* from the classification of rocks that he presented as a third chapter, following his taxonomies for animals and plants. The first "species," *Quartzum aqueum* (transparent quartz) includes ordinary glasslike quartz; the second, *Quartzum album* (white quartz) encompasses less valued, opaquely water-worn quartz pebbles; the third, *Quartzum tinctum* (colored quartz) gathers together the colored varieties that mimic more-valuable gemstones (Linnaeus calls them false topaz, ruby, and sapphire, for example); and the fourth, *Quartzum opacum* (opaque quartz) describes the even less useful and less transparent flintstones.

But the nature of quartz, and the basis of relationships among minerals in general, defies the required logic of causality for any system legitimately described in Linnaean binomial terms. The members of *Quartzum aqueum,* for

VITRESCENTES. 149

2. QVARTZUM. FRAGMENTIS INDETERMINATIS, PELLUCIDIS, SOLIDIS, ANGULATIS, ACUTIS.

1. QVARTZUM aqueum. Kieſel.

Quartzum aquei coloris. *Syſt. nat.* Quartz.
Quartzum ſolidum pellucidum. *W.* 103. Quartzum vulgare.
Locus ubique in rupium fiſſuris.
Natum ex aqua in rupibus detenta.
Paraſiticum ſemper fuit, licet ſæpe diſperſum.
Scintillas dat omne quartzum cum chalybe vivaciſſimas.
Uſus: pro Vitro artificiali, Metallis fundendis pro vehiculo.

2. QVARTZUM album. wäſſerichter Quarz.

Quartzum aqueo-album. *Syſt. nat.* Caillou Quartzum vulgare vel potius Silex.
Locus rupes, minus tamen frequens.
Natum ex aqua & marmore.
Vitreſcit facillime. Fragmenta ſæpe ſubfarinacea.
Uſus: Metallurgis cupri maxime expetitum.

3. QVARTZUM tinctum.

Quartzum ſolidum opacum coloratum. *W.* 104.
Locus plerumque ad fodinas vel mineras.
Natum ex quartzo metallo tincto.

Variat colore		
luteo.	♄	unächter Topas.
rubro.	♂	unächter Rubin.
purpureo.	♂, ♀	Amethyſt.
coeruleo.	♀	Sapphier.
viridi.	♀	Schmaragd.

4. QVARTZUM opacum. (Kattflinta.)

Quartzum opacum fragile & rigidum. *W.* 102.
Locus fere ubique in rupibus & terra.

5. QVARTZUM ſubcotaceum. (Saltſlag.)

Quartzum granulatum cohærens. *W.* 104.
Locus: Fahlunæ fodina.
Grana conglomerata inſtar muriæ.

K 3 3. SI-

A page from the mineralogical chapter of the 1748 edition of Linnaeus's Systema Naturae, *showing that he tried to apply his method of binomial nomenclature to rocks, as well as to organisms.*

example, do not hang together as a set of mutually closest historical relatives, all physically derived in continuity from a common ancestor that generated no other offspring. Rather, the specimens of this false species look alike because simple rules of chemistry and physics dictate that transparent quartz will form whenever silicon and oxygen ions come together under certain conditions of temperature, pressure, and composition. The members of this "species" can claim no historical or genealogical coherence. One specimen might have originated half a billion years ago from a cooling magma in Africa, and another just fifty years ago in a bomb crater in Nevada. Minerals must be classified according to their own causes of order, a set of rules distinctly different from the evolutionary and genealogical principles that build the interrelationships among organisms.

Linnaeus clearly overreached in supposing that he had discovered the one true system for all natural objects. In the twelfth and final edition of *Systema Naturae* (1766), he included a section entitled *Imperium Naturae,* dedicated to extolling his hierarchical and binomial method as universally valid. God made all things, Linnaeus argues, and he must have used a single and universal method, now discovered by his most obedient (and successful) servant. Linnaeus writes: *"Omnes res creatae sunt divinae sapientiae et potentiae testes"* (all created things are witnesses of divine knowledge and power). Using a common classical metaphor (the thread of Ariadne that led Perseus out of the labyrinth after he had killed the Minotaur), Linnaeus praises himself as the code cracker of this universal order: "knowledge of nature begins with our understanding of her methods by means of a systematic nomenclature that works like Ariadne's thread, permitting us to follow nature's meanders with accuracy and confidence."

Ironically, however, Linnaeus had succeeded (in a truly ample, albeit not universal, domain of nature) precisely because he had constructed a logic that correctly followed the causes of order in the organic world, but could not, for the same reasons, be extended to cover inorganic objects *not* built and interrelated by ties of genealogical continuity and evolutionary transformation. The strength of any great system shines most brightly in the light of limits that give sharp and clear definition to the large domain of its non-universal action! By understanding why the Linnaean system works for organisms and *not* for rocks, we gain our best insight into the importance of his achievements in specifying the *varied* nature of *disparate* causes for nature's order among her many realms.

On the same theme of power in exceptions, and to make a somewhat ironic point in closing, Linnaeus's hope that he had discovered a fully universal basis

(God's own rules of creation) for classifying all natural objects has recently suffered another fascinating blow—but this time from inside (that is, from the world of organisms). Science had already denied Linnaeus's universality more than two hundred years ago by accepting his procedures only for historically generated genealogical systems based on a geometry of branching (the evolution of organisms as a primary example), and rejecting his binomial schemes for rocks, diseases, and other systems based upon different theoretical foundations of order. (In fact, Linnaeus did try to establish a binomial classification of human diseases as well—in one of his least successful treatises, *Genera morborum,* published in 1736 and immediately forgotten!) But now, one of the most important biological discoveries of our age has also challenged the universal application of Linnaean taxonomy in the world of organisms as well.

We need not fret for fat, furry, multicellular creatures—the plants, fungi, and animals of our three great multicellular and macroscopic kingdoms of life. For evolution, in this visible world of complex creatures, does follow the Linnaean topology nearly all the time. That is, the basic structural rule that validates the binomial system works quite satisfactorily at this level—for branches never join once they have separated, and each species therefore becomes a permanently independent lineage, forging no further combinations with others after its origin. Evolution cannot make a nifty new species of mammal by mixing half a dolphin with half a bat to generate an all-purpose flyer and swimmer.

Until a few years ago, we thought that this rule of permanent separation also applied to the world of unicellular bacteria—the true dominators of earth and rulers of life in my opinion (see my book *Full House,* 1996). In other words, we assumed that the bacterial foundation of the tree of life grew in a fully Linnaean manner, just like the multicellular section. (Actually, and to emphasize the importance of the discovery described below, the bacterial domain occupies most of life's tree because the three multicellular kingdoms sprout as three terminal twigs on just one of the tree's three great limbs, the other two being entirely bacterial.)

And, apparently, we were wrong. By a set of processes collectively called "lateral gene transfer" (LGT for short), individual genes and short sequences of genes can move from one bacterial species to another. For two reasons, these transfers may challenge Linnaean logic in a serious way. First, LGT does not seem to respect taxonomic separation. That is, genes from genealogically distant bacterial species seem to enter a host species with no greater difficulty than genes from closely related species. Second, the process occurs frequently enough to preclude dismissal as a rare and peculiar exception to the prevailing

Linnaean rule of strict branching with no subsequent amalgamation. (If only a percent or two of bacterial genomes represented imports from distant species by LGT, we could view the phenomenon as a fascinating anomaly that does not degrade the primary signal of Linnaean reality. But, at least for some species, LGT may be sufficiently common to flash a primary signal of its own. In *E. coli,* the familiar bacillus found in all human guts, for example, 755 of 4,288 genetic units [about 18 percent of the entire genome] record at least 234 events of lateral genetic transfer during the last 100 million years.)

Professional evolutionary biologists have been puzzled and excited by these discoveries about LGT. But the word has hardly filtered through to the interested public—an odd situation given the status of LGT as a challenge to one of our most basic assumptions about the nature and fundamental topology of evolution itself, not to mention the foundation of Linnaean logic as well! Perhaps most of us just don't care about invisible bacteria, whereas we would sit up and take notice if we heard that LGT played a major role in the evolution of animals. Or perhaps the issue strikes most people as too abstract to command the same level of attention that we heap and hype upon such events of minimal theoretical interest as the discovery of a new carnivorous dinosaur larger than *Tyrannosaurus.* But I would not so disrespect the concerns of public understanding. Properly explained, the theoretical challenge of LGT to some truly fundamental views about the nature of evolution and classification should be fascinating to all people interested in science and natural history.

To put the matter baldly, if LGT plays a large enough role in bacterial evolution to overcome the Linnaean signal of conventional branching without subsequent joining, then binomial logic really doesn't work. An honest diagnosis could not then recommend a simple remedy of some minor repairs or small plaster patches—for the Linnaean system would truly be broken by the collapse of this central theoretical prerequisite. The hierarchical basis of Linnaean logic demands that life's history develop as a tree, without amalgamation of branches once a lineage originates in independence. But if LGT dominates the composition of bacterial genomes, then trees cannot express the topology of evolutionary relationships, because the pathways of life then form a meshwork, as bacteria evolve by importing genes from any position, no matter how evolutionarily distant, on the genealogical net.

But don't just trust the words of this expert on land snails (in the realm of fat, furry things) and diligent essayist on subjects beyond his genuine expertise. Consider these measured words from a technical article by the leading researcher on the subject (W. Ford Doolittle of Dalhousie University, Halifax, writing in the June 25, 1999, issue of *Science,* a special report on evolution from

America's leading journal for professional scientists). Doolittle wrote, in an article titled "Phylogenetic classification and the universal tree":

> If "lateral gene transfer" can't be dismissed as trivial in extent or limited to species categories of genes, then no hierarchical universal classification can be taken as natural. Molecular phylogeneticists will have failed to find the "true tree," not because their methods are inadequate or because they have chosen the wrong genes, but because the history of life cannot properly be represented as a tree.

Do not lament for the spirit of Linnaeus. Yes, his dreams about the discovery of a universal system suffered two sequential blows—first, soon after his death, when scientists recognized that his logic could only work for organisms, and not for rocks and all the rest of the natural world. And, second, as discovered only in the last decade, when Linnaean taxonomy encountered a strong biological challenge from the frequency of lateral gene transfer, the ultimate tree-buster, in the substantial domain of bacteria, albeit not so strongly expressed in our own world of multicellular life (although the sequence of the human genome, published in February 2001, does reveal some important bacterial "immigrants" as well).

As a truly great scientist, Linnaeaus understood the central principle that honorable error, through overextension of exciting ideas, "comes with the territory"—and that theories gain both strength and better definition by principled limitations upon their realm of legitimate operation. Moreover, as the modern founder of the truly noble science of taxonomy, Linnaeaus also understood that all classifications must embody passionate human choices about the causes of order—in short, theories that must be subject to continuous revision and correction of error—and cannot only be conceived as passive descriptions of objective nature on the philatelic model.

Thus, taxonomies must express both concepts and percepts—and must therefore teach us as much about ourselves and our mental modes as about the structure of external nature.* Surely Linnaeaus, of all people, comprehended this fundamental and ineluctable interrelationship of mind and nature, for

*I would like to dedicate this essay to Ernst Mayr, the greatest taxonomist of the twentieth (and twenty-first) century, who remains as intellectually active as ever at age ninety-six, and who taught me, through his writing and the human contact of personal friendship, the central principle of our science (and of this essay): that taxonomies are active theories about the causes of natural order, not objective, unchanging, and preexisting stamp albums for housing nature's obvious facts.

when he composed, at the very beginning of *Systema Naturae,* his formal description of his newly crowned species, *Homo sapiens,* he linked us (in various editions)—in only one case correctly, as we now know—with three other mammals: monkeys, sloths, and bats. For each of these three, Linnaeus penned a conventional and objective description in terms of hairiness, body size, and number of fingers and toes. But for *Homo sapiens,* he chose the path of terseness and wrote just the three Latin words of another familiar motto. Not *natura non facit saltum* this time, but the foremost intellectual challenge of classical wisdom: *Nosce te ipsum* — Know thyself.

22

Abscheulich! (Atrocious)

REVOLUTIONS CANNOT BE KIND TO PROMINENT AND UNRE-constructed survivors of a superseded age. But the insight and dignity of vanquished warriors, after enough time has elapsed to quell the immediate passions of revolt, often inspire a reversal of fortune in the judgment of posterity. (Even the most unabashed northerner seems to prefer Robert E. Lee to George McClellan these days.)

This essay details a poignant little drama in the lives of three great central European scientists caught in the intellectual storm of Darwin's *Origin of Species,* published in 1859. This tale, dormant for a century, has just achieved a vigorous second life, based largely on historical misapprehension and creationist misuse. Ironically, once we disentangle the fallacies and supply a proper context for understanding, our admiration must flow to Darwin's two most prominent opponents from a dispersed and defeated conceptual world: the Estonian (but ethnic German) embryologist and general naturalist

Karl Ernst von Baer (1792–1876), who spent the last forty years of his life teaching in Russia, and the Swiss zoologist, geologist, and paleontologist Louis Agassiz (1807–1873), who decamped to America in the 1840s and founded Harvard's Museum of Comparative Zoology, where I now reside as curator of the collection of fossil invertebrates that he began.

Meanwhile, our justified criticism must fall upon the third man in this topsy-turvy drama, the would-be hero of the new world order: the primary enthusiast and popularizer of Darwin's great innovation, the German naturalist Ernst Haeckel (1834–1919). Haeckel's forceful, eminently comprehensible, if not always accurate, books appeared in all major languages and surely exerted more influence than the works of any other scientist, including Darwin and Huxley (by Huxley's own frank admission), in convincing people throughout the world about the validity of evolution.

I willingly confess to hero-worship for the raw intellectual breadth and power of three great men: Darwin, who constructed my world; Lavoisier, because the clarity of his mind leaves me awestruck every time I read his work; and Karl Ernst von Baer, who lived too long and became too isolated to win the proper plaudits of posterity. But T. H. Huxley, who ranks fourth on my personal list, regarded von Baer as the greatest pre-Darwinian naturalist of Europe, and I doubt that any expert with the detailed knowledge to render judgment about general brilliance and specific accomplishments would disagree.

As the leading embryologist of the early nineteenth century, von Baer discovered the mammalian egg cell in 1827 and then, in 1828, published the greatest monograph in the history of the field: *Entwickelungsgeschichte der Thiere* (The developmental history of animals). He then suffered a mental breakdown and never returned to the field of embryology. Instead he moved to St. Petersburg in 1834 (a common pattern for central European scientists, as Russia, lacking a system of modern education, imported many of its leading professors in scientific subjects from abroad). There he enjoyed a long and splendid second career as an Arctic explorer, a founder of Russian anthropology, and a geomorphologist credited with discovering an important law relating the erosion of riverbanks to the earth's rotation.

Von Baer's theories of natural history allowed for limited evolution among closely related forms, but not for substantial transformation between major groups. Moreover, he held no sympathy for Darwin's mechanistic views of evolutionary causality. Darwin's book shook the aged von Baer from decades of inactivity in his old zoological realm—and this great man, whom Agassiz, in his last (and posthumously published) article of 1874, would call "the aged

Nestor of the science of Embryology," came roaring back with a major critique entitled *Über Darwins Lehre* (On Darwin's Theory).

In a second article written to criticize a brave new world that often disparaged or even entirely forgot the discoveries of previous generations, von Baer made a rueful comment in 1866 that deserves enshrinement as one of the great aphorisms in the history of science. Invoking Louis Agassiz, his younger friend and boon companion in rejecting the new theory of mechanistic evolution, von Baer wrote:

> Agassiz says that when a new doctrine is presented, it must go through three stages. First, people say that it isn't true, then that it is against religion, and, in the third stage, that it has long been known [my translation from the German original].

Ernst Haeckel (1834–1919), with his characteristic mixture of gusto and bluster, fancied himself a Darwinian general, embattled in Agassiz's first two stages and unfurling the new evolutionary banner not only for biological truth, but for righteousness of all stripes. In 1874 he wrote in his most popular book, *Anthropogenie* (The Evolution of Man):

> On one side spiritual freedom and truth, reason and culture, evolution and progress stand under the bright banner of science; on the other side, under the black flag of hierarchy, stand spiritual slavery and falsehood, irrationality and barbarism, superstition and retrogression. . . . Evolution is the heavy artillery in the struggle for truth.

Men of large vision often display outsized foibles as well. No character in the early days of Darwinism can match Haeckel for enigmatic contrast of the admirable and the dubious. No one could equal his energy and the volume of his output—most of high quality, including volumes of technical taxonomic description (concentrating on microscopic radiolarians, and on jellyfishes and their allies), not only theoretical effusions. But no major figure took so much consistent liberty in imposing his theoretical beliefs upon nature's observable factuality.

I won't even discuss Haeckel's misuse of Darwinian notions in the service of a strident German nationalism based on claims of cultural, and even biological, superiority—a set of ideas that became enormously popular and did provide later fodder for Nazi propagandists (obviously not Haeckel's direct fault, although scholars must bear some responsibility for exaggerated, but not distorted, uses of their arguments—see D. Gasman, *The Scientific Origins of National Socialism: Social Darwinism in Ernst Haeckel and the German Monist League* [London: MacDonald, 1972]). Let's consider only his drawings of organisms, supposedly a far more restricted subject, imbued with far less opportunity for any "play" beyond sober description.

I do dislike the common phrase "artistic license," especially for its parochially smug connotation (when used by scientists) that creative humanists care little for empirical accuracy. (After all, the best artistic "distortions" record great skill and conscious intent, applied for definite and fully appropriate purposes; moreover, when great artists choose to depict external nature as seen through our eyes, they have done so with stunning accuracy.) But I don't know how else to describe Haeckel, who was, by the way, a skilled artist and far more than a Sunday painter.

Haeckel published books at the explicit interface of art and science—and here he stated no claim for pure fidelity to nature. His *Kunstformen der Natur* (Artforms of Nature), published in 1904 and still the finest work ever printed in this genre, contains one hundred plates of organisms crowded into intricate geometric arrangements. One can identify the creatures, but their invariably curved and swirling forms so closely follow the reigning conventions of art nouveau (called *Jugendstil* in Germany) that one cannot say whether the plates should be labeled as illustrations of actual organisms or primers for a popular artistic style.

But Haeckel also prepared his own illustrations for his technical monographs and scientific books—and here he did claim, while standard practice and legitimate convention also required, no conscious departure from fidelity to nature. Yet Haeckel's critics recognized from the start that this master naturalist, and more than competent artist, took systematic license in "improving" his specimens to make them more symmetrical or more beautiful. In particular, the gorgeous plates for his technical monograph on the taxonomy of radiolarians (intricate and delicate skeletons of single-celled planktonic organisms) often "enhanced" the actual appearances (already stunningly complex and remarkably symmetrical) by inventing structures with perfect geometric regularity.

This practice cannot be defended in any sense, but distortions in technical monographs cause minimal damage because such publications rarely receive attention from readers without enough professional knowledge to recognize the fabrications. "Improved" illustrations masquerading as accurate drawings spell much more trouble in popular books intended for general audiences lacking the expertise to separate a misleading idealization from a genuine signal from nature. And here, in depicting vertebrate embryos in several of his most popular books, Haeckel took a license that subjected him to harsh criticism in his own day and, in a fierce brouhaha (or rather a tempest in a teapot), has resurfaced in the last two years to haunt him (and us) again, and even to give some false comfort to creationists.

We must first understand Haeckel's own motivations—not as any justification for his actions, but as a guide to a context that has been sadly missing from most recent commentary, thereby leading to magnification and distortion of this fascinating incident in the history of science. Haeckel remains most famous today as the chief architect and propagandist for a famous argument that science disproved long ago, but that popular culture has never fully abandoned, if only because the standard description sounds so wonderfully arcane and mellifluous—"ontogeny recapitulates phylogeny," otherwise known as the theory of recapitulation or, roughly, the claim that organisms retrace their evolutionary history, or "climb their own family tree" to cite an old catchphrase, during their embryological development. Thus the gill slits of the early human embryo supposedly repeat our distant ancestral past as a fish, while the transient embryonic tail, developing just afterward, marks the later reptilian phase of our evolutionary ascent. (My first technical book, *Ontogeny and Phylogeny* [Harvard University Press, 1977], includes a detailed account of the history of recapitulation—an evolutionary notion exceeded only by natural selection itself for impact upon popular culture. See essay 8 for a specific, and unusual, expression of this influence in the very different field of psychoanalysis.)

As primary support for his theory of recapitulation, and to advance the argument that all vertebrates may be traced to a common ancestor, Haeckel frequently published striking drawings, showing parallel stages in the development of diverse vertebrates, including fishes, chickens, and several species of mammals from cows to humans. The accompanying figure (on page 310) comes from a late, inexpensive, popular English translation, published in 1903, of his most famous book, *The Evolution of Man.* Note how the latest depicted stages (bottom row) have already developed the distinctive features of adulthood (the tortoise's shell, the chick's beak). But Haeckel draws the earliest stages of the first row, showing tails below and gill slits just under the primordial head, as virtually identical for all embryos, whatever their adult destination. Haeckel could thus claim that this near identity marked the common ancestry of all vertebrates—for, under the theory of recapitulation, embryos pass through a series of stages representing successive adult forms of their evolutionary history. An identical embryonic stage can only imply a single common ancestor.

To cut to the quick of this drama: Haeckel exaggerated the similarities by idealizations and omissions. He also, in some cases—in a procedure that can only be called fraudulent—simply copied the same figure over and over again. At certain stages in early development, vertebrate embryos do look more alike, at least in gross anatomical features easily observed with human eyes, than do the adult tortoises, chickens, cows, and humans that will develop from them. But these early embryos also differ far more substantially, one from the other, than Haeckel's figures show. Moreover, Haeckel's drawings never fooled expert embryologists, who recognized his fudgings right from the start.

At this point a relatively straightforward factual story, blessed with a simple moral message as well, becomes considerably more complex, given the foibles and practices of the oddest primate of all. Haeckel's drawings, despite their noted inaccuracies, entered into the most impenetrable and permanent of all quasi-scientific literatures: standard student textbooks of biology. I do not know how the transfer occurred in this particular case, but the general (and highly troubling) principles can be easily identified. Authors of textbooks can-

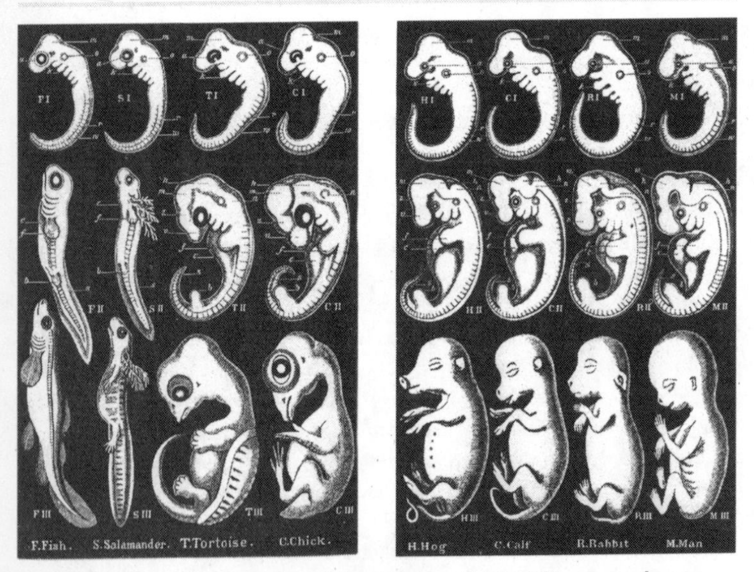

A famous chart of the embryological development of eight different vertebrates, as drawn by Haeckel for a 1903 edition of his popular book, The Evolution of Man. *Haeckel exaggerated the similarities of the earliest stages (top row) with gill slits and tails.*

not be experts in all subdisciplines of their subject. They should be more careful, and they should rely more on primary literature and the testimony of expert colleagues. But shortcuts tempt us all, particularly in the midst of elaborate projects under tight deadlines.

Therefore, textbook authors often follow two suboptimal routes that usually yield adequate results, but can also engender serious trouble: they copy from previous textbooks, and they borrow from the most widely available popular sources. No one ever surpassed Haeckel in fame and availability as a Darwinian spokesman who also held high professional credentials as a noted professor at the University of Jena. So textbook authors borrowed his famous drawings of embryonic development, probably quite unaware of their noted inaccuracies and outright falsifications—or (to be honest about dirty laundry too often kept hidden) perhaps well enough aware, but then rationalized with the ever-tempting and ever-dangerous argument, "Oh well, it's close enough to reality for student consumption, and it does illustrate a general truth with permissible idealization." (I am a generous realist on most matters of human foibles. But I confess to raging fundamentalism on this issue. The smallest compromise in dumbing down by inaccuracy destroys integrity and places an author upon a slippery slope of no return.)

Once ensconced in textbooks, misinformation becomes cocooned and effectively permanent because, as stated above, textbooks copy from previous texts. (I have written two previous essays on this lamentable practice—one on the amusingly perennial description of the "dawn horse" eophippus as "fox terrier" in size, even though most authors, including yours truly, have no idea about the dimensions or appearance of this breed; and the other on the persistent claim that elongating giraffe necks provide our best illustration of Darwinian natural selection versus Lamarckian use and disuse, when, in fact, no meaningful data exist on the evolution of this justly celebrated structure.)

We should therefore not be surprised that Haeckel's drawings entered nineteenth-century textbooks. But we do, I think, have the right to be both astonished and ashamed by the century of mindless recycling that has led to the persistence of these drawings in a large number, if not a majority, of modern textbooks! Michael Richardson of the St. George Hospital Medical School in London, a colleague who deserves nothing but praise for directing attention to this old issue, wrote to me (letter of August 16, 1999):

> If so many historians knew all about the old controversy [over Haeckel's falsified drawings] then why did they not communicate this information to the numerous contemporary authors who use

> the Haeckel drawings in their books? I know of at least fifty recent biology texts which use the drawings uncritically. I think this is the most important question to come out of the whole story.

The recent flap over this more-than-twice-told tale—an almost comical manifestation of the famous dictum that those unfamiliar with history (or simply careless in reporting) must be condemned to repeat the past—began with an excellent technical paper by Richardson and six other colleagues in 1997 ("There is no highly conserved embryonic stage in the vertebrates: implications for current theories of evolution and development," *Anatomy and Embryology,* volume 196: 91–106; following a 1995 article by Richardson alone in *Developmental Biology,* volume 172: 412–21). In these articles, Richardson and colleagues discussed the original Haeckel drawings, briefly noted the contemporary recognition of their inaccuracies, properly criticized their persistent appearance in modern textbooks, and then presented evidence (discussed below) of the differences in early vertebrate embryos that Haeckel's tactics had covered up, and that later biologists had therefore forgotten. Richardson invoked this historical tale in order to make an important point, also mentioned below, about exciting modern work in the genetics of development.

From this excellent and accurate beginning, the reassertion of Haeckel's old skullduggery soon spiraled into an abyss of careless reporting and self-serving utility. The news report in *Science* magazine by Elizabeth Pennisi (September 5, 1997) told the story well, under an accurate headline ("Haeckel's embryos: fraud rediscovered") and a textual acknowledgment of "Haeckel's work, first found to be flawed more than a century ago." But the shorter squib in Britain's *New Scientist* (September 6, 1997) began the downward spiral by implying that Richardson had discovered Haeckel's misdeed for the first time.

As so often happens, this ersatz version, so eminently more newsworthy than the truth, opened the floodgates to the following sensationalist (and nonsensical) account: a primary pillar of Darwinism, and of evolution in general, has been revealed as fraudulent after more than a century of continuous and unchallenged centrality in biological theory. If evolution rests upon such flimsy support, perhaps we should question the entire enterprise and give creationists, who have always flubbed their day in court, their day in the classroom.

Michael Behe, a Lehigh University biologist who has tried to resuscitate the most ancient and tired canard in the creationist arsenal (Paley's "argument from design," based on the supposed "irreducible complexity" of intricate biological structures, a claim well refuted by Darwin himself in his famous discussion of transitional forms in the evolution of complex eyes), reached the nadir in an op-ed piece for the *New York Times* (August 13, 1999), commenting on the Kansas School Board's decision to make instruction in evolution optional within the state's science curriculum—an antediluvian move, fortunately reversed in February 2001, following the political successes of scientists, activists, and folks of goodwill (and judgment) in general, in voting fundamentalists off the state school board and replacing them with elected members committed to principles of good education and respect for nature's factuality. (In fairness, I liked Behe's general argument in this piece, for he stayed away from irrelevant religious issues and attacked the Kansas decision by saying that he would never get a chance to present his supposed refutations if students didn't study evolution at all.)

As his putatively strongest refutation of Darwinism, Behe cites the ersatz version of Richardson's work on Haeckel's drawings. (Behe presents only two other arguments, one that he accepted as true [the evolution of antibiotic resistance by several bacterial strains], the second judged as "unsupported by current evidence" [the "classic" case of industrial melanism in moths], with only this third point—the tale of Haeckel's drawings—declared "downright false." So if this piece represents Behe's best shot, I doubt that creationists will receive much of a boost from their latest academic poster boy.) Behe writes:

> The story of the embryos is an object lesson in seeing what you want to see. Sketches of vertebrate embryos were first made in the last 19th century by Ernst Haeckel, an admirer of Darwin. In the intervening years, apparently nobody verified the accuracy of Haeckel's drawings . . . If supposedly identical embryos were once touted as strong evidence for evolution, does the recent demonstration of variation in embryos now count as evidence against evolution?

From this acme of media hype and public confusion, we should step back and reassert the two crucial points that accurately site Haeckel's drawings as a poignant and fascinating historical tale and a cautionary warning about scientific carelessness (particularly in the canonical and indefensible practices of textbook writing)—but not, in any way, as an argument against evolution, or a sign of weakness in Darwinian theory. Moreover, as a testament to greatness of intellect and love of science, whatever the ultimate validity of an underlying worldview honorably supported by men of such stature, we may look to the work of von Baer and Agassiz, Darwin's most valiant opponents in his own day, for our best illustrations of these two clarifying points.

1. *Haeckel's forgeries as old news.* Tales of scientific fraud excite the imagination for good reason. Getting away with this academic equivalent of murder for generations, and then being outed a century after your misdeeds, makes even better copy. Richardson reexamined Haeckel's drawings for good reasons, and he never claimed that he had uncovered the fraud. But press commentary then invented and promulgated this phony version—and these particular chickens came home to a creationist roost (do pardon this rather mixed metaphor!).

Haeckel's expert contemporaries recognized what he had done, and said so in print. For example, a famous 1894 article by Cambridge University zoologist Adam Sedgwick ("On the law of development commonly known as von Baer's law") included the following withering footnote of classical Victorian understatement:

> I do not feel called upon to characterise the accuracy of the drawings of embryos of different classes of Vertebrata given by Haeckel in his popular works. . . . As a sample of their accuracy, I may refer the reader to the varied position of the auditory sac in the drawings of the younger embryos.

I must confess to a personal reason, emotional as well as intellectual, for long and special interest in this tidbit of history. Some twenty years ago I found, on the open stacks of our Museum's library, Louis Agassiz's personal copy of the first (1868) edition of Haeckel's *Natürliche Schöpfungsgeschichte* (The Natural History of Creation). After his death, Agassiz's library passed into the museum's general collection, where indifferent librarianship (before the present generation) led to open access, through nonrecognition, for such priceless treasures.

I noted, with the thrill that circumstances vouchsafe to an active scholar only a few times in a full career, that Agassiz had penciled copious marginal notes—some forty pages' worth in typed transcription—into this copy. But I couldn't read his scribblings. Agassiz, a typical Swiss polyglot, annotated books in the language of their composition. Moreover, when he wrote marginalia into a German book published in Roman type, he composed the notes in Roman script (which I can read and translate). But when he read a German book printed in old, but easily decipherable, *Fraktur* type (as in Haeckel's 1868 edition), he wrote his annotations in the corresponding (and now extinct) *Sütterlin* script (which I cannot read at all). Fortuna, the Roman goddess, then smiled upon me, for my secretary, Agnes Pilot, had been educated in Germany just before the Second World War—and she, *Gott sei Dank,* could still read this archaic script. So she transliterated Agassiz's squiggles into readable German in Roman type, and I could finally sense Agassiz's deep anger and distress.

In 1868, Agassiz at age sixty-one, and physically broken by an arduous expedition to Brazil, felt old, feeble, and bypassed—especially in the light of his continued opposition to evolution. (His own graduate students had all "rebelled" and embraced the new Darwinian model.) He particularly disliked Haeckel for his crass materialism, his scientifically irrelevant and vicious swipes at religion, and his haughty dismissal of earlier work (which he often shamelessly "borrowed" without attribution). And yet, in reading through Agassiz's extensive marginalia, I sensed something noble about the quality of his opposition, however ill-founded in the light of later knowledge.

To be sure, Agassiz waxes bitter at Haeckel's excesses, as in the accompanying figure of his final note appended to the closing flourish of Haeckel's book, including the author's gratuitous attack on conventional religion as "the dark beliefs and secrets of a priestly class." Agassiz writes sardonically: *Gegeben im Jahre I der neuen Weltordnung. E. Haeckel.* (Given in year one of the new world order. E. Haeckel.) But Agassiz generally sticks to the high road, despite ample provocation, by marshaling the facts of his greatest disciplinary expertise (in geology, paleontology, and zoology) to refute Haeckel's frequent exaggerations and rhetorical inconsistencies. Agassiz may have been exhausted and discouraged, but he could still put up a whale of a fight, even if only in private. (See my previous 1979 publication for details: "Agassiz's later, private thoughts on evolution: his marginalia in Haeckel's *Natürliche Schöpfungsgeschichte (1868),*" in Cecil Schneer, ed., *History of Geology* [University of New Hampshire: The University Press of New England, 277–82].)

Agassiz proceeded in generally measured prose until he came to page 240, where he encountered Haeckel's falsified drawings of vertebrate embryology—a subject of extensive personal research and writing on Agassiz's part. He immediately recognized what Haeckel had done, and he exploded in fully justified rage. Above the nearly identical pictures of dog and human embryos, Agassiz wrote: *Woher copiert? Gekünstelte Ähnlichkeit mit Ungenauigkeit verbunden, z.b. Coloboma, Nabel, etc.* (Where were these copied from? [They include] artistically crafted similarities mixed with inaccuracies, for example, the eye slit, umbilicus, etc.)

At least these two drawings displayed some minor differences. But when Agassiz came to page 248, he noticed that Haeckel had simply copied the same exact figure three times in supposedly illustrating a still earlier embryonic stage of a dog (left), a chicken (middle), and a tortoise (right). He wrote above this figure: *Woher sind diese Figuren entnommen? Es gibt sowas in der ganzen*

Litteratur nicht. Diese Identität ist nicht wahr. (Where were these figures taken from? Nothing like this exists in the entire literature. This identity is not true.)

Finally, on the next page, he writes his angriest note next to Haeckel's textual affirmation of this threefold identity. Haeckel stated: "If you take the young embryos of a dog, a chicken and a tortoise, you cannot discover a single difference among them." And Agassiz sarcastically replied: *Natürlich — da diese Figuren nicht nach der Natur gezeichnet, sondern eine von der andern copiert ist! Abscheulich.* (Naturally—because these figures were not drawn from nature, but rather copied one from the other! Atrocious.)

2. *Haeckel's forgeries as irrelevant to the validity of evolution or Darwinian mechanisms.* From the very beginning of this frenzied discussion two years ago, I have been thoroughly mystified as to what, beyond simple ignorance or self-serving design, could ever have inspired the creators of the sensationalized version to claim that Haeckel's exposure challenges Darwinian theory, or even evolution itself. After all, Haeckel used these drawings to support his theory of recapitulation—the claim that embryos repeat successive adult stages of their ancestry. For reasons elaborated at excruciating length in my book *Ontogeny and Phylogeny,* Darwinian science conclusively disproved and abandoned this idea by 1910 or so, despite its persistence in popular culture. Obviously, neither evolution nor Darwinian theory needs the support of a doctrine so conclusively disconfirmed from within.

I do not deny, however, that the notion of greater embryonic similarity, followed by increasing differentiation toward the adult stages of related forms, has continued to play an important, but scarcely defining, role in biological the-

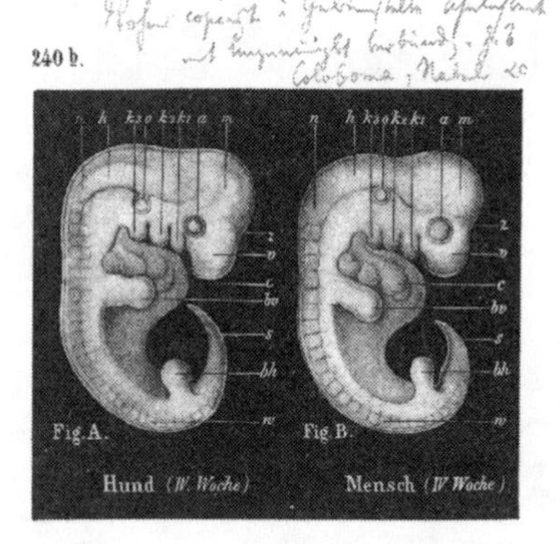

Early embryonic stages of dog (left) and human (right), as drawn by Haeckel for an 1868 book, but clearly "fudged" in exaggerating and even making up some of the similarities. Louis Agassiz's copy, with his angry words of commentary at top.

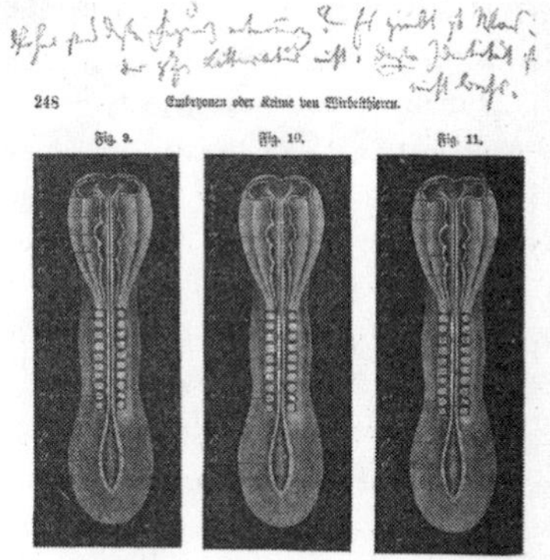

Agassiz's angry comments (written above) upon Haeckel's false figure of similarities in the early embryonic stages of dog, chicken, and tortoise. Haeckel simply copied the same drawing three times.

Diese fünf Hirnblasen sind ursprünglich bei allen Wirbelthieren, die überhaupt ein Gehirn besitzen, gleichmäßig angelegt, und bilden sich erst allmählich bei den verschiedenen Gruppen so verschiedenartig aus, daß es nachher sehr schwierig ist, in den ganz entwickelten Gehirnen die gleichen Theile wieder zu erkennen. Wenn Sie die jungen Embryonen des Hundes, des Huhns und der Schildkröte in Fig. 9, 10 und 11 vergleichen, werden Sie nicht im Stande sein, einen Unterschied wahrzunehmen. Wenn Sie dagegen die viel weiter entwickelten Embryonen in Fig. C—F mit einander vergleichen, werden Sie schon deutlich die ungleichartige Ausbildung erkennen, und namentlich wahr-

Agassiz's angry comments on Haeckel's falsified drawings—ending with the judgment used as the title for this essay: "Abscheulich!"

ory—but through the later evolutionary version of another interpretation first proposed by von Baer in his 1828 treatise, one of the greatest works ever published in the history of science. In a pre-evolutionary context, von Baer argued that development, as a universal pattern, must proceed by a process of differentiation from the general to the specific. Therefore, the most general features of all vertebrates will arise first in embryology, followed by a successive appearance of ever more specific characters of particular groups.

In other words, you can first tell that an embryo will become a vertebrate rather than an arthropod, then a mammal rather than a fish, then a carnivore rather than a rodent, and finally good old Rover rather than Ms. Tabby. Under von Baer's reading, a human embryo grows gill slits not because we evolved from an adult fish (Haeckel's recapitulatory explanation), but because all vertebrates begin their embryological lives with gills. Fish, as "primitive" vertebrates, depart least from this basic condition in their later development, whereas mammals, as most "advanced," lose their gills and grow lungs during their maximal embryological departure from the initial and most generalized vertebrate form.

Von Baer's Law—as biologists soon christened this principle of differentiation—received an easy and obvious evolutionary interpretation from Darwin's hand. The intricacies of early development, when so many complex organs differentiate and interconnect in so short a time, allow little leeway for substantial change, whereas later stages with fewer crucial connections to the central machinery of organic function permit greater latitude for evolutionary change. (In rough analogy, you can always paint your car a different color, but you had better not mess with basic features of the internal combustion engine as your future vehicle rolls down the early stages of the assembly line.)

The evolutionary version of von Baer's law suggests that embryos may give us better clues about ancestry than adults—but not because they represent ancestral adults in miniature, as Haeckel and the recapitulationists believed. Rather, embryos indicate ancestry because generalized features of large groups offer better clues than do specialized traits of more-restricted lineages. In a standard example, some parasites become so anatomically degenerate as adults that they retain no distinctive traits of their larger affiliation—as, for example, in the parasitic barnacle *Sacculina* that, as an adult, becomes little more than an amorphous bag of feeding and reproductive tissue within the body of its crab host. But the larval stages that must seek and penetrate a crab can hardly be distinguished from the early stages of ordinary barnacles. Darwin made the key point succinctly when he stated in *The Origin of Species* that "community in embryonic structure reveals community of descent."

Von Baer's law makes good sense—but nothing in Darwinian theory implies or requires its validity, while evolution itself clearly permits embryology to proceed in either direction (or in no linearized manner at all)—either from embryonic similarity to adult discordance (as in groups that follow von Baer's principle), or from larval discordance to adult likeness (as in several invertebrate groups, notably some closely related sea urchin species, where larvae have adapted to highly different lifestyles of planktonic floating versus development from yolk-filled eggs that remain on the sea floor, while the highly similar adults of both species continue to live and function like ordinary sea urchins).

The "bottom line"—to use a popular phrase from another walk of human life—may now be simply stated: the validity and relative frequency of von Baer's law remains an open, empirical question within evolutionary theory, an issue that can only be resolved by observational evidence from a wide variety of organisms. Moreover, this issue has become quite important in the light of current excitement over recent advances in genetics that have finally allowed us to identify and trace the genes regulating early development. In this crucial and valid context, Richardson wisely chose to reevaluate our complacency about the probable validity of von Baer's law.

Richardson realized that the continuing republication of Haeckel's fraudulent figures might be tipping our beliefs in von Baer's favor for indefensible reasons of inherited and unquestioned tradition (based on falsified drawings, to boot), rather than by good observational evidence. He therefore called attention to this likely source of unrecognized bias as he marshaled several colleagues to make the basic observations that could resolve a truly open question, falsely regarded by many colleagues as an issue decided long ago, partly on the basis of Haeckel's doctored evidence.

The jury will be out for some time as they debate, and actively research, this important issue, too long neglected, in the sciences of natural history. But the 1997 paper of Richardson and six colleagues has already poked some important holes in the old and (as we now learn) poorly documented belief in early embryonic similarity among related lineages, followed by increasing disparity toward adulthood. The early embryonic stages of vertebrates are not nearly so similar as Haeckel's phony drawings had led us to believe. For example, at the stage that Haeckel chose for maximal similarity, the somite count (number of vertebral segments) of actual vertebrate embryos ranges from eleven for a Puerto Rican tree frog to sixty for a "blind worm" (the common name for an unfamiliar group of limbless amphibians with a basically snakelike adult form). Moreover, although Haeckel drew his embryos as identical in both size and form, actual vertebrate embryos at their stage of maximal anatomical similarity span a tenfold range in body size.

In short, the work of Richardson and colleagues goes by a simple and treasured name in my trade: "good science." The flap over Haeckel's doctored drawings should leave us feeling ashamed about the partial basis of a widely shared bias now properly exposed and already subjected to exciting new research. But Haeckel's high Victorian (or should I say Bismarckian) misdeeds provide no fodder to foes of Darwin, or of evolution—although we should feel sheepish (and well instructed) about our belated obedience to a grand old motto: Physician, heal thyself.

In other words, to give von Baer and Agassiz a final due, we need not fear the first and second stages of a scientific revolution because we will fight like hell (perhaps unwisely and too well, but at least with gusto) so long as we regard a new idea as either ridiculous or opposed to "religion" (that is, to conventional belief). But we must beware the dreaded third stage—for when we capitulate and then smugly state that we knew it all along, we easily fall into the greatest danger of all—arrogant complacency—because we have ceased to question and observe. And no situation in science could possibly be more *abscheulich*—atrocious!

23

Tales of a Feathered Tail

One fine day, or so the legend proclaims, Joseph Stalin received a telegram from his exiled archrival, Leon Trotsky. Overjoyed by the apparent content, Stalin rounded up the citizenry of Moscow for an impromptu rally in Red Square. He then addressed the crowd below: "I have received the following message of contrition from Comrade Trotsky, who has obviously been using his Mexican retreat for beneficial reflection: 'Comrade Stalin: You are right! I was wrong! You are the leader of the Russian people!'"

But as waves of involuntary applause rolled through the square, a Jewish tailor in the front row—Trotsky's old school chum from yeshiva days—bravely mounted the platform, tapped Stalin on the shoulder, and took the microphone to address the crowd. "Excuse me, Comrade Stalin," he said. "The words, you got them right; but the meaning, I'm not so sure." Then the tailor read the telegram again, this time with the intended intonation of disgust and the

Velociraptor*, a ground-running, meat-eating dinosaur from the Gobi, may well have been covered in feathers.*

rising inflection of inquiry: '"Comrade Stalin: *You* are right?? *I* was wrong?? *You* are the leader of the Russian people??'"

I have never been able to regard this joke with equanimity, because I can't help wondering what happened to the poor tailor, who undoubtedly suffered far more than Trotsky, albeit anonymously. But I value the tale as a lesson about the importance of context. We may get every word right but, in a pungent military acronym (the superlative degree of SNAFU, I have been assured by army grammarians), still get the meaning FUBAR (defined by the dictionary, in genteel terms, as "fouled [euphemism] up beyond all recognition"). As an unfortunate, yet eminently understandable, consequence of its central status in biology and its challenging implications for our view of human origins and history, evolution probably exceeds all other scientific subjects in featuring straightforward facts enshrouded in difficult or ambiguous meanings.

The popular understanding of evolution includes at least two false assumptions, so widely shared and so deeply (if unconsciously) embedded in the context of conventional explanations that many plain facts, easily grasped at a superficial level of overt recitation, almost always enter the public discourse of newspapers, films, and magazines in a highly confused form that "science writers" either mistake for the actual opinions of scientists or, more cynically, choose to present as the literary equivalent of "easy listening" for succor in drive-time traffic jams.

In the picture conveyed by these two related fallacies, evolution becomes, first of all, the transformation of one kind of entity into another, body and soul. So fish evolve into amphibians in a "conquest" of the land, and apes leave the safety of trees, eventually to become human by facing the dangers of terra firma with a weapon in their liberated hand and a fresh twinkle of insight emanating from an enlarged organ behind their eye. In the second component of this transformational view, descendants win victory from the heart of their valor in the face of natural selection—for "later" must mean "better," as the land yields to explorational metaphors of conquest or colonization while the African savannas, for the first time in planetary history, ring with sounds of progress now expressed in the voice of real language.

But evolution proceeds by the branching of bushes, not by the morphing of one form into another, with the old disappearing into the triumph of the new. Novelties begin as little branches on old trees, not as butterflies of Michael Jordan refashioned from the caterpillar components of Joe Airball. Moreover, most novelties, at least at their origin, grow as tiny twigs of addition to persisting and vigorous bushes, not as higher realizations of ancestors that literally gave their all to a transcendence of their former grubby selves.

Amphibians and all their descendants have done well enough on land, but fins beat feet on the vertebrate bush, where the majority of twigs (species) sprout among fishes. I do not deny the transient success, and interesting novelties, of humans. But *Homo sapiens* occupies only one twig on a modest primate bush of some two hundred species, and even our most distantly related subgroups, in both evolutionary and geographic terms (say, the San of southern Africa and the Sami of northern Finland), show very little genetic divergence, whereas two populations of the same species of chimpanzee, separated by only a few hundred miles of African real estate, have evolved many more genetic differences, one from the other. (This initially surprising fact makes evident sense once we recast our conceptions in properly bushy terms. All living humans descended from common ancestors who lived in Africa less than 200,000 years ago—despite our subsequent spread throughout the world. The two chimpanzee populations may have remained in geographical proximity, but they split from a common ancestor far longer ago, thus providing much more time for the evolution of genetic differences in the separated groups.)

Finally, and at the broadest scale, we will grasp the principle that novelty arises by branching and not by the wholesale transformation of all ancestors into better descendants only when we recognize that bacteria still constitute most of life's tree—including the entire basal trunk that they built by themselves at life's cellular origin—and that all multicellular kingdoms occupy just a few, if admittedly quite healthy, branches at the terminus of a single bough.

Many of my essays stress this theme of mentally liberating bushes versus constraining ladders because I believe that no other misconception so skews public understanding of evolution. I have treated a variety of topics under this rubric: why the air bladders of fishes evolved from lungs and not vice versa, as nearly everyone assumes (including Darwin himself in this case); why the cramming of primates into a halfway corner, and not at a triumphant terminus, of a linear walk through the hall of fossil mammals in New York's Museum of Natural History makes such revolutionary sense; and why the "out of Africa" theory (on the origin of all modern humans from a recent population of African ancestors) and not the multiregional theory (of our threefold parallel origin from ancestral *Homo erectus* populations in Europe, Africa, and Asia) represents conventional evolutionary thought based on origin by branching and not the iconoclastic shock featured in most press reports, which have also misconstrued the truly peculiar and theoretically unlikely multiregional theory as transformational orthodoxy.

But I have heretofore desisted in applying this favorite theme to serious public misunderstandings of the apparently accurate claim that birds descended from dinosaurs—probably because I don't like to attack generalities head-on but prefer the path of insinuation by small but fascinating tidbits and also because dinosaurs really are just a tad overexposed and scarcely need more publicity from this student of snails. But my tidbit just arrived in the professional literature, thus permitting a tale about the bushy reform of avian origins at two levels: first the dreaded generality and then the tidbit.

1. *The basic relationship of birds and dinosaurs.* I don't mean to toss any cold water upon the almost surely valid claim, and one of the most interesting conclusions of late-twentieth-century paleontology, that birds descended from a lineage of small-bodied, bipedal dinosaurs. But the conventional interpretive "take" on this accurately stated fact could benefit from a flat-out dousing, if only because the gain in general understanding of evolution might more than compensate the loss of a charming but truly misleading characterization of a fact.

I should point out, first of all, that the basic claim does not justify the feelings of surprise or weirdness conveyed by most popular accounts. Birds did not evolve from massive sauropods or antediluvian, tanklike ankylosaurs or even from the large tyrannosaurs (which do, in fact, lie fairly close to birds on the dinosaur bush). Rather, birds branched off from a lineage of small, two-legged, meat-eating, running dinosaurs—full members of the group by proper criteria of descent (hence the validity of the one-liner that "birds evolved from dinosaurs"), but scarcely a version calculated to evoke either the fear or the power associated with our usual icon of dinosaurian immensity.

Moreover, the assertion of this evolutionary linkage during the past twenty years does not mark a stunningly new and utterly surprising discovery, but rather reaffirms an old idea that seemed patently obvious to many paleontologists in Darwin's day (notably to T. H. Huxley, who defended the argument in several publications), but then fell from popularity for a good reason based on an honest error.

The detailed anatomical correspondence between *Archaeopteryx,* the earliest toothed bird of Late Jurassic times, and the small running dinosaurs now known as maniraptors (and including such popularized forms as *Deinonychus* and *Troodon*) can hardly fail to suggest a close genealogical relationship. But following Huxley's initial assertion of an evolutionary link, paleontologists reached the false conclusion that all dinosaurs had lost their clavicles, or collarbones—a prominent component of bird skeletons, where the clavicles enlarge and fuse to form the furcula, or wishbone. Since complex anatomical structures, coded by numerous genes working through intricate pathways of development, cannot reevolve after a complete loss, the apparent absence of clavicles in dinosaurs seemed to preclude a directly ancestral status to birds, though most paleontologists continued to assert a relationship of close cousinhood.

The recent discovery of clavicles in several dinosaurs—including the forms closest to birds—immediately reinstated Huxley's old hypothesis of direct evolutionary descent. I don't mean to downplay the significance of a firm resolution for the evolutionary relationship between birds and dinosaurs, but for sheer psychological punch, the revivification of an old and eminently sensible idea cannot match the impact of discovering a truly pristine and almost shockingly unexpected item in nature's factual arsenal.

But I throw my iciest pitcher of water (a device to open eyes wide shut, not an intended punishment) into the face of a foolish claim almost invariably featured—usually as a headline—in any popular article on the origin of birds: "Dinosaurs didn't die out after all; they remain among us still, more numerous than ever, but now twittering in trees rather than eating lawyers off john seats."

This knee-jerk formulation sounds right in the superficial sense that buttresses most misunderstandings in science, for the majority of our errors reflect false conventionalities of hidebound thinking (conceptual locks) rather than failures to find the information (factual lacks) that could resolve an issue in purely observational terms. After all, if birds evolved from dinosaurs (as they did) and if all remaining dinosaurs perished in a mass extinction triggered by an impacting comet or asteroid 65 million years ago—well, then, we must have been wrong about dinosaurian death and incompetence, for our latter-day tyrannosaurs in the trees continually chirp the New Age message of *Jurassic Park:* life finds a way. (In fact, as I write this paragraph, a mourning dove mocks my mammalian pretensions in a minor key, from a nest beneath my air conditioner. *Sic non transit gloria mundi!*)

Only our largely unconscious bias for conceiving evolution as a total transformation of one entity into a new and improved model could buttress the common belief that canonical dinosaurs—the really big guys, in all their brontosaurian bulk or tyrannosaurian terror—live on as hawks and hummingbirds. For we do understand that most species of dinosaurs just died, plain and simple, without leaving any direct descendants. Under a transformational model, however, any ancestral bird carries the legacy of all dinosaurs at the heart of its courageous persistence, just as the baton in a relay race embodies all the efforts of those who ran before.

However, under a corrected branching model of evolution, birds didn't descend from a mystical totality but only from the particular little bough that generated an actual avian branch. The dinosaurian ancestors of birds lie among the smallest bipedal carnivores (think of little *Compsognathus,* tragically mistaken for a cute pet in the sequel to *Jurassic Park*)—creatures that may be "all dinosaur" by genealogy but that do not seem so functionally incongruous as progenitors of birds. Ducks as direct descendants of *Diplodocus* (a sauropod dinosaur of maximal length) would strain my credulity, but ostriches as later offshoots from a dinosaurian line that began with little *Oviraptor* (a small, lithe carnivore of less than human, and much less than ostrich, height) hardly strains my limited imagination.

As a second clarification offered by the branching model of evolution, we must distinguish similarity of form from continuity in descent: two important concepts of very different meaning and far too frequent confusion. The fact of avian descent from dinosaurs (continuity) does not imply the persistence of the functional and anatomical lifestyle of our culture's canonical dinosaurs. Evolution does mean change, after all, and our linguistic conventions honor the results of sufficiently extensive changes with new names. I don't call my dainty poodle a wolf or my car a horse-drawn carriage, despite the undoubted ties of genealogical continuity.

To draw a more complex but precise analogy in evolutionary terms: Mammals evolved from pelycosaurs, the "popular" group of sail-backed reptiles often mistaken for dinosaurs in series of stamps or sets of plastic monsters from the past. But I would never make the mistake of claiming that *Dimetrodon* (the most familiar and carnivorous of pelycosaurian reptiles) still exists because I am now typing its name, while whales swim in the sea and mice munch in my kitchen. In their descent from pelycosaurs, mammals evolved into such different creatures that the ancestral name, defined for a particular set of anatomical forms and functions, no longer describes the altered descendants. Moreover, and to reemphasize the theme of branching, pelycosaurs included three major subgroups, only two bearing sails on their backs. Mammals probably evolved as a branch of the third, sailless group. So even if we erroneously stated that pelycosaurs still lived because mammals now exist, we could not grant this status to a canonical sail-backed form, any more than we could argue for brontosaurian persistence because birds descended from a very different lineage of dinosaurs.

2. *A tidbit with feathers.* If birds evolved from small running dinosaurs, and if feathers could provide no aerodynamic benefit in an initial state of rudimentary size and limited distribution over the body, then feathers (which, by long-standing professional consensus and clear factual documentation, evolved from reptilian scales) must have performed some other function at their first appearance. A thermodynamic role has long been favored for the first feathers on the small-bodied and highly active ancestors of birds. Therefore, despite some initial skepticism, abetted by a few outlandish and speculative reconstructions in popular films and fiction, the hypothesis of feathered dinosaurs as avian ancestors gained considerable favor. Then, in June 1998, Ji Qiang and three North American and Chinese colleagues reported the discovery of two feathered dinosaurs from Late Jurassic or Early Cretaceous rocks in China ("Two Feathered Dinosaurs from Northeastern China," *Nature* 393, 25 June 1998).

The subject has since exploded in both discovery and controversy, unfortunately intensified by the reality of potential profits previously beyond the contemplation of impoverished Chinese farmers—a touchy situation compounded by the lethal combination of artfully confected hoaxes and enthusiastically wealthy, but scientifically naïve, collectors. At least one fake (the so-called *Archaeoraptor*) has been exposed, to the embarrassment of *National Geographic,* while many wonderful and genuine specimens languish in the vaults of profiteers.

But standards have begun to coagulate, and at least one genus—*Caudipteryx* ("feather-tailed," by etymology and actuality)—holds undoubted status as a feathered runner that could not fly. And so at least until the initiating tidbit for this essay appeared in the August 17, 2000, issue of *Nature,* one running dinosaur with utterly unambiguous feathers on its tail and forearms seemed to stand forth as an ensign of Huxley's intellectual triumph and the branching of birds within the evolutionary tree of ground-dwelling dinosaurs. But the new article makes a strong, if unproven, case for an inverted evolutionary sequence, with *Caudipteryx* interpreted as a descendant of flying birds, secondarily readapted to a running lifestyle on terra firma, and not as a dinosaur in a lineage of exclusively ground-dwelling forms (T. D. Jones, J. O. Farlow, J. A. Ruben, D. M. Henderson, and W. J. Hillenius, "Cursoriality in Bipedal Archosaurs," *Nature* 406 [17 August 2000]).

The case for secondary loss of flight rests upon a set of anatomical features that *Caudipteryx* shares with modern ground birds that evolved from flying ancestors—a common trend in several independent lineages, including ostriches, rheas, cassowaries, kiwis, moas, and others. By contrast, lineages of exclusively ground-dwelling forms, including all groups of dinosaurs suggested as potential ancestors of birds, evolved different shapes and proportions for the same features. In particular, as the accompanying illustration shows, ground-running and secondarily flightless birds—in comparison with small dinosaurs of fully terrestrial lineages—tend to have relatively shorter tails, relatively longer legs, and a center of gravity located in a more forward (headward) position. By all three criteria, the skeleton of *Caudipteryx* falls into the domain of flightless birds rather than the space of cursorial (running) dinosaurs.

Jones and colleagues have presented an interesting hypothesis demanding further testing and consideration, but scarcely (by their own acknowledgment) a firm proof or even a compelling probability. Paleontologists have unearthed only a few specimens of *Caudipteryx,* none complete. Moreover, we do not know the full potential for ranges of anatomical variation in the two relevant lifestyles.

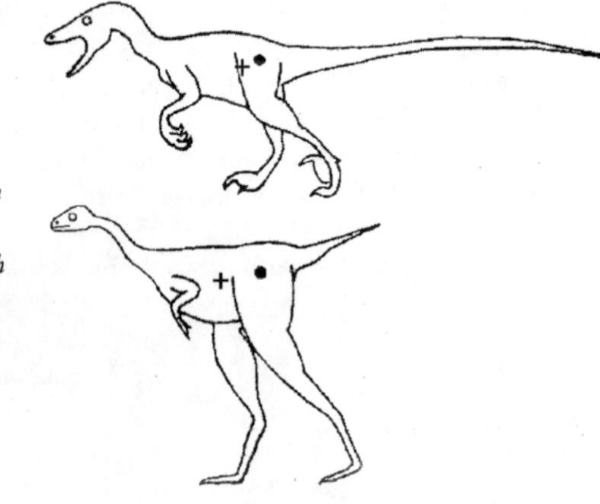

The flightless Cretaceous bird Caudipteryx *(bottom) was closer in proportions to a modern emu or ostrich than to a bipedal dinosaur (top). The plus signs mark the torsos' centers of gravity, and hip joints are indicated by dots.*

Perhaps *Caudipteryx* belonged to a fully terrestrial lineage of dinosaurs that developed birdlike proportions for reasons unrelated to any needs or actualities of flight.

I do not raise this issue here to vent any preference (for I remain neutral in a debate well beyond my own expertise, and I do regard the existence of other genera of truly feathered dinosaurs as highly probable, if not effectively proved).* Nor do I regard the status of *Caudipteryx* as crucial to the largely settled question about the dinosaurian ancestry of birds. If *Caudipteryx* belongs to a fully terrestrial lineage of dinosaurs, then its feathers provide striking confirmation for the hypothesis, well supported by several other arguments, that this defining feature of birds originated in a running ancestor for reasons unrelated to flight. But if *Caudipteryx* is a secondarily flightless bird, the general hypothesis of dinosaurian ancestry suffers no blow, though *Caudipteryx* itself loses its potential role as an avian ancestor (while gaining an equally interesting status as the first known bird to renounce flight).

Rather, I mention this tidbit to close my essay because the large volume of press commentary unleashed by the hypothesis of Jones and his colleagues showed me yet again—this time for the microcosm of *Caudipteryx* rather than for the macrocosm of avian origins in general—just how strongly our transformational biases and our failure to grasp the reality of evolution as a branching bush distort our interpretations of factual claims easily understood by all. In short, I was astonished to note that virtually all these press commentaries reported the claim for secondary loss of flight in *Caudipteryx* as deeply paradoxical and stunningly surprising (even while noting the factual arguments supporting the assertion with accuracy and understanding).

In utter contrast, the hypothesis of secondary loss of flight in *Caudipteryx* struck me as interesting and eminently worthy of further consideration but also as entirely plausible. After all, numerous lineages of modern birds have lost their ability to fly and have evolved excellent adaptations for running in a rapid and sustained manner on the ground. If flightlessness has evolved in so many independent lineages of modern birds, why should a similar event surprise us merely by occurring so soon after the origin of birds? (I might even speculate that Cretaceous birds exceeded modern birds in potential for evolutionary loss of flight, for birds in the time of *Caudipteryx* had only recently evolved as flying forms from running ancestors. Perhaps these early birds still retained

---

*An article by Q. Ti and several colleagues, published soon after this essay appeared (*Nature,* 26 April 2001), elegantly proves the existence of feathered dinosaurs that could not have used these structures for flight.

Originally interpreted as a feathered dinosaur, Caudipteryx *(left) may be a secondarily flightless bird.* Oviraptor *(right) was named "egg thief" because its skeleton was found atop fossilized eggs. Later studies showed it was probably protecting its own nest.*

enough features of their terrestrial ancestry to facilitate a readaptation to ground life in appropriate ecological circumstances.) Moreover, on the question of timing in our admittedly spotty fossil record, *Archaeopteryx* (the first known bird) lived in Late Jurassic times, while *Caudipteryx* probably arose at the beginning of the subsequent Cretaceous period—plenty of time for a flying lineage to redeploy one of its species as a ground-dwelling branch.

After considerable puzzlement, I think that I finally understand the reason for such a stark contrast between my lack of surprise and the sense of deep paradox conveyed by most press reports. As an implication of my view (expressing a professional consensus) that evolutionary novelty arises by a process of branching, the discovery of an earlier "first time" for a common and repeated event—loss of flight and secondary adaptation to effective ground running—surely attracts interest as a lovely nugget of discovery but scarcely evokes any theoretical surprise.

But in the usual public misconception of evolution as a story of wholesale transformation into something better, such an early "falling off" from "the program" seems almost perverse. After all, birds had just taken to the air a few tens of millions of years (at most) before the appearance of *Caudipteryx.* Why would a lineage fall out of step so early in the game? Once the program rolls to a full and triumphant completion, then evolution might permit an ostrich or two to slip off the main line and pursue its own bohemian path in a now strange but once ancestral land. But such events surely cannot occur in the vigorous youth of a lineage that has just snatched winged victory from the jaws of terrestrial dinosaurian death.

Perhaps I have treated a garden-variety error with unfair disdain in the sarcasm of the preceding paragraph. But the fallacy behind this common feeling of surprise, evoked by the eminently plausible hypothesis of *Caudipteryx* as a flightless bird, originates in a pervasive bias that renders much of the fascination of evolution inaccessible to millions of genuinely interested students and lovers of science.

The vigorous branching of life's tree, and not the accumulating valor of mythical marches to progress, lies behind the persistence and expansion of organic diversity in our tough and constantly stressful world. And if we do not grasp the fundamental nature of branching as the key to life's passage across the geological stage, we will never understand evolution aright. Tennyson caught the essence of life's challenge when he personified nature's unforgiving geological ways, as expressed in the fossil record of extinction:

> From scarped cliff and quarried stone
> She cries, "A thousand types are gone:
> I care for nothing, all shall go."

Yes, all shall eventually go, but some shall branch, thus permitting life to persist. To cite a sardonic song of self-mockery in leftist circles: "Trotsky got the ice pick [the weapon used by his murderers] . . . and so say all of us." And I do shudder even to contemplate the fate of the poor tailor. But the totality of life feints, dodges, and branches—and therefore, above all, hangs on in beauty and fascination. Psalm 1 invokes the right picture for a different purpose: "And he shall be like a tree planted by the rivers of water . . . his leaf also shall not wither; and whatsoever he doeth shall prosper." And Darwin employed the same image, both as metaphor and as literal topology this time, in the final words of the focal chapter in *The Origin of Species*—chapter 4, titled "Natural Selection," with its closing literary flourish on extinction and branching as the motors of evolution's tree and life's glory:

As buds give rise by growth to fresh buds and these, if vigorous, branch out and overtop on all sides many a feebler branch, so by generation I believe it has been with the great Tree of Life, which fills with its dead and broken branches the crust of the earth, and covers the surface with its ever branching and beautiful ramifications.

VII

Natural Worth

24

An Evolutionary Perspective on the Concept of Native Plants

AN IMPORTANT, BUT WIDELY UNAPPRECIATED, CONCEPT in evolutionary biology draws a clear and careful distinction between the historical origin and current utility of organic features. Feathers, for example, could not have originated for flight because five percent of a wing in an evolutionary intermediate between small running dinosaurs and birds could not have served any aerodynamic function (though feathers, derived from reptilian scales, provide important thermodynamic benefits right away). But feathers were later co-opted to keep birds aloft in a most exemplary fashion (see essay 23 for a detailed discussion of this subject). In a similar manner, our large brains could not have evolved to permit modern descendants to read and write, though these much later functions now define an important aspect of the modern utility of consciousness.

Similarly, the later use of an argument, often in a context foreign or even opposite to the intent of originators, must be separated from the validity and purposes of initial formulations. Thus, for example, Darwin's theory of natural selection cannot be diminished, either morally or scientifically, because later racists and warmongers perverted the concept of a "struggle for existence" into a rationale for genocide. However, we must admit a crucial difference between the origin and later use of a biological feature, and the origin and later use of an idea. The first, or anatomical, case involves no conscious action and cannot be submitted to any moral judgment. But ideas originate by explicit intent for overt purposes, and we have some ethical responsibility for the consequences of our deeds. An inventor may be fully exonerated for true perversions of his purposes (Hitler's use of Darwin), but unfair extensions consistent with the logic of original motivations do entail some moral demerit (academic racists of the nineteenth century did not envision or intend the Holocaust, but some of their ideas did fuel the "final solution").

I want to examine the concept of "native plants" within this framework—for this complex concept includes a remarkable mixture of sound biology, invalid ideas, false extensions, ethical implications, and political usages both intended and unanticipated. Clearly, Nazi ideologues provided the most chilling uses (see articles of J. Wolschke-Bulmahn and G. Groening, cited in the bibliography to this essay, for example). In advocating native plants along the *Reichsautobahnen,* Nazi architects of the Reich's motor highways explicitly compared their proposed restriction to Aryan purification of people. By this procedure, Reinhold Tüxen hoped "to cleanse the German landscape of unharmonious foreign substance." In 1942 a team of German botanists made the analogy explicit in calling for the extirpation of *Impatiens parviflora,* a supposed interloper: "As with the fight against Bolshevism, our entire Occidental culture is at stake, so with the fight against this Mongolian invader, an essential element of this culture, namely the beauty of our home forest, is at stake."

At the other extreme of kindly romanticism, gentle arguments for native plants have stressed their natural "rightness" in maximally harmonious integration of organism and environment, a modern invocation of the old doctrine of *genius loci.* Consider this statement, for example, from a 1982 book by C. A. Smyser and others, titled *Nature's Designs: A Practical Guide to Natural Landscaping:*

> Man makes mistakes; nature doesn't. Plants growing in their natural habitat look fit and therefore beautiful. In any undeveloped area you can find a miraculously appropriate assortment of plants, each one contributing to the overall appearance of a unified natural landscape. The balance is preserved by the ecological conditions of the place, and the introduction of an alien plant could destroy this balance.

In other words, these authors claim, evolution has produced a harmony that contrived gardens can only defy.

Or consider these words from former President Clinton (though I doubt that he wrote the text personally), in an April 26, 1994, memorandum "for the heads of executive departments and agencies" on "environmentally and economically beneficial practices on federal landscaped grounds": "The use of native plants not only protects our natural heritage and provides wildlife habitat, but also can reduce fertilizer, pesticide, and irrigation demands and their associated costs because native plants are suited to the local environment and climate."

This general argument boasts a long pedigree, as well illustrated in Jens Jensen's remark in *Our Native Landscape,* published in his famous 1939 book, *Siftings:*

> It is often remarked, "native plants are coarse." How humiliating to hear an American speak so of plants with which the Great Master has decorated his land! To me no plant is more refined than that which belongs. There is no comparison between native plants and those imported from foreign shores which are, and shall always remain so, novelties.

Yet the slippery slope from this benevolent version toward dangerous *Volkist* nationalism may be discerned, and quite dramatically, in another statement from the same Jens Jensen—this time published in a German magazine in 1937.

> The gardens that I created myself shall . . . be in harmony with their landscape environment and the racial characteristics of its inhabitants. They shall express the spirit of America and therefore shall be free of foreign character as far as possible. The Latin and

> the Oriental crept and creeps more and more over our land, coming from the South, which is settled by Latin people, and also from other centers of mixed masses of immigrants. The Germanic character of our cities and settlements was overgrown. . . . Latin spirit has spoiled a lot and still spoils things every day.

How tenuous the space between *genius loci* (and respect for all the other spirits in their proper places as well)—and "my *locus* is best, while others must be uprooted, either as threats or as unredeemable inferiors." How easy the fallacious transition between a biological argument and a political campaign.

When biologically based claims engender such a range of political usages (however dubious, and however unfairly), we encounter a special responsibility to examine the scientific validity of the underlying arguments, if only to acquire weapons to guard against usages that properly inspire our ethical opposition. Any claim for preferring native plants must rest upon some construction of evolutionary theory—a difficult proposition to defend (as I shall argue) because evolution has been so widely misconstrued and, when properly understood, so difficult to utilize for the defense of intrinsic native superiority. This difficulty did not exist in pre-Darwinian creationist biology, because the old paradigm of "natural theology" held that God displayed both his existence and his attributes of benevolence and omniscience in the optimal design of organic form and the maximal harmony of local ecosystems. Native must therefore be right and best because God made each creature to dwell in its proper place.

But evolutionary theory fractured this equation of existence with optimality by introducing the revolutionary idea that all anatomies and interactions arise as transient products of complex history, not as created optimalities. Evolutionary defenses of native plants rest upon two quite distinct aspects of the revolutionary paradigm that Darwin introduced. (I shall argue that neither provides an unambiguous rationale, and that many defenders of native plants have mixed up these two distinct arguments, therefore rendering their defense incoherent.)

The Functional Argument Based on Adaptation

Popular impression regards Darwin's principle of natural selection as an optimizing force, leading to the same end of local perfection that God had supplied directly in older views of natural theology. If natural selection works for the best forms and most balanced interactions that could possibly exist in any one spot, then native must be best—for native has been honed to optimality in the refiner's fire of Darwinian competition. (In critiquing horticulturists for this misuse of natural selection, I am not singling out any group for an unusual or particularly naïve misinterpretation. This misreading of natural selection has become pervasive in our culture, and also records a primary fallacy of much professional thinking as well.)

In *Siftings,* Jens Jensen expressed this common viewpoint with particular force:

> There are trees that belong to low grounds and those that have adapted themselves to highlands. They always thrive best amid the conditions they have chosen for themselves through many years of selection and elimination. They tell us that they love to grow here, and only here will they speak in their fullest measure. . . . I have often marvelled at the friendliness of certain plants for each other, which, through thousands of years of selection, have lived in harmonious relation.

But natural selection does not preferentially generate plants that humans happen to regard as attractive. Nor do natural systems always yield rich associations of numerous, well-balanced species. Plants that we label "weeds" will dominate in many circumstances, however transiently (where "transient" can mean more than a human lifetime on the natural time scales of botanical succession). Such weeds cannot be called less "native"—in the sense of evolving indigenously—than plants of much more restricted habitat and geography. Moreover, weeds often grow as virtual monocultures, choking out more-diverse assemblages that human intervention could maintain. C. A. Smyser, in his 1982 book previously cited in this article, admits the point, but does not seem to grasp the logical threat thus entailed against an equation of "natural" with "right" or "preferable." Smyser states:

> You may have heard of homeowners who simply stopped mowing or weeding and now call their landscapes "natural." The truth is that these so-called no-work, natural gardens will be long dominated by exotic weed species, most of which are pests and look downright ugly. Eventually, in fifty to one hundred years, native plants will establish themselves and begin to create an attractive environment.

But not all "weed" species can be called "exotic" in the sense of being artificially imported from other geographic areas. Weeds must often be classified as indigenous, even though their geographic ranges tend to be large, and their means of natural transport efficient and well developed.

The evolutionary fallacy in equating native with optimally adapted can be explicated most clearly by specifying the central theme of natural selection as a causal principle. As Darwin recognized, and stated so forcefully, natural selection generates adaptation to changing local environments—and nothing else. The Darwinian mechanism includes no concept of general progress or universal betterment. The "struggle for existence" can only yield local appropriateness. Moreover, and even more important for debates about superiority of native plants, natural selection is only a "better than" principle, not an optimizing device. That is, natural selection can only transcend the local standard and cannot work toward universal "improvement"—for once a species prevails over others at a location, no pressure of natural selection need arise to promote further adaptation. (Competition within species will continue to eliminate truly defective individuals and may promote some refinement by selection of fortuitous variants with still more advantageous traits, but the great majority of successful species are highly stable in form and behavior over long periods of geological time.)

For this reason, many native plants, evolved by natural selection as adaptive to their regions, fare poorly against introduced species that never experienced the local habitat. If natural selection produced optimality, this very common situation could never arise, for native forms would prevail in any competition against intruders. But most Australian marsupials succumb to placentals imported from other continents, despite tens of millions of years of isolation—surely more than enough time for Australian natives to attain irreplaceable incumbency, if natural selection generated optimality. And *Homo sapiens,* after arising in Africa, seems able to prevail in any exotic bit of real estate, almost anywhere in the world!

Thus the primary rationale so often stated for preferring native plants—that, as locally evolved, such species must be best adapted—cannot be sustained. I strongly suspect that a large majority of well-adapted natives could be supplanted by some exotic form that has never experienced the immediate habitat. In Darwinian terms, this exotic would be better adapted than the native—though we may well, on defensible aesthetic or even ethical grounds, prefer the natives (for nature's factuality can never enjoin our moral decisions).

We should, I think, grant the legitimacy of only one limited point from evolutionary biology on the subject of adaptation in native plants. At least we *do know* that well-established natives must be adequately adapted, and we can observe their empirical balances with other local species. We *cannot know* what an exotic species will do—and we can all cite numerous, and tragic, stories of exotics imported for a restricted and benevolent reason that then grew like kudzu to everyone's disgust and detriment. We also know that natives grow appropriately—though not necessarily optimally—in their environment, while exotics may not fit without massive human "reconstruction" of habitat, an intervention that many ecologically minded people deplore. I confess that nothing strikes me as more vulgar or inappropriate than a bright green lawn in front of a mansion in the Arizona desert, an artificial construction sucking up precious water that already must be imported from elsewhere. A preference for natives does foster humility and does counteract human arrogance—for such preference does provide the only sure protection against our profound ignorance of potential consequences when we import exotics. But the standard argument—that natives should be preferred as best adapted—must be firmly rejected as simply false within Darwinian theory.

The Geographic Argument Based on Appropriate Place

This second argument, harder to formulate and less clearly linked to a Darwinian postulate, somehow seems even more deeply embedded (as a fallacy) into conventional arguments for preferring native plants. This argument holds that plants occupy their natural geographic ranges for reasons of maximal appropriateness. Why, after all, would a plant live only in a particular region of five hundred square kilometers unless this domain constituted its "natural" home—the place where this species, and no other, fits best? Smyser, for example, writes: "In any area there is always a type of vegetation that would exist without being planted or protected. This native vegetation consists of specific groups of plants that adapted to specific environmental conditions." But the deepest principle of evolutionary biology—the construction of all current biological phenomena as outcomes of contingent history, rather than optimally manufactured situations—exposes this belief as nonsense.

Organisms do not necessarily, or even generally, inhabit the geographic area best suited to their attributes. Since organisms (and their areas of habitation) originate as products of a history laced with chaos, contingency, and genuine randomness, current patterns (although obviously workable) will rarely express anything close to an optimum, or even a "best possible on this earth now"—

whereas the pre-Darwinian theory of natural theology, with direct creation of best solutions, and no appreciable history thereafter (or ever), could have validated an idea of native as best. Consequently, although native plants must be adequate for their environments, evolutionary theory grants us no license for viewing natives as the best-adapted inhabitants conceivable, or even as the best available among all species on the planet.

An enormous literature in evolutionary biology documents the various, and often peculiar, mechanisms whereby organisms achieve fortuitous transport as species spread to regions beyond their initial point of origin. Darwin himself took particular interest in this subject. During the 1850s, in the years just before publication of the *Origin of Species* in 1859, he wrote several papers on the survival of seeds in salt water. (How long could seeds float without sinking? Would they still germinate after such a long bath?) He determined that many seeds could survive long enough to reach distant continents by floating across oceans—and that patterns of colonization therefore reflected historical accidents of available pathways, and not a set of optimal environments.

Darwin then studied a large range of "rarely efficient" means of transport beyond simple carriage by waves—natural rafts of intertwined logs (often found floating in the ocean hundreds of miles from river mouths), mud caked on birds' feet, residence in the gut of birds with later passage in feces. In his usually thorough and obsessive way, Darwin assiduously collected information and found more than enough means of fortuitous transport. He wrote to a sailor who had been shipwrecked on Kerguelen Island to find out if he remembered any seeds or plants growing from driftwood on the beach. He asked an inhabitant of Hudson Bay if seeds might be carried on ice floes. He studied the contents of ducks' stomachs. He received, in the mail and with delight, a pair of partridges' feet caked with mud; he rooted through bird droppings. He even followed a suggestion of his eight-year-old son that they float a dead and well-fed bird. Darwin wrote, in a letter, that "a pigeon has floated for thirty days in salt water with seeds in crop and they have grown splendidly." In the end, Darwin found more than enough mechanisms to move viable seeds.

"Natives," in short, are the species that happened to find their way (or evolve *in situ*), not the best conceivable species for a spot. As in my first argument about adaptation, the proof that current incumbency as "native" does not imply superiority against potential competitors exists in abundance among hundreds of imported interlopers that have displaced natives throughout the world—eucalypts in California, kudzu in the American Southeast, rabbits and other placental mammals in Australia, and humans just about everywhere.

"Native" species can only be defined as those forms that first happened to gain and keep a footing. We rightly decry the elitist and parochial claims of American northeastern WASPs to the title of native. But, however "politically incorrect" the point, the fashionable status of "Indians" (so called by Columbus's error) as "Native Americans" makes just as little sense in biological terms. "Native Americans" arrived in a geological yesterday, some twenty thousand years ago (perhaps a bit earlier), probably on the geographic fortuity of a pathway across the Bering Strait, perhaps by some equivalent of *Kon-Tiki*. The first Americans were no more intrinsically suited than any other people to New World real estate. They just happened to arrive first.

In this context, the only conceivable rationale for the moral or practical superiority of "natives" (read first-comers) must lie in a romanticized notion that old inhabitants learn to live in ecological harmony with surroundings, while later interlopers tend to become exploiters. But this notion, however popular among "new agers," must be dismissed as romantic drivel. People are people, whatever their technological status; some learn to live harmoniously for their own good—and others don't, to their own detriment or destruction. Preindustrial people have been just as rapacious (though not so quickly, perhaps, for lack of tools) as the worst modern clearcutters. The Maori people of New Zealand wiped out a rich fauna of some twenty moa (giant bird) species within a few hundred years. The "native" Polynesians of Easter Island wiped out everything edible or usable (and, in the end, could find no more logs to build boats or to raise their famous statues), and finally turned to self-destruction.

In summary of my entire argument from evolutionary theory, "native" plants cannot be deemed biologically best in any justifiable way. "Natives" are only the plants that happened to arrive first and be able to flourish (the evolutionary argument based on geography and history)—while their capacity for flourishing only indicates a status as "better than" others available, not as optimal or globally "best suited" (the evolutionary argument based on adaptation and natural selection).

Speaking biologically, the only general defense that I can concoct for native plants lies in protection thus afforded against our overweening arrogance. At least we know what natives will do in an unchanged environment—for they have generally inhabited an area for a long time, and have therefore stabilized and adapted. We never know for sure what an imported interloper will do—and our consciously planted exotics have "escaped" to disastrous spread and extirpation of natives (the kudzu model) as often as they have supplied the intended horticultural or agricultural benefits.

As a final ethical point (and I raise this issue as a concerned human being, not as a scientist, for my profession can offer no direct moral insight), I do understand the appeal of the ethical argument that we should leave nature alone and preserve as much as we can of the life that existed and developed before our very recent geological appearance. Like all evolutionary biologists, I treasure nature's bounteous diversity of species (the thought of half a million described species of beetles—and so many more, yet undescribed—fills me with an awe I can only call reverent). And I do understand that much of this variety lies in geographic diversity (different organisms evolved in similar habitats in many places on our planet, as a result of limits and accidents of access). I would certainly be horrified to watch the botanical equivalent of McDonald's uniform architecture and cuisine wiping out every local diner in America. Cherishing native plants does allow us to defend and preserve a maximal amount of local variety.

But we must also acknowledge that the argument for strict "nativism" has an ethical downside inherent in the notion that "natural" must be right and best—for such an attitude easily slides from the philistinism of denying any role to human intelligence and good taste, thence to the foolish romanticism of viewing all that humans might accomplish in nature as "bad" (and how then must we judge Olmsted's Central Park), and even, in an ugly perversion—but realized in our time by Nazi invocation of nativist doctrine—to the claim that my "native" is best, and yours fit only for extirpation.

The best defense against all these misuses, from mild to virulent, lies in a profoundly humanistic notion as old as Plato—one that we often advance in sheepish apology, but should rather honor and cherish: the idea that "art" should be defined as the caring, tasteful, and intelligent *modification* of nature for respectful human utility. If we can practice this art in partnership with nature, rather than by exploitation (and if we also set aside large areas for rigidly minimal disturbance, so that we never forget, and may continue to enjoy, what nature accomplished during nearly all of her history without us), then we may achieve optimal balance.

People of goodwill may differ on the best botanical way to capture the "spirit of democracy"—from one end of maximal "respect" for nature by using only her unadorned and locally indigenous ("native") products, to the other of maximal use of human intelligence and aesthetic feeling in sensitive and "respectful" mixing of natives and exotics, just as our human populations have so benefited from imported diversity. Jens Jensen extolled the first view:

> When we are willing to give each plant a chance fully to develop its beauty, so as to give us all it possesses without any interference,

> then, and only then, shall we enjoy ideal landscapes made by man. And is not this the true spirit of democracy? Can a democrat cripple and misuse a plant for the sake of show and pretense?

But is all cultivation—hedgerows? topiary?—crippling and misuse? The loaded nature of ethical language lies exposed in Jensen's false claim, quoted just above. Let us consider, in closing, another and opposite definition of democracy that can certainly claim the sanction of ancient usage. In a 1992 article, J. Wolschke-Bulmahn and G. Groening cite a stirring and poignant argument made by Rudolf Borchardt, a Jew who later died by Nazi hands, against the nativist doctrine as perverted by Nazi horticulturists:

> If this kind of garden owning barbarian became the rule, then neither a gillyflower nor a rosemary, neither a peach-tree nor a myrtle sapling nor a tea-rose would ever have crossed the Alps. Gardens connect people, times and latitudes. If these barbarians ruled, the great historic process of acclimatization would never have begun and today we would horticulturally still subsist on acorns. . . . The garden of humanity is a huge democracy.

I cannot state a preference in this wide sweep of opinions, from pure hands-off romanticism to thorough overmanagement (though I trust that most of us would condemn both extremes). Absolute answers to such ethical and aesthetic questions do not exist in any case. But we will not achieve clarity on this issue if we advocate a knee-jerk equation of "native" with morally best, and fail to recognize the ethical power of a contrary view, supporting a sensitive cultivation of all plants, whatever their geographic origin, that can enhance nature and bring both delight and utility to humans. Do we become more "democratic" when we respect organisms only in their natural places (how then, could any non-African human respect himself), or shall we persevere in the great experiment of harmonious and mutually reinforcing geographic proximity—as the prophet Isaiah sought in his wondrous vision of a place where the wolf might dwell with the lamb and such nonnatives as the calf and the lion might feed together—where "they shall not hurt nor destroy in all my holy mountain."

Bibliography

Clinton, W. J. 1994. *Memorandum for the heads of executive departments and agencies.* Office of the Press Secretary, 26 April 1994.

Druse, K., and M. Roach. 1994. *The Natural Habitat Garden.* New York: Clarkson Potter.

Gould, S. J. 1991. Exaptation: A crucial tool for an evolutionary psychology. *Journal of Social Issues* 47(3):43–65.

Gould, S. J., and R. C. Lewontin. 1979. The spandrels of San Marco and the Panglossian paradigm: A critique of the adaptationist programme. *Proceedings of the Royal Society of London B* 205:581–98.

Groening, G., and J. Wolschke-Bulmahn. 1992. Some notes on the mania for native plants in Germany. *Landscape Journal* 11(2):116–26.

Jensen, J. 1956. *Siftings,* the major portion of *The Clearing and collected writings.* Chicago: Ralph Fletcher Seymour.

Paley, W. 1802. *Natural Theology.* London: R. Faulder.

Smyser, C. A. 1982. *Nature's Design: A Practical Guide to Natural Landscaping.* Emmaus, Pa.: Rodale Press.

Wolschke-Bulmahn, J. 1995. Political landscapes and technology: Nazi Germany and the landscape design of the *Reichsautobahnen* (Reich Motor Highways). Selected CELA Annual Conference Papers, vol. 7. Nature and Technology. Iowa State University, 9–12 September 1995.

Wolschke-Bulmahn, J., and G. Groening. 1992. The ideology of the nature garden: Nationalistic trends in garden design in Germany during the early twentieth century. *Journal of Garden History* 12(1):73–80.

25

Age-Old Fallacies of Thinking and Stinking

WE SHUDDER AT THE THOUGHT OF REPEATING THE INItial sins of our species. Thus, Hamlet's uncle bewails his act of fratricide by recalling Cain's slaying of Abel:

> O! my offense is rank, it smells to heaven;
> It hath the primal eldest curse upon 't;
> A brother's murder!

Such metaphors of unsavory odor seem especially powerful because our sense of smell lies so deep in our evolutionary construction, yet remains (perhaps for this reason) so undervalued and often unmentioned in our culture. A later seventeenth-century English writer recognized this potency and particularly warned his readers against using olfactory metaphors because common people will take them literally:

> Metaphorical expression did often proceed into a literal construction; but was fraudulent. . . . How dangerous it is in sensible things to use metaphorical expressions unto the people, and what absurd conceits they will swallow in their literals.

This quotation appears in the 1646 work of Sir Thomas Browne: *Pseudodoxia Epidemica: or, Enquiries into Very Many Received Tenents* [*sic*], *and Commonly Presumed Truths.* Browne, a physician from Norwich, remains better known for his wonderful and still widely read work of 1642, the part autobiographical, part philosophical, and part whimsical *Religio Medici,* or "Religion of a Doctor." The *Pseudodoxia Epidemica* (his Latinized title for a plethora of false truths) became the granddaddy of a most honorable genre still vigorously pursued—exposés of common errors and popular ignorance, particularly the false beliefs most likely to cause social harm.

I cited Browne's statement from the one chapter (among more than a hundred) sure to send shudders down the spine of modern readers—his debunking of the common belief "that Jews stink." Browne, although almost maximally philo-Semitic by the standards of his century, was not free of all prejudicial feelings against Jews. He attributed the origin of the canard about Jewish malodor—hence, my earlier quotation—to a falsely literal reading of a metaphor legitimately applied (or so he thought) to the descendants of people who had advocated the crucifixion of Jesus. Browne wrote: "Now the ground that begat or propagated this assertion, might be the distasteful averseness of the Christian from the Jew, upon the villainy of that fact, which made them abominable and stink in the nostrils of all men."

As a rationale for debunking a compendium of common errors, Browne correctly notes that false beliefs arise from incorrect theories about nature and therefore serve as active impediments to knowledge, not just as laughable signs of primitivity: "To purchase a clear and warrantable body of truth, we must forget and part with much we know." Moreover, Browne notes, truth is hard to ascertain and ignorance is far more common than accuracy. Writing in the mid-seventeenth century, Browne uses "America" as a metaphor for domains of uncharted ignorance, and he bewails our failure to use good tools of reason as guides through this *terra incognita:* "We find no open tract . . . in this labyrinth; but are oft-times fain to wander in the America and untravelled parts of truth."

The *Pseudodoxia Epidemica,* Browne's peregrination through the maze of human ignorance, contains 113 chapters gathered into seven books on such general topics as mineral and vegetable bodies, animals, humans, Bible tales, and geographical and historical myths. Browne debunks quite an array of common opinions, including claims that elephants have no joints, that the legs of badgers are shorter on one side than the other, and that ostriches can digest iron.

As an example of his style of argument, consider book 3, chapter 4: "That a bever [sic] to escape the hunter, bites off his testicles or stones"—a harsh tactic that, according to legend, either distracts the pursuer or persuades him to settle for a meal smaller than an entire body. Browne labels this belief as "a tenet very ancient; and hath had thereby advantages of propagation. . . . The Egyptians also failed in the ground of their hieroglyphick, when they expressed the punishment of adultery by the bever depriving himself of his testicles, which was amongst them the penalty of such incontinency."

Browne prided himself on using a mixture of reason and observation to achieve his debunking. He begins by trying to identify the source of error—in this case a false etymological inference from the beaver's Latin name, *Castor,* which does not share the same root with "castration" (as the legend had assumed), but derives ultimately from a Sanskrit world for "musk"; and an incorrect interpretation of purposeful mutilation from the internal position, and therefore near invisibility, of the beaver's testicles. He then cites the factual evidence of intact males, and the reasoned argument that a beaver couldn't reach his own testicles even if he wanted to bite them off (and thus, cleverly, the source of common error—the external invisibility of the testicles—becomes the proof of falsity!).

> The testicles properly so called, are of a lesser magnitude, and seated inwardly upon the loins: and therefore it were not only a fruitless attempt, but impossible act, to eunuchate or castrate themselves: and might be an hazardous practice of art, if at all attempted by others.

Book 7, chapter 2 debunks the legend "that a man hath one rib less than a woman"—"a common conceit derived from the history of Genesis, wherein it stands delivered, that Eve was framed out of a rib of Adam." (I regret to report that this bit of nonsense still commands some support. I recently appeared on a nationally televised call-in show for high school students, where one young woman, a creationist, cited this "well-known fact" as proof of the Bible's inerrancy and evolution's falsity.) Again, Browne opts for a mixture of logic and observation in stating: "this will not consist with reason or inspection." A simple count on skeletons affirms equality of number between sexes (Browne, after all, maintained his "day job" as a physician and should have known). Moreover, reason provides no argument for assuming that Adam's single loss would be propagated to future members of his sex:

> Although we concede there wanted one rib in the sceleton of Adam, yet were it repugnant unto reason and common observation, that his posterity should want the same [in the old meaning of "want" as "lack"]. For we observe that mutilations are not transmitted from father unto son; the blind begetting such as can see, men with one eye children with two, and cripples mutilate in their own persons do come out perfect in their generations.

Book 4, chapter 10—"That Jews Stink"—is one of the longest, and clearly held special importance for Dr. Browne. He invokes more-elaborate arguments, but follows the same procedure used to dispel less noxious myths—citation of contravening facts interlaced with more general support from logic and reason.

Browne begins with a statement of the fallacy: "That Jews stink naturally, that is, that in their race and nation there is an evil savor, is a received opinion." Browne then allows that species may have distinctive odors, and that individuals surely do: "Aristotle says no animal smells sweet save the pard. We confess that beside the smell of the species, there may be individual odors, and every man may have a proper and peculiar savor; which although not perceptible unto man, who hath this sense but weak, is yet sensible unto dogs, who hereby can single out their masters in the dark."

In principle, then, discrete groups of humans might carry distinctive odors, but reason and observation permit no such attribution to Jews as a group: "That an unsavory odor is gentilitous or national unto the Jews, if rightly understood, we cannot well concede, nor will the information of Reason or Sense induce it."

On factual grounds, Browne asserts, direct experience has provided no evidence for this noxious legend: "This offensive odor is no way discoverable in their Synagogues where many are, and by reason of their number could not be concealed: nor is the same discernible in commerce or conversation with such as are cleanly in apparel, and decent in their houses." The "test case" of Jewish converts to Christianity proves the point, for even the worst bigots do not accuse such people of smelling bad: "Unto converted Jews who are of the same seed, no man imputeth this unsavory odor; as though aromatized by their conversion, they lost their scent with their religion, and smelt no longer." If people of

Jewish lineage could be identified by smell, the Inquisition would greatly benefit from a sure-fire guide for identifying insincere converts: "There are at present many thousand Jews in Spain . . . and some dispensed withal even to the degree of Priesthood; it is a matter very considerable, and could they be smelled out, would much advantage, not only the Church of Christ, but also the Coffers of Princes."

Turning to arguments from reason, foul odors might arise among groups of people from unhealthy habits of diet or hygiene. But Jewish dietary laws guarantee moderation and good sense, while drinking habits tend to abstemiousness—"seldom offending in ebriety or excess of drink, nor erring in gulosity or superfluity of meats; whereby they prevent indigestion and crudities, and consequently putrescence of humors."

If no reason can therefore be found in Jewish habits of life, the only conceivable rationale for a noxious racial odor would lie in a divine "curse derived upon them by Christ . . . as a badge or brand of a generation that crucified their Salvator." But Browne rejects this proposal even more forcefully as a "conceit without all warrant; and an easie way to take off dispute in what point of obscurity soever." The invocation of miraculous agency, when no natural explanation can be found, is a coward's or lazy man's escape from failure. (Browne does not object to heavenly intervention for truly great events like Noah's flood or the parting of the Red Sea, but a reliance upon miracles for small items, like the putative racial odor of unfairly stigmatized people, makes a mockery of divine grandeur. Browne then heaps similar ridicule on the legend that Ireland has no snakes because Saint Patrick cast them out with his rod. Such inappropriate claims for a myriad of minor miracles only stifles discussion about the nature of phenomena and the workings of genuine causes.)

But Browne then caps his case against the proposition "that Jews stink" with an even stronger argument based on reason. The entire subject, he argues, makes no sense because the category in question—the Jewish people—does not represent the kind of entity that could bear such properties as a distinctive national odor.

Among the major fallacies of human reason, such "category mistakes" are especially common in the identification of groups and the definition of their characters—problems of special concern to taxonomists like myself. Much of Browne's text is archaic, and strangely fascinating, therefore, as a kind of conceptual fossil. But his struggle with errors of categories in debunking the proposition "that Jews stink" interleaves a layer of modern relevance, and uncovers a different kind of reason for contemporary interest in the arguments of *Pseudodoxia Epidemica*.

Browne begins by noting that traits of individuals can't automatically be extended to properties of groups. We do not doubt that individuals have distinctive odors, but groups might span the full range of individual differences, and thereby fail to maintain any special identity. What kind of group might therefore qualify as a good candidate for such distinctive properties?

Browne argues that such a group would have to be tightly defined, either by strict criteria of genealogy (so that members might share properties by heredity of unique descent) or by common habits and modes of life not followed by other people (but Browne had already shown that Jewish lifestyles of moderation and hygiene disproved any claim for unsavory national odor).

Browne then clinches his case by arguing that the Jewish people do not represent a strict genealogical group. Jews have been dispersed throughout the world, reviled and despised, expelled and excluded. Many subgroups have been lost by assimilation, others diluted by extensive intermarriage. Most nations, in fact, are strongly commingled and therefore do not represent discrete groups by genealogical definition; this common tendency has been exaggerated among the Jewish people. Jews are not a distinct hereditary group, and therefore cannot maintain such properties as a national odor:

> There will be found no easie assurance to fasten a material or temperamental propriety upon any nation; . . . much more will it be difficult to make out this affection in the Jews; whose race however pretended to be pure, must needs have suffered inseparable commixtures with nations of all sorts. . . . It being therefore acknowledged that some [Jews] are lost, evident that others are mixed, and not assured that any are distinct, it will be hard to establish this quality [of national odor] upon the Jews.

In many years of pondering over fallacious theories of biological determinism, and noting their extraordinary persistence and tendency to reemerge after presumed extirpation, I have been struck by a property that I call "surrogacy." Specific arguments raise a definite charge against a particular group—that Jews stink, that Irishmen drink, that women love mink, that Africans can't think—but each specific claim acts as a surrogate for any other. The general form of argument remains perennially the same, always permeated by identical fallacies over the centuries. Scratch the argument that women, by their biological nature, cannot be effective as heads of state and you will uncover the same structure of false inference underlying someone else's claim that African Americans will never form a high percentage of the pool of Ph.D. candidates.

Thus, Browne's old refutation of the myth "that Jews stink" continues to be relevant for our modern struggle, since the form of his argument applies to our current devaluings of people for supposedly inborn and unalterable defects of intelligence or moral vision. Fortunately (since I belong to the group), Jews are not taking much heat these days (though I need hardly mention the searing events of my parents' generation to remind everyone that current acceptance should breed no complacency). Following Browne's strategy, any particular version of this general claim can be debunked with a mixture of factual citation and logical argument. I shall not go through the full exercise here, lest this essay become a book. But I do wish to emphasize that Browne's crowning point in refuting the legend "that Jews stink"—his explication of category mistakes in defining Jews as a biological group—also undermines the modern myth of black intellectual inferiority, from Jensen and Shockley in the 1960s to Murray and Herrnstein and *The Bell Curve* in the 1990s.

The African American population of the United States today cannot be identified as a genealogical unit in the same sense that Browne's Jews lacked inclusive definition by descent. As a legacy of our ugly history of racism, anyone with a visually evident component of African ancestry belongs to the category of "black" even though many persons so designated can trace their roots to substantial, often majoritarian, Caucasian sources as well. (An old "trick" question for baseball aficionados asks: "What Italian American player hit more than forty home runs for the Brooklyn Dodgers in 1953"? The answer is "Roy Campanella," who had a Caucasian Italian father and a black mother, but who, by our social conventions, became identified as black.)

(As a footnote on the theme of surrogacy, explanations of the same category mistake for blacks and Jews often follow the same prejudicial form of blaming the victim. Browne, though generally and refreshingly free of anti-Jewish bias, cites a particularly ugly argument in explaining high rates of miscegenation between Jews and Christians—the supposed lasciviousness of Jewish women and their preference for blond Christian men over swarthy and unattractive Jews. Browne writes: "Nor are fornications infrequent between them both [Jewish women and Christian men]; there commonly passing opinions of invitement, that their women desire copulation with them rather than their own nation, and affect Christian carnality above circumcised venery." American racists often made the same claim during slavery days—a particularly disgraceful lie in this case, for the argument functions primarily to excuse rapists by blaming the truly powerless. For example, Louis Agassiz wrote in 1863: "As soon as the sexual desires are awakening in the young men of the South, they find it easy to gratify themselves by the readiness with which they are met by colored [half-breed] house servants. . . . This blunts his better instincts in that direction and leads him gradually to seek more spicy partners, as I have heard the full blacks called by fast young men.")

Obviously, we cannot make a coherent claim for "blacks" being innately anything by heredity if the people so categorized do not form a distinctive genealogical grouping. But the category mistake goes far, far deeper than dilution by extensive intermixture with other populations. The most exciting and still emerging discovery in modern paleoanthroplogy and human genetics will force us to rethink the entire question of human categories in a radical way. We shall be compelled to recognize that "African black" cannot rank as a racial group with such conventional populations as "Native American," "European Caucasian," or "East Asian," but must be viewed as something more inclusive than all the others combined, not really definable as a discrete group, and therefore not available for such canards as "Africans are less intelligent" or "Africans sure can play basketball."

The past decade of anthropology has featured a lively debate about the origin of the only living human species, *Homo sapiens.* Did our species emerge separately on three continents (Africa, Europe, and Asia) from precursor populations of *Homo erectus* inhabiting all these areas—the so-called multiregionalist view? Or did *Homo sapiens* arise in one place, probably Africa, from just one of these *Homo erectus* populations, and then spread out later to cover the globe—the so-called out-of-Africa view?

The tides of argument have swung back and forth, but recent evidence seems to be cascading rapidly toward Out of Africa. As more and more genes are sequenced and analyzed for their variation among human racial groups, and as we reconstruct genealogical trees based upon these genetic differences, the same strong signal and pattern seem to be emerging: *Homo sapiens* arose in Africa; the migration into the rest of the world did not begin until about 100,000 years ago.

In other words, *all* non-African racial diversity—whites, yellows, reds, everyone from the Hopi to the Norwegians to the Fijians—may not be much older than one hundred thousand years. By contrast, *Homo sapiens* has lived in Africa for a far longer time. Consequently, since genetic diversity roughly correlates with time available for evolutionary change, genetic variety among Africans alone exceeds the sum total of genetic diversity for everyone else in the rest of the world combined! How, therefore, can we lump "African blacks" together as a single group, and imbue them with traits either favorable or unfavorable, when they represent more evolutionary space and more genetic variety than we find in all non-African people in all the rest of the world? Africa includes most of humanity by any proper genealogical definition; all the rest of us occupy a branch within the African tree. This non-African branch has surely flourished, but can never be topologically more than a subsection within an African structure.

We will need many years, and much pondering, to assimilate the theoretical, conceptual, and iconographic implications of this startling reorientation in our views about the nature and meaning of human diversity. For starters, though, I suggest that we finally abandon such senseless statements as "African blacks have more rhythm, less intelligence, greater athleticism." Such claims, apart from their social perniciousness, have no meaning if Africans cannot be construed as a coherent group because they represent more diversity than all the rest of the world put together.

Our greatest intellectual adventures often occur within ourselves—not in the restless search for new facts and new objects on the earth or in the stars, but from a need to expunge old prejudices and build new conceptual structures. No hunt can promise a sweeter reward, a more admirable goal, than the excitement of thoroughly revised understanding—the inward journey that thrills real scholars and scares the bejesus out of the rest of us. We need to make such an internal expedition in reconceptualizing our views of human genealogy and the meaning of evolutionary diversity. Thomas Browne—for we must award him the last word—praised such inward adventures above all other intellectual excitement. Interestingly, in the same passage, he also invoked Africa as a metaphor for unknown wonder. He could not have known the uncanny literal accuracy of his words (from *Religio Medici,* book 1, section 15):

> I could never content my contemplation with those general pieces of wonder, the flux and reflux of the sea, the increase of the Nile, the conversion of the [compass] needle to the north; and have studied to match and parallel those in the more obvious and neglected pieces of nature, which without further travel I can do in the cosmography of myself; we carry with us the wonders we seek without us: there is all Africa and her prodigies in us; we are that bold and adventurous piece of nature.

26

The Geometer of Race

INTERESTING STORIES OFTEN LIE ENCODED IN NAMES THAT seem either capricious or misconstrued. Why, for example, are political radicals called "left" and their conservative counterparts "right"? In most European legislatures, maximally distinguished members sat at the chairman's right, following a custom of courtesy as old as our prejudices for favoring the dominant hand of most people. (These biases run deep, extending well beyond can openers and writing desks to language itself, where *dextrous* comes from the Latin for "right," and *sinister* for "left.") Since these distinguished nobles and moguls tended to espouse conservative views, the right and left wings of the legislature came to define a geometry of political views.

Among such apparently capricious names in my own field of biology and evolution, none seems more curious, and none elicits more inquiry from correspondents and questioners after lectures, than the official designation of light-skinned people from Europe, western Asia, and North Africa as Caucasian. Why should this most common racial group of the Western world be named for a range of mountains in Russia? J. F. Blumenbach (1752–1840), the German naturalist who established the most influential of all racial classifications, invented this name in 1795, in the third edition of his seminal work, *De generis humani varietate nativa* (On the Natural Variety of Mankind). Blumenbach's original definition cites two reasons for his choice—the maximal beauty of people from this small region, and the probability that humans had first been created in this area. Blumenbach wrote:

> *Caucasian variety.* I have taken the name of this variety from Mount Caucasus, both because its neighborhood, and especially its southern slope, produces the most beautiful race of men, and because . . . in that region, if anywhere, we ought with the greatest probability to place the autochthones [original forms] of mankind.

Blumenbach, one of the greatest and most honored naturalists of the Enlightenment, spent his entire career as a professor at the University of Göttingen in Germany. He first presented his work *De generis humani varietate nativa* as a doctoral dissertation to the medical faculty of Göttingen in 1775, as the minutemen of Lexington and Concord began the American Revolution. He then republished the text for general distribution in 1776, as a fateful meeting in Philadelphia proclaimed our independence. The coincidence of three great documents in 1776—Jefferson's Declaration of Independence (on the politics of liberty), Adam Smith's *Wealth of Nations* (on the economics of individualism), and Blumenbach's treatise on racial classification (on the science of human diversity)—records the social ferment of these decades, and sets the wider context that makes Blumenbach's taxonomy, and his decision to call the European race Caucasian, so important for our history and current concerns.

The solution to big puzzles often hinges upon tiny curiosities, easy to miss or to pass over. I suggest that the key to understanding Blumenbach's classification, the foundation of so much that continues to influence and disturb us today, lies in a peculiar criterion that he invoked to name the European race Caucasian—the supposed maximal beauty of people from this region. Why, first of all, should anyone attach such importance to an evidently subjective assessment; and why, secondly, should an aesthetic criterion become the basis for a scientific judgment about place of origin? To answer these questions, we must turn to Blumenbach's original formulation of 1775, and then move to the changes he introduced in 1795, when Caucasians received their name.

Blumenbach's final taxonomy of 1795 divided all humans into five groups defined by both geography and appearance—in his order, the "Caucasian variety" for light-skinned people of Europe and adjacent areas; the "Mongolian variety" for inhabitants of eastern Asia, including China and Japan; the "Ethiopian variety" for dark-skinned people of Africa; the "American variety" for native populations of the New World; and the "Malay variety" for Polynesians and Melanesians of Pacific islands, and for the aborigines of Australia. But Blumenbach's original classification of 1775 recognized only the first four of these five, and united members of the "Malay variety" with the other people of Asia, later named "Mongolian" by Blumenbach.

We now encounter the paradox of Blumenbach's reputation as the inventor of modern racial classification. The original four-race system, as I shall illustrate in a moment, did not arise from Blumenbach's observations or theorizing, but only represents, as Blumenbach readily admits, the classification adopted and promoted by his guru Carolus Linnaeus in the founding document of taxonomy, the *Systema Naturae* of 1758. Therefore, the later addition of a "Malay variety" for some Pacific peoples originally included in a broader Asian group represents Blumenbach's only original contribution to racial classification. This change seems so minor. Why, then, do we credit Blumenbach, rather than Linnaeus, as the founder of racial classification? (One might prefer to say "discredit," as the enterprise does not, for good reason, enjoy high repute these days.) I wish to argue that Blumenbach's apparently small change actually records a theoretical shift that could not have been broader, or more portentous, in scope. This change has been missed or misconstrued in most commentaries because later scientists have not grasped the vital historical and philosophical principle that theories should be construed as models subject to visual representation, usually in clearly definable geometric terms.

By moving from the Linnaean four-race system to his own five-race scheme, Blumenbach radically changed the geometry of human order from a geographically based model without explicit ranking to a double hierarchy of worth, oddly based upon perceived beauty, and fanning out in two directions from a Caucasian ideal. The addition of a Malay category, as we shall see, provided the focus for this geometric reformulation—and Blumenbach's "minor" change between 1775 and 1795 therefore becomes the key to a conceptual transformation rather than a simple refinement of factual information within an old scheme.

Blumenbach idolized his teacher Linnaeus. On the first page of the 1795 edition of his racial classification, Blumenbach hailed "the immortal Linnaeus, a man quite created for investigating the characteristics of the works of nature, and arranging them in systematic order." Blumenbach also acknowledged Linnaeus as the source of his original fourfold classification: "I have followed Linnaeus in the number, but have defined my varieties by other boundaries" (1775 edition). Later, in adding his "Malay variety," Blumenbach identified his change as a departure from his old guru Linnaeus: "It became very clear that the Linnaean division of mankind could no longer be adhered to; for which reason I, in this little work, ceased like others to follow that illustrious man."

Linnaeus divided his species *Homo sapiens* into four varieties, defined primarily by geography and secondarily by appearance and supposed behavior (Linnaeus also included two other false or fanciful varieties within *Homo sapiens—ferus* for "wild boys" occasionally discovered in the woods and possibly raised by animals [most turned out to be retarded or mentally ill youngsters abandoned by their parents]; and *monstrosus* for travelers' tales of hairy people with tails, and other assorted fables.)

Linnaeus then presented the four major varieties arranged by geography and, interestingly, *not* in the ranked order favored by most Europeans in the racist tradition. He discussed, in sequence, *Americanus, Europeus, Asiaticus,* and *Afer* (or African). In so doing, Linnaeus presented nothing at all original, but merely mapped humans onto the four geographic regions of conventional cartography.

In the first line of his descriptions, Linnaeus characterized each group by three words for color, temperament, and posture in that order. Again, none of these three categories implies any ranking by worth. Moreover, Linnaeus again bowed to classical taxonomic theories rather than his own observations in making these decisions. For example, his separations by temperament (or "humor") record the ancient medical theory that a person's mood arises from a balance of four fluids (*humor,* in Latin, means "moisture")—blood, phlegm, choler (or yellow bile), and melancholy (or black bile). One of the four substances may dominate, and a person therefore becomes sanguine (the cheerful realm of blood), phlegmatic (sluggish), choleric (prone to anger), or melancholic (sad). Four geographic regions, four humors, four races.

For the American variety, Linnaeus wrote *"rufus, cholericus, rectus"* (red, choleric, upright); for the European, *"albus, sanguineus, torosus"* (white, sanguine, muscular); for the Asian, *"luridus, melancholicus, rigidus"* (pale yellow, melancholy, stiff); and for the African, *"niger, phlegmaticus, laxus"* (black, phlegmatic, relaxed).

I don't mean to deny that Linnaeus held conventional beliefs about the superiority of his own European variety over all others. He surely maintained the almost universal racism of his time—and being sanguine and muscular as a

European surely sounds better than being melancholy and stiff as an Asian. Moreover, Linnaeus included a more overtly racist label in his last line of description for each variety. Here he tries to epitomize supposed behavior in a single word following the statement *regitur* (ruled)—for the American, *consuetudine* (by habit); for the European, *ritibus* (by custom); for the Asian, *opinionibus* (by belief); and for the African, *arbitrio* (by caprice). Surely, regulation by established and considered custom beats the unthinking rule of habit or belief, and caprice can only represent the least desirable among the four criteria, thus leading to the implied and conventional racist ranking of Europeans first, Asians and Americans in the middle, and Africans at the bottom.

Nonetheless, and despite these implications, the overt geometry of Linnaeus's model is neither linear nor hierarchical. When we epitomize his scheme as an essential picture in our mind, we see a map of the world divided into four regions, with the people in each region characterized by a list of different traits. In short, Linnaeus uses cartography as a primary principle for human ordering; if he had wished to advocate linear ranking as the essential picture of human variety, he would surely have listed Europeans first and Africans last, but he started with Native Americans instead.

The shift from a geographic to a hierarchical ordering of human diversity marks a fateful transition in the history of Western science—for what, short of railroads and nuclear bombs, has generated more practical impact, in this case almost entirely negative, upon our collective lives and nationalities? Ironically, J. F. Blumenbach became the primary author of this shift—for his five-race scheme became canonical, as he changed the geometry of human order from Linnaean cartography to linear ranking by putative worth.

I say ironic because Blumenbach surely deserves plaudits as the least racist, most egalitarian, and most genial of all Enlightenment writers on the subject of human diversity. How peculiar that the man most committed to human unity, and to inconsequential moral and intellectual differences among groups, should have changed the mental geometry of human order to a scheme that has promoted conventional racism ever since. Yet, on second thought, this situation should not be deemed so peculiar or unusual—for most scientists have always been unaware of the mental machinery, and particularly of the visual or geometric implications, behind their particular theorizing (and underlying all human thought in general).

An old tradition in science proclaims that changes in theory must be driven by observation. Since most scientists believe this simplistic formula, they assume that their own shifts in interpretation only record their better understanding of novel facts. Scientists therefore tend to be unaware of their own mental impositions upon the world's messy and ambiguous factuality. Such mental manipulations arise from a variety of sources, including psychological predisposition and social context. Blumenbach lived in an age when ideas of progress, and of the cultural superiority of European life, dominated the political and social world of his contemporaries. Implicit and loosely formulated (or even unconscious) notions of racial ranking fit well with such a worldview. In changing the geometry of human order to a system of ranking by worth, I doubt that Blumenbach operated consciously in the overt service of racism. I think that he only, and largely passively, recorded the pervasive social view of his time. But ideas have consequences, whatever the motives or intentions of their promoters.

Blumenbach certainly thought that his switch from the Linnaean four-race system to his own five-race scheme—the basis for his fateful geometric shift, as we shall see, from cartography to hierarchy—arose only from his improved understanding of nature's factuality. He so stated in the second (1781) edition of his treatise, when he announced his change: "Formerly in the first edition of this work, I divided all mankind into four varieties; but after I had more actively investigated the different nations of Eastern Asia and America, and, so to speak, looked at them more closely, I was compelled to give up that division, and to place in its stead the following five varieties, as more consonant to nature." And, in the preface to the third edition of 1795, Blumenbach states that he gave up the Linnaean scheme in order to arrange "the varieties of man according to the truth of nature." When scientists adopt the myth that theories arise solely from observation, and do not scrutinize the personal and social influences emerging from their own psyches, they not only misunderstand the causes of their changed opinions, but may also fail to comprehend the deep and pervasive mental shift encoded by their own new theory.

Blumenbach strongly upheld the unity of the human species against an alternative view, then growing in popularity (and surely more conducive to conventional forms of racism), that each major race had been separately created. He ended the third edition of his treatise by writing: "No doubt can any longer remain but that we are with great probability right in referring all varieties of man . . . to one and the same species."

As his major argument for unity, Blumenbach notes that all supposed racial characters grade continuously from one people to another, and cannot define any separate and bounded group.

> For although there seems to be so great a difference between widely separate nations, that you might easily take the inhabitants

> of the Cape of Good Hope, the Greenlanders, and the Circassians for so many different species of man, yet when the matter is thoroughly considered, you see that all do so run into one another, and that one variety of mankind does so sensibly pass into the other, that you cannot mark out the limits between them.

He particularly refutes the common claim that black Africans, as lowest on the conventional racist ladder, bear unique features of their inferiority: "There is no single character so peculiar and so universal among the Ethiopians, but what it may be observed on the one hand everywhere in other varieties of men."

Blumenbach believed that *Homo sapiens* had been created in a single region and had then spread out over the globe. Our racial diversity, he then argued, arose as a result of our movement to other climates and topographies, and our consequent adoption of different habits and modes of life in these various regions. Following the terminology of his time, Blumenbach referred to these changes as "degenerations"—not intending, by this word, the modern sense of deterioration, but the literal meaning of departure from an initial form of humanity at the creation (*de* means "from," and *genus* refers to our original stock).

Most of these degenerations, Blumenbach argues, arise directly from differences in climate—ranging from such broad patterns as the correlation of dark skin with tropical environments, to more particular (and fanciful) attributions, including a speculation that the narrow eye slits of some Australian people may have arisen as a response to "constant clouds of gnats . . . contracting the natural face of the inhabitants." Other changes then originate as a consequence of varying modes of life adopted in these different regions. For example, nations that compress the heads of babies by swaddling boards or papoose carriers end up with relatively long skulls. Blumenbach holds that "almost all the diversity of the form of the head in different nations is to be attributed to the mode of life and to art."

Blumenbach does not deny that such changes, promoted over many generations, may eventually become hereditary (by a process generally called "Lamarckism," or "inheritance of acquired characters" today, but serving as the folk wisdom of the late eighteenth century, and not as a peculiarity of Lamarck's biology, as Blumenbach's support illustrates). "With the progress of time," Blumenbach writes, "art may degenerate into a second nature."

But Blumenbach strongly held that most racial variation, as superficial impositions of climate and mode of life, could be easily altered or reversed by moving to a new region or by adopting new styles of behavior. White Europeans living for generations in the tropics may become dark-skinned, while Africans trans-

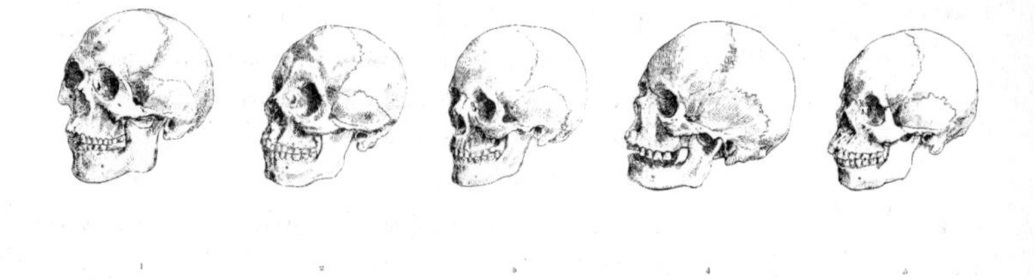

Blumenbach's new geometry of racial ranking in two lines of degeneration from a Caucasian center (middle skull), with one line (to the left) moving from American to Mongolian, and the other (to the right) from Malay to African. From Blumenbach's treatise of 1795.

ported as slaves to high latitudes may eventually become white: "Color, whatever be its cause, be it bile, or the influence of the sun, the air, or the climate, is, at all events, an adventitious and easily changeable thing, and can never constitute a diversity of species."

Backed by these views on the superficiality of racial variation, Blumenbach stoutly defended the mental and moral unity of all peoples. He held particularly strong opinions on the equal status of black Africans and white Europeans—perhaps because Africans had been most stigmatized by conventional racist beliefs.

Blumenbach established a special library in his house devoted exclusively to writings by black authors. We may regard him today as patronizing in praising "the good disposition and faculties of these our black brethren," but paternalism surely trumps contempt. He campaigned for the abolition of slavery when such views did not enjoy widespread assent, and he asserted the moral superiority of slaves to their captors, speaking of a "natural tenderness of heart, which has never been benumbed or extirpated on board the transport vessels or on the West India sugar plantations by the brutality of their white executioners."

Blumenbach affirmed "the perfectibility of the mental faculties and the talents of the Negro," and he listed the fine works of his library, offering special praise for the poetry of Phillis Wheatley, a Boston slave whose writings have only recently been rediscovered and reprinted in America: "I possess English, Dutch, and Latin poems by several [black authors], amongst which however above all, those of Phillis Wheatley of Boston, who is justly famous for them, deserves mention here." Finally, Blumenbach noted that many Caucasian nations could not boast so fine a set of authors and scholars as black Africa has produced under the most depressing circumstances of prejudice and slavery: "It would not be difficult to mention entire well-known provinces of Europe, from out of which you would not easily expect to obtain off-hand such good authors, poets, philosophers, and correspondents of the Paris Academy."

Nonetheless, when Blumenbach presented his implied mental picture of human diversity—his transposition from Linnaean geography to hierarchical ranking—he chose to identify a central group as closest to the created ideal, and then to characterize other groups by relative degrees of departure from this archetypal standard. He therefore devised a system (see the accompanying illustration from his treatise) that placed a single race at the pinnacle of closest approach to the original creation, and then envisioned two symmetrical lines of departure from this ideal toward greater and greater degeneration.

We may now return to the riddle of the name Caucasian, and to the significance of Blumenbach's addition of a fifth race, the Malay variety. Blumenbach chose to regard his own European variety as closest to the created ideal, and he then searched within the diversity of Europeans for a smaller group of greatest perfection—the highest of the highest, so to speak. As we have seen, he identified the people around Mount Caucasus as the closest embodiments of an original ideal, and he then named the entire European race for their finest representatives.

But Blumenbach now faced a dilemma. He had already affirmed the mental and moral equality of all peoples. He therefore could not use these conventional standards of racist ranking to establish degrees of relative departure from the Caucasian ideal. Instead, and however subjective (and even risible) we may view the criterion today, Blumenbach chose physical beauty as his guide to ranking. He simply affirmed Europeans as most beautiful, with people of the Caucasus on the highest pinnacle of comeliness (hence his linking, in the quotation presented at the beginning of this essay, of maximal beauty with place of human origin—for Blumenbach viewed all subsequent variation as departure from a created ideal, and the most beautiful people must therefore live closest to our primal home).

Blumenbach's descriptions make continual reference to his personal sense of relative beauty, presented as an objective and quantifiable property, not subject to doubt or disagreement. He describes a Georgian female skull (from closest to Mount Caucasus) in his collection as "really the most beautiful form of skull which . . . always of itself attracts every eye, however little observant." He then defends his European standard on aesthetic grounds:

> In the first place, that stock displays . . . the most beautiful form of the skull, from which, as from a mean and primeval type, the others diverge by most easy gradations. . . . Besides, it is white in color, which we may fairly assume to have been the primitive color of mankind, since . . . it is very easy for that to degenerate into brown, but very much more difficult for dark to become white.

Blumenbach then presented all human variety on two lines of successive departure from this Caucasian ideal, ending in the two most degenerate (least attractive, not morally unworthy or mentally obtuse) forms of humanity—Asians on one side, and Africans on the other. But Blumenbach also wanted to designate intermediary forms between ideal and most degenerate—especially since he advocated an even gradation as his primary argument for human unity. In his original four-race system, he could identify Native Americans as intermediary between Europeans and Asians, but who would serve as the transitional form between Europeans and Africans?

The four-race system included no appropriate group, and could therefore not be transformed into the new geometry of a pinnacle with two symmetrical limbs leading to maximal departure from ideal form. But invention of a fifth racial category for forms intermediate between Europeans and Africans would complete the new geometry—and Blumenbach therefore added the Malay race, not as a minor factual refinement, but as the enabler of a thorough geometric transformation in theories (mental pictures) about human diversity. As an intermediary between Europeans and Africans, the Malay variety provided crucial symmetry for Blumenbach's hierarchical taxonomy. This Malay addition therefore completed the geometric transformation from an unranked geographic model to the conventional hierarchy of implied worth that has fostered so much social grief ever since. Blumenbach epitomized his system in this geometric manner, and explicitly defended the necessary role of his Malay addition:

> I have allotted the first place to the Caucasian . . . which makes me esteem it the primeval one. This diverges in both directions into two, most remote and very different from each other; on the one side, namely, into the Ethiopian, and on the other into the Mongolian. The remaining two occupy the intermediate positions between that primeval one and these two extreme varieties; that is, the American between the Caucasian and Mongolian; the Malay between the same Caucasian and Ethiopian.

Scholars often suppose that academic ideas must remain, at worst, harmless and, at best, mildly amusing or even instructive. But ideas do not reside in the ivory tower of our usual metaphor about academic irrelevancy. In his famous vision of strength and weakness, Pascal epitomized humans as thinking reeds—and ideas do motivate human history. Would Hitler have flourished without racism, America without liberty? Blumenbach lived as a cloistered professor all his life, but his ideas reverberate through our wars, our conquests, our sufferings, and our hopes. I therefore end by returning to the coincidence of 1776, as Jefferson wrote the Declaration of Independence while Blumenbach published the first edition of his treatise in Latin. Consider the words of Lord Acton on the power of ideas to propel history, as illustrated by potential passage from Latin to action:

> It was from America that . . . ideas long locked in the breast of solitary thinkers, and hidden among Latin folios—burst forth like a conqueror upon the world they were destined to transform, under the title of the Rights of Man.

27

The Great Physiologist of Heidelberg

IF YOU SUSPEND BOTH REASON AND KNOWLEDGE, AND then gaze upon the ruins of the medieval castle on the hill, lit so softly at night and visible from all points in the city below; if you then recall the lively drinking songs from Sigmund Romberg's *Student Prince,* and conjure up an image of dashing young men purposely scarring their faces in frivolous duels—then the usual image of Heidelberg as a primary symbol of European romanticism and carefree charm might pass muster. But when you trace the tales of internecine destruction that created these sets, then the visions become fiction, and a gritty historical reality emerges from gentle mythology.

Heidelberg boasts an ancient pedigree, for the town's name first appears in a document written in 1196, while the university, founded in 1386, ranks as Germany's oldest. But only one or two medieval buildings still stand (while the castle lies in ruins), because the city suffered an architectural equivalent of genocide—"devoured [unto] the foundations thereof" (Lamentations 4:11)—in several disastrous religious and political wars of the seventeenth century. The Thirty Years War (1618–48) had wrought enough destruction, but when the Protestant elector (ruler) of the Rhineland Palatinate (with Heidelberg as capital) married his daughter to the brother of France's Catholic king Louis XIV, he only courted further trouble—for the elector's son died without heir in 1685, and Louis then laid claim to the territory. French armies destroyed Heidelberg in 1689, and the few remnants then succumbed to fire in 1693.

If our all-too-human tendencies toward xenophobia and anathematization of differences can place such closely allied and ethnically similar people on warpaths of total destruction, what hope can we maintain for toleration or decency toward people of more different appearance and cultural background? A sad chapter in the history of science must chronicle the support provided by supposedly "factual" arguments for the designation of different people as inferior beings. Science, to be fair, did not invent the concept of an inherent gradation in worth, with the promulgator's own group on top and his immediate enemies and more distant prospects for conquest on the bottom. But the doctrine of racism—the claim for intrinsically biological and therefore ineradicable differences in intellectual or moral status among peoples—has built a powerful buttress for our ancient inclinations toward xenophobia.

During the heyday of European colonialism in the nineteenth century, scarcely any Western scientist denied such gradations of worth—either as ordained by divine or natural law in versions favored before Darwin's discoveries, or as developed by the workings of evolution in explanations that triumphed in the closing decades of Victoria's reign. Black Africans received especially short shrift in these racist classifications.

Such opinions flowed with particular ease from basically conservative scientists, like the great French anatomist Georges Cuvier (1769–1832), who also favored strict divisions among social classes back home. Cuvier wrote in 1817:

> The Negro race is confined to the area south of the Atlas Mountains. With its small cranium, its flattened nose, its protruding jaw, and its large lips, this race clearly resembles the monkeys. The people belonging to it have always remained barbarians.

But even scientists of more egalitarian bent at home, including such passionate abolitionists as Charles Darwin, did not challenge the general consensus. In his most striking statement (from *The Descent of Man,* 1871), Darwin argues that a gap between two closely related living species does not disprove evolution because the intermediary stages, linking both forms to a common ancestor, died out long ago. The large gap that now separates the highest ape and the lowest man, Darwin asserts, will grow even wider as extinctions continue:

> The civilized races of man will almost certainly exterminate and replace throughout the world the savage races. At the same time the anthropomorphous apes . . . will no doubt be exterminated. The break will then be rendered wider, for it will intervene between man in a more civilized state, as we may hope, than the Caucasian, and some ape as low as a baboon, instead of at present between the negro or Australian and the gorilla.

The few "egalitarians" of those times—defined in this context as scientists who denied inherent differences in intellect or morality among races—limited their views to abstract potentials, and did not challenge conventional opinions about gradation in actual achievements. Alfred Russel Wallace, for example, strongly supported inherent equality (or at least minimal difference), but he did not doubt that English society had reached a pinnacle of realization, while African savages remained mired in barbarity: "Savage languages," he wrote, "contain no words for abstract conceptions . . . The singing of savages is a more or less monotonous howling."

Even J. F. Blumenbach (1752–1840), the great Enlightenment thinker (see essay 26) who devised the classification of races that became standard in nineteenth-century science, stoutly defended intellectual equality, while never doubting gradational differences in inherent beauty, with his own Caucasian race on a pinnacle obvious to all. Blumenbach devised the term Caucasian (still employed today) for the white races of Europe because he regarded the people living near Mount Caucasus as the best among the comeliest—"really the most beautiful form of skull," he writes, "which always of itself attracts every eye, however little observant."

I have, during the quarter-century of this series of essays, written about most of these few egalitarians, if only because iconoclasm always attracts me, while moral rectitude (at least by the preferences of most people today) always inspires admiration. But I have never treated the single most remarkable document in this small tradition, probably because its largely unknown author never extended his anthropological research beyond this lone foray into a subject (the status of races) and a language (English) otherwise absent from his extensive and highly valued work.

Perhaps Friedrich Tiedemann (1781–1861) had learned a sad lesson about the fruits of xenophobia from the history of his own adopted city, for "the great physiologist of Heidelberg"—an accolade for Tiedemann from the pen of England's leading anatomist, Richard Owen—served as professor of anatomy, physiology, and zoology at the University (from 1816 until his retirement in 1849), where the ruined castle, perched on a hill above his lecture hall, stood as a mute testimony to human folly and venality.

Following a common pattern among the intellectual elite of his generation (his father served as a professor of Greek and classical literature), Tiedemann wandered among many European universities to study with the greatest teachers of his time. Thus he learned philosophy from Schelling at Würzburg; anatomy from Franz Joseph Gall (the founder of phrenology) in Marburg; zoology from Cuvier in Paris; and anthropology from Blumenbach in Göttingen. Although he did not publish any work on human races until 1836, near the end of his active career in science, he must have internalized the core of this debate during these youthful *Wanderjahren*.

Tiedemann may have chosen his professors for other reasons, but he studied with the most prominent scholars of both persuasions—with the leading egalitarians Blumenbach and Gall (who used phrenology to advocate the material basis of consciousness, and who favored multiple organs of intelligence, expressed in bumps and other features of cranial architecture, largely because each person would then excel in some specific faculty, while no measure could then rank people or groups in a linear order of "general" worth); and with such eminent supporters of racial ranking as Cuvier, and the medical anatomist S. T. Soemmerring, who landed Tiedemann his first job in 1807. (Interestingly, in Tiedemann's 1836 article on race, the focus of this essay, he quotes both Soemmerring and Cuvier in a strong critique on the opening page, but then praises both Gall and Blumenbach later. Tiedemann also dedicated his 1816 book on the comparative anatomy and embryology of brains, the second document discussed in this essay, to Blumenbach. Obviously, Tiedemann remembered the lessons of his youth—and then developed his own critiques and preferences. What more could a teacher desire?

Tiedemann's career began on a fast track, with a textbook on zoology and anatomical dissertations on fish hearts (1809), large reptiles (1811 and 1817), and the lymphatic and respiratory organs of birds. He did not neglect invertebrates, either, winning a prize in 1816 from the Académie des Sciences in Paris for a treatise on the anatomy of echinoderms. He then turned his attention to the first of two major projects in his career—a remarkable study, published in 1816, on the embryology of the human brain compared with the anatomy of adult brains throughout the Vertebrata.

When Tiedemann took up his position as professor in Heidelberg, his interests switched to physiology—hence the nature of Owen's accolade, used as the title of this essay—largely because he met the remarkable young chemist Leopold Gmelin, and the two men recognized that a combination of anatomical and chemical expertise could resolve some outstanding issues in the mechanics and functioning of human organs. Thus, in the second major project of his career, Tiedemann collaborated with Gmelin on a series of remarkable discoveries about human digestion—recognizing, for example, that the intestines and other organs participate in the process, not only the stomach (as previously believed); that digestion involves chemical transformation (the conversion of starch into glucose, for example), not merely dissolution; and that hydrochloric acid works as a powerful agent of digestion in the stomach.

Then, in 1836, Tiedemann published a stunning article in English in the *Philosophical Transactions of the Royal Society of London,* Britain's most prestigious scientific journal (then and now): "On the Brain of the Negro, compared with that of the European and the Orang-Outang" (pages 497–527). Why such a shift in focus of research, and why such a foray into a language not his own and never before used to express his research? I do not know the full answer to this intriguing question (for biographical materials on Tiedemann are, to say the least, sparse). But a consideration of his life and work, combined with an exegesis of his two leading publications, provides a satisfactory beginning.

Tiedemann's unusual paper of 1836 states the egalitarian argument pure and simple—with no ifs, ands, or buts about inferior culture or suboptimal beauty. He does follow Blumenbach in accepting European definitions as universal aesthetic norms—a claim that can only strike our modern sensibilities as almost naïvely humorous. But, unlike Blumenbach, he then holds that Africans measure up to Caucasian standards of beauty. Of Africans living freely in the continent's interior, untouched by slavery, Tiedemann writes:

> Their skin is not so black as that of the Negroes on the coast of Guinea, and their black hair is not so woolly, but long, soft, and silky. They have neither flat noses, thick lips, nor prominent cheekbones; sloping contracted forehead, nor a skull compressed from both sides, which most naturalists consider as the universal characteristics of a Negro. Most of them have well-formed skulls, long faces, handsome, even Roman or aquiline noses, thin lips, and

> agreeable features. The Negresses of these nations are as finely formed as the men, and are, with the exception of their color, as handsome as European women.

(This remarkable statement illustrates the vintage of conventional prejudice, as the nineteenth century's firmest egalitarian scientist never doubts the "obviously" greater beauty of light skin, straight hair, thin lips, and "Roman or aquiline noses"!)

Tiedemann then argues that the false impression of African ugliness arose from limited studies of people suppressed by slavery and living at the coast:

> The mistaken notion of these naturalists arose from [study] . . . of a few skulls of Negroes living on the coasts, who, according to credible travellers, are the lowest and most demoralized of all the Negro tribes; the miserable remains of an enslaved people, bodily and spiritually lowered and degraded by slavery and ill treatment.

The technical argument of Tiedemann's paper follows a clear and simple logic to an equally firm conclusion—an exemplar of scientific reasoning, so long as the data hold up to scrutiny and withstand the light of new findings. Tiedemann develops two sources of information to reach the same conclusion. He first uses his anatomical expertise to search for distinctions among the brains of Caucasians (both males and females), black Africans, and orangutans. And he finds no structural differences among humans of different races and sexes. He begins with the cerebrum, the traditional "seat" of intelligence, and concludes: "In the internal structure of the brain of the Negro I did not observe any difference between it and that of the European." He then studies any other part used by scientific colleagues to assert differences in rank—particularly to test the claim that blacks have thicker nerves than whites. Again, he finds no distinction: "Hence there is no remarkable difference between the medulla oblongata and spinal cord of the Negro and that of the European, except the difference arising from the different size of the body."

Tiedemann then moves on to a second and clinching argument based on size, for some colleagues had accepted the conclusion of no structural difference, but had then defended racial ranking on supposed grounds of "more is better," arguing that Caucasians possessed the largest brains among human races, and African blacks the smallest. Tiedemann understood the complexity of this subject, and the consequent need for statistical analysis. He recognized that weighing a brain or two could not decide the issue because brains grow in correlation with bodies, and larger bodies therefore house bigger brains, quite independently of any hypothetical differences caused by racial inequalities. Tiedemann understood, for example, that small brains of women only reflected their smaller bodies—and that appropriate corrections for size might put women ahead. He therefore writes, controverting the greatest and most ancient authority of all:

> Although Aristotle has remarked that the female brain is absolutely smaller than the male, it is nevertheless not relatively smaller compared with the body; for the female body is in general lighter than that of the male. The female brain is for the most part even larger than the male, compared with the size of the body.

Tiedemann also recognized that brain sizes varied greatly among adults of any individual race. Therefore, a prejudiced observer could tout any desired view merely by choosing a single skull to fit his preferences, no matter how unrepresentative such a specimen might be as surrogate for an entire group. Thus, Tiedemann noted, many anthropologists had simply chosen the smallest-brained and biggest-jawed African skull they could find, and then presented a single drawing as "proof" of what every (Caucasian) observer already "knew" in any case! Tiedemann therefore labored to produce the largest compilation of data ever assembled, with all items based entirely on his own measurements for skulls of all races. (He followed the crude, but consistent, method of weighing the skull, filling the cavity with "dry millet-seed," weighing again, and finally expressing the capacity of the brain case as the weight of the skull filled with seed minus the weight of the empty skull.)

From his extensive tables (38 male African and 101 male Caucasian skulls, for example—see my later comments on his methods and results), Tiedemann concluded that no differences in size of the brain can distinguish human races. In one of the most important conclusions of nineteenth-century anthropology—a statement, based on extensive data, that at least placed a brake upon an otherwise unchallenged consensus in the opposite direction—Tiedemann wrote:

> We can also prove, by measuring the cavity of the skull in Negroes and the men of the Caucasian, Mongolian, American, and Malayan races, that the brain of the Negro is as large as that of the European and other nations. . . . Many naturalists have incorrectly asserted that Europeans have larger brains than Negroes.

Finally, Tiedemann closed the logical circle to clinch his argument by invoking the materialist belief of his teacher, F. J. Gall: brain stuff engenders thought, and brain size must therefore correlate at least roughly with intellectual capacity. If the brains of all races differ in neither size nor anatomy, we cannot assert a biological basis for differences in intellect among groups and must, on the contrary, embrace the opposite hypothesis of equality, unless some valid argument, not based on size or structure, can be advanced (an unlikely prospect for scientific materialists like Tiedemann and Gall). Tiedemann writes:

> The brain is undoubtedly the organ of the mind. . . . In this organ we think, reason, desire, and will. In short, the brain is the instrument by which all the operations called intellectual are carried on. . . . An intimate connection between the structure of the brain and the intellectual faculties in the animal kingdom cannot be doubted. As the facts which we have advanced plainly prove that there are no well-marked and essential differences between the brain of the Negro and the European, we must conclude that no innate difference in the intellectual faculties can be admitted to exist between them.

Claims for African inferiority have almost always been based on prejudiced observation of people degraded by the European imposition of slavery:

> Very little value can be attached to those researches, when we consider that they have been made for the most part on poor and

Tiedemann's Data

Sample	Number of Skulls	Smallest	Largest	Average
		weight in ounces		
Caucasian (all)	101	28	57	40.08
European Caucasian	77	33	57	41.34
Asian Caucasian	24	28	42	36.04
Malayan	38	31	49	39.84
American	24	26	59	39.33
Mongolian	18	25	49	38.94
Ethiopian	38	32	54	37.84

> unfortunate Negroes in the Colonies, who have been torn from their native country and families, and carried into the West Indies, and doomed there to a perpetual slavery and hard labor. . . . The original and good character of the Negro tribes on the Western Coast of Africa has been corrupted and ruined by the horrors of the slave trade, since they have unfortunately become acquainted with Europeans.

I have now done my ordinary duty as an essayist: I have told the forgotten story of the admirable Tiedemann with sufficient detail, and in his own words, to render the flavor of his concerns, the compelling logic of his argument, and the careful documentation of his empirical research. But, curiously, rather than feeling satisfaction for a job adequately done, I am left with a feeling of paradox based on a puzzle that I cannot fully resolve, but that raises an interesting issue in the social and intellectual practice of science—thus giving this essay a fighting chance to move beyond the conventional.

The paradox arises from internal evidence of Tiedemann's strong *predisposition* toward belief in the equality of races. I should state explicitly that I refer, in stating this claim, not to the logic and data so well presented in his published work (as discussed above), but to evidence based on idiosyncrasies of presentation and gaps in stated arguments, that Tiedemann undertook the research for his 1836 article with his mind *already set* (or at least strongly inclined) to a verdict of equality. Such preferences—especially for judgments generally regarded as so morally honorable—might not be deemed surprising, except that the ethos of science, both then and now, discourages such *a priori* convictions as barriers to objectivity. An old proverb teaches that "to err is human." To have strong preferences before a study may be equally human—but scientists are supposed to disregard such biases if their existence be recognized, or at least to remain so unconscious of their sway that a heartfelt and genuine belief in objectivity persists in the face of contrary practice.

Tiedemann's preferences for racial equality do not arise from either of the ordinary sources for such a predisposition. First, I could find no evidence that Tiedemann's political or social beliefs inclined him in a "liberal" or "radical" direction toward an uncommon belief of egalitarianism. Tiedemann grew up in an intellectually elite and culturally conservative family. He particularly valued stable government, and he strongly opposed popular uprisings. Three of his sons served as army officers, and his eldest was executed under martial law imposed by a temporarily successful revolt (while his two other military sons fled to exile) during the revolutions of 1848. When peace and conventional order returned, the discouraged Tiedemann retired from the university, and published little more (largely because his eyesight had become so poor) beyond a final book (in 1854) titled *The History of Tobacco and Other Similar Means of Enjoyment.*

Second, some scientists tend, by temperament, to embrace boldly hypothetical pronouncement, and to publish exciting ideas before adequate documentation can affirm their veracity. But Tiedemann built a well-earned reputation for the exactly opposite behavior of careful and meticulous documentation, combined with extreme caution in expressing beliefs that could not be validated by copious data. The definitive eleventh edition of the *Encyclopaedia Britannica* (1910–11), in its single short paragraph on Tiedemann, includes this sole assessment of his basic scientific approach: "He maintained the claims of patient and sober anatomical research against the prevalent speculations of the school of Lorenz Oken, whose foremost antagonist he was long reckoned." (Oken led the oracular movement known as *Naturphilosophie.* He served as a sort of antihero in my first book, *Ontogeny and Phylogeny,* published in 1977—so I have a long acquaintance, entirely in Tiedemann's favor, with his primary adversary.)

I have found, in each of Tiedemann's two major publications on brains and races, a striking indication—rooted in information that he does *not* use, or data that he does *not* present (for surprising or illogical absences often speak more loudly than vociferous assertions)—of his predisposition toward racial equality.

1. *Creating the standard argument, and then refraining from the usual interpretation: Tiedemann's masterpiece of 1816.*

By customary criteria of new discovery, copious documentation, and profound theoretical overview, Tiedemann's 1816 treatise on the embryology of the human brain, as contrasted with adult brains in all vertebrates (fish to mammals), has always been judged as his masterpiece (I quote from my copy of the French translation of 1823). As a central question in pre-Darwinian biology, scientists of Tiedemann's generation yearned to know whether all developmental processes followed a single general law, or if each pursued an independent path. Two processes stood out for evident study: the growth of organs in the embryology of "higher" animals, and the sequence of structural advance (in created order, not by evolutionary descent) in a classification of animals from "lowest" to "highest" along the chain of being.

In rough terms, both sequences seemed to move from small, simple, and homogeneous beginnings to larger, more-complex, and more-differentiated endpoints. But how similar might these two sequences be? Could adults of lower animals really be compared with transitory stages in the embryology of higher creatures? If so, then a single law of development might pervade nature to reveal the order and intent of the universe and its creator. This heady prospect drove a substantial amount of biological research during the late eighteenth and early nineteenth centuries. Tiedemann, beguiled by the prospect of discovering such a universal pattern, wrote that the "two routes" to such knowledge "are those of comparative anatomy and the anatomy of the fetus, and these shall become, for us, a veritable thread of Ariadne." (When Ariadne led Theseus through the labyrinth to the Minotaur, she unwound a thread along the path, so that Theseus could find his way out after his noble deed of bovicide. The "thread of Ariadne" thus became a standard metaphor for a path to the solution of a particularly difficult problem.)

Tiedemann's densely documented treatise announced a positive outcome for this grand hope of unification: the two sequences of human fetal development and comparative anatomy of brains from fish to mammals coincide perfectly. He wrote in triumph:

> I therefore publish here the research that I have done for several years on the brain of the [human] fetus. . . . I then present an exposition of the comparative anatomy of the structure of the brain in the four classes of vertebrate animals [fish, reptiles, birds, and mammals in his taxonomy]—all in order to prove that the formation of this organ in the [human] fetus, followed from month to month during its development, passes through the major stages of organization reached by the [vertebrate] animals in their complexity. We therefore cannot doubt that nature follows a uniform plan in the creation and development of the brain in both the human fetus and the sequence of vertebrate animals.

Thus, Tiedemann had reached one of the most important and most widely cited conclusions of early-nineteenth-century zoology. Yet he never extended this notion, the proudest discovery of his life, to establish a sequence of human races as well—although virtually all other scientists did. Nearly every major defense of conventional racial ranking in the nineteenth century expanded Tiedemann's argument from embryology and comparative anatomy to variation within a sequence of human races as well—by arguing that a supposedly linear order from African to Asian to European expresses the same universal law of progressive development.

Even the racial "liberals" of nineteenth-century biology invoked the argument of "Tiedemann's line" when the doctrine suited their purposes. T. H. Huxley, for example, proposed a linear order of races to fill the gap between apes and humans as an argument for evolutionary intermediacy: "The difference in weight of brain between the highest and the lowest man is far greater, both relatively and absolutely, than that between the lowest man and the highest ape."

But Tiedemann himself, the inventor of the basic argument, would not extend his doctrine into a claim that variation within a species (distinctions among human races in this case) must follow the same linear order as differences among related species. I can only assume that he demurred (as logic surely permits, and as later research has confirmed, for variations within and among species represent quite different biological phenomena) because he did not wish to use his argument as a defense for racial ranking. At least we know that one of his eminent colleagues read his silence in exactly this light—For Richard Owen, refuting Huxley's claim, cited Tiedemann with the accolade used as a title to this essay:

> Although in most cases the Negro's brain is less than that of the European, I have observed individuals of the Negro race in whom the brain was as large as the average one of Caucasians; and I concur with the great physiologist of Heidelberg, who has recorded similar observations, in connecting with such cerebral development the fact that there has been no province of intellectual activity in which individuals of the pure Negro race have not distinguished themselves.

2. *Developing the first major data set and then failing to notice an evident conclusion not in your favor (even if not particularly damaging, either).*

When I wrote *The Mismeasure of Man,* published in 1981, I discovered that most of the major data sets presented in the name of racial ranking contained evident errors that should have been noted by their authors, and would have reversed their conclusions, or at least strongly compromised the apparent strength of their arguments. Even more interestingly, I found that those scientists usually published the raw data that allowed me to correct their errors. I therefore had to conclude that these men had not based their conclusions upon conscious fraud—for fakers try to cover up the tracks of their machinations. Rather, their errors had arisen from unconscious biases so strong and so unquestioned (or even unquestionable in their system of beliefs and values) that information now evident to us remained invisible to them.

Fair is fair. The same phenomenon of unconscious bias must also be exposed in folks we admire for the sagacity, even the moral virtue, of their courageous and iconoclastic conclusions—for only then can we extend an exposé about beliefs we oppose into a more interesting statement about the psychology and sociology of scientific practice in general.

I have just discovered an interesting instance of nonreporting in the tables that Tiedemann compiled to develop his case for equality in brain sizes among human races. (To my shame, I never thought about pursuing this exercise when I wrote *The Mismeasure of Man,* even though I reported Paul Broca's valid critiques of different claims in Tiedemann's data to show that Broca often criticized others when their conclusions denied his own preferences, but did not apply the same standards to "happier" data of his own construction.)

Tiedemann's tables, the most extensive quantitative study of variation available in 1836, provides raw data for 320 male skulls in all five of Blumenbach's major races, including 101 "Caucasians" and 38 "Ethiopians" (African blacks). But Tiedemann only lists each skull individually (in old apothecaries' weights of ounces and drams), and presents no summary statistics for groups—no ranges, no averages. But these figures can easily be calculated from Tiedemann's raw data, and I have done so in the appended table.

Tiedemann bases his argument entirely on the overlapping ranges of smallest to largest skulls in each race—and we can scarcely deny his correct conclusion that no difference exists between Ethiopians (32 to 54 ounces among 38 skulls) and Europeans (28 to 57 ounces for a larger sample of 101 skulls). But as I scanned his charts of raw data. I suspected that I might find some interesting differences among the means for each racial group—the obvious summary statistic (even in Tiedemann's day) for describing some notion of an average.

Indeed, as my table and chart show, Tiedemann's uncalculated mean values do differ—and in the traditional order advocated by his opponents, with gradation from a largest Caucasian average, through intermediary values for Malayans, Americans (so-called "Indians," not European immigrants), and Mongolians, to lowest measures for his Ethiopian group. The situation becomes even more complicated when we recognize that these mean differences do not challenge Tiedemann's conclusion, even though an advocate for the other side could certainly advertise this information in such a false manner. (Did Tiedemann calculate these means and not publish them because he sensed the confusion that would then be generated—a procedure that I would have to label as indefensible, however understandable? Or did he never calculate them because he got what he wanted from the more obvious data on ranges, and then never proceeded further—the more common situation of failure to recognize

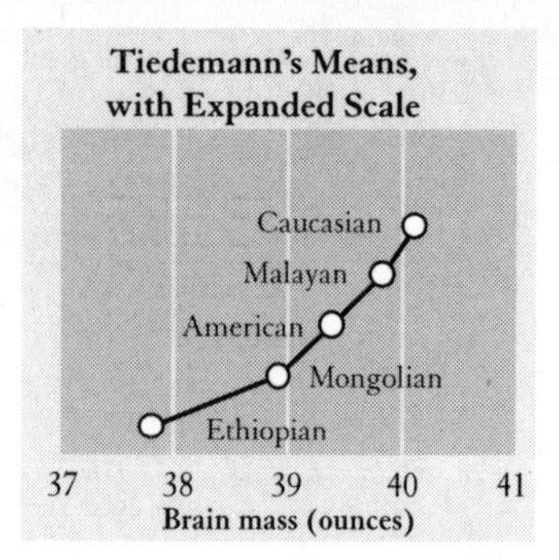

My compilation of average values of brain mass for each race, as calculated from raw data that Tiedemann presented but did not use for such a calculation.

potential interpretations as a consequence of unconscious bias. I rather suspect the second scenario, as more consistent with Tiedemann's personal procedures and the actual norms, as opposed to the stated desirabilities, of scientific study in general, but I cannot disprove the first conjecture.)

My appended graph of Tiedemann's uncalculated data does validate his position. The ranges are large and fully overlapping for the crucial comparison of Caucasians and Ethiopians (with the substantially larger Caucasian sample including the smallest and the largest single skull for the entire sample of both groups, as expected). The differences in mean values are tiny compared with the ranges, and, for this reason, probably of no significance in the judgment of intelligence. Moreover, the small variation among means probably reflects differences in body size rather than any stable distinction among races—and Tiedemann had documented the positive correlation of brain and body size in asserting the equality of brains in men and women (as previously cited). Tiedemann's own data indicate the probable control of mean differences in brain weight by body size. He divides his Caucasian chart into two parts by geography—for Europeans and Asians (mostly East Indians). He also notes that Caucasian males from Asia tend to be quite small in body size. Note that his mean brain size for these (presumably smallest-bodied) Caucasians from Asia stands at 36.04 ounces, the minimal value of his entire chart, lying well below the Ethiopian mean of 37.84 ounces.

But data can be "massaged" to advance almost any desired point, even when nothing "technically" inaccurate mars the presentation. For example, the mean differences in Tiedemann's data look trivial when properly scaled against the large ranges of each sample. But if I expand the scale, amalgamate the European and Asian Caucasians into one sample (Tiedemann kept them separate), omit the ranges, and plot only the mean values in conventional order of nineteenth-

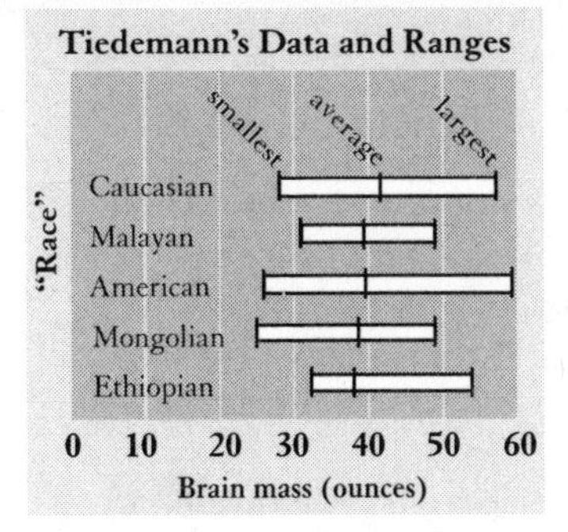

A different presentation of both ranges and average values for the brain mass of each major race, calculated by me from Tiedemann's data. Note that the mean values (the vertical black line in the middle of each horizontal white bar, representing the full ranges for each race) scarcely differ in comparison with the full overlap of ranges, thus validating Tiedemann's general conclusion.

century racial rankings, the distinctions can be made to seem quite large, and an unsophisticated observer might well conclude that significant differences in intrinsic mental capacity had been documented.

In conclusion, since Tiedemann clearly approached his study of racial differences with a predisposition toward egalitarian conclusions, and since he differed from nearly all his scientific colleagues in promoting this result, we must seek the source for his defense largely outside the quality and persuasive character of his data. Indeed, and scarcely surprising for an issue so salient in Tiedemann's time and so continually troubling and tragic ever since, he based his judgment on a moral question that, as he well understood, empirical data might illuminate but could never resolve: the social evils of racism, and particularly of slavery.

Tiedemann recognized that scientific data about facts of nature could not validate moral judgments about the evils of slavery—as conquerors could always invent other justifications for enslaving people judged equal to themselves in mental might, while many abolitionists accepted the inferiority of black Africans, but argued all the more strongly for freedom because decency requires special kindness toward those not so well suited for success. But Tiedemann also appreciated a social reality that blurred the logical separation of facts and morals: in practice, most supporters of slavery promoted inferiority as an argument for tolerating an institution that would otherwise be hard to justify under a rubric of supposedly "Christian" values: if "they" are not like "us," and if "they" are too benighted to govern themselves in the complexities of modern living, then "we" gain the right to take over. If scientific facts pointed to equality of intellectual capacity, then many conventional arguments for slavery would fall.

Modern scientific journals generally insist upon the exclusion of overt moral arguments from ostensibly factual accounts of natural phenomena. But the more literary standards and interdisciplinary preferences of Tiedemann's time permitted far more license, even in leading scientific journals like the *Philosophical Transactions* (an appropriate, if now slightly archaic, name used by this great scientific journal since its foundation in the seventeenth century). Tiedemann could therefore state his extrascientific reasons literally "up front"—for the first paragraph of his article announces *both* his scientific and ethical motives, and also resolves the puzzle of his decision to publish in English:

> I take the liberty of presenting to the Royal Society a paper on a subject which appears to me to be of great importance in the natural history, anatomy, and physiology of Man; interesting also in a political and legislative point of view. Celebrated naturalists . . . look upon the Negroes as a race inferior to the European in organization and intellectual powers, having much resemblance with the Monkey. . . . Were it proved to be correct, the Negro would occupy a different situation in society from that which has been so lately given him by the noble British Government.

In short, Tiedemann published his paper in English to honor and commemorate the abolition of the slave trade in Great Britain. The process had been long and tortuous (also torturous). Under the vigorous prodding of such passionate abolitionists as William Wilberforce, Britain had abolished the West Indian slave trade in 1807, but had not freed those already enslaved. (Wilberforce's son, Bishop Samuel, aka "Soapy Sam," Wilberforce became an equally passionate anti-Darwinian—for what goes around admirably can come around ridiculously, and history often repeats itself by Marx's motto, "the first time as tragedy, the second as farce.") Full manumission, with complete abolition, did not occur until 1834—a great event in human history that Tiedemann chose to celebrate in the most useful manner he could devise in his role as a professional scientist: by writing a technical article to promote a true argument that, he hoped, would do some moral good as well.

I cited Tiedemann's opening paragraph to praise his wise mixture of factual information and moral concern, and to resolve the puzzle of his publication in a foreign tongue. I can only end with his closing paragraph, an even more forceful statement of the moral theme, and a testimony to a most admirable man, whom history has forgotten, but who did his portion of good with the tools that his values, his intellectual gifts, and his sense of purpose had provided:

The principal result of my researches on the brain of the Negro is, that neither anatomy nor physiology can justify our placing them beneath the Europeans in a moral or intellectual point of view. How is it possible, then, to deny that the Ethiopian race is capable of civilization? This is just as false as it would have been in the time of Julius Caesar to have considered the Germans, Britons, Helvetians, and Batavians incapable of civilization. The slave trade was the proximate and remote reason of the innumerable evils which retarded the civilization of the African tribes. Great Britain has achieved a noble and splendid act of national justice in abolishing the slave trade. The chain which bound Africa to the dust, and prevented the success of every effort that was made to raise her, is broken.

VIII

Triumph and Tragedy on the Exact Centennial of *I Have Landed*, September 11, 2001

Introductory Statement

THE FOUR SHORT PIECES IN THIS FINAL SECTION, ADDED for obvious and tragic reasons after the completion of this book in its original form, chronicle an odyssey of fact and feeling during the month following an epochal moment that may well be named, in history's archives, simply by its date rather than its cardinal event—not D-Day, not the day of JFK's assassination, but simply as "September 11th." I would not have been able to bypass the subject in any case, but I simply couldn't leave this transformation of our lives and sensibilities unaddressed in this book, because my focus and title—*I Have Landed*—memorializes the beginning of my family in America at my grandfather's arrival on September 11, 1901, exactly, in the most eerie coincidence that I have ever viscerally experienced, 100 years to the day before our recent tragedy, centered less than a mile from my New York home. The four pieces in this set consciously treat the same theme, and build upon each other by carrying the central thought, and some actual phrases as well, from one piece to the next in sequence—a kind of repetition that I usually shun with rigor in essay collections, but that seems right, even required, in this singular circumstance. For I have felt such a strong need, experienced emotionally almost as a duty, to emphasize a vital but largely invisible theme of true redemption, so readily lost in the surrounding tragedy, but flowing from an evolutionary biologist's professional view of complex systems in general, and human propensities in particular: the overwhelming predominance of simple decency and goodness, a central aspect of our being as a species, yet so easily obscured by the efficacy of rare acts of spectacularly destructive evil. Thus, one might call this section "four changes rung on the same theme of tough hope and steadfast human nature"—as my chronology moves from some first thoughts in "exile" in Halifax, to first impressions upon returning home to Ground Zero, to musings on the stunning coincidence at the centennial of *I Have Landed,* to more general reflections, at a bit more emotional distance, of a lifelong New Yorker upon the significance of his city's great buildings and their symbolic meaning for human hope and transcendence.

28

The Good People of Halifax*

IMAGES OF DIVISION AND ENMITY MARKED MY FIRST CONtact, albeit indirect, with Nova Scotia—the common experience of so many American schoolchildren grappling with the unpopular assignment of Longfellow's epic poem *Evangeline,* centered upon the expulsion of the Acadians in 1755. My first actual encounter with Maritime Canada, as a teenager on a family motor trip in the mid-1950s, sparked nothing but pleasure and fascination, as I figured out the illusion of Moncton's magnetic hill, marveled at the tidal phenomena of the Bay of Fundy (especially the reversing rapids of Saint John and the tidal bore of Moncton), found peace of spirit at Peggy's Cove, and learned some history in the old streets of Halifax.

---

*This piece appeared in Canada's national newspaper, *The Globe and Mail,* on September 20, 2001. The last paragraph paraphrases several lines from the Canadian national anthem. Mr. Sukanen truly existed (as does the town of Moose Jaw in Saskatchewan). Evangeline belongs to Mr. Longfellow's artistry, but their link, as Canadians at home vs. those now resident in the United States, reinforces the theme of this small tribute.

I have been back, always with eagerness and fulfillment, a few times since, for reasons both recreational and professional—a second family trip, one generation later, and now as a father with two sons aged 3 and in utero; a lecture at Dalhousie; or some geological field work in Newfoundland. My latest visit among you, however, was entirely involuntary and maximally stressful. I live in lower Manhattan, just one mile from the burial ground of the Twin Towers. As they fell victim to evil and insanity on Tuesday, September 11, during the morning after my sixtieth birthday, my wife and I, en route from Milan to New York, flew over the *Titanic*'s resting place and then followed the route of her recovered dead to Halifax. We sat on the tarmac for eight hours, and eventually proceeded to the cots of Dartmouth's sports complex, then upgraded to the adjacent Holiday Inn. On Friday, at three o'clock in the morning, Alitalia brought us back to the airport, only to inform us that their plane would return to Milan. We rented one of the last two cars available and drove, with an intense mixture of grief and relief, back home.

The general argument of this piece, amidst the most horrific specifics of any event in our lifetime, does not state the views of a naively optimistic Pollyanna, but rather, and precisely to the contrary, attempts to record one of the deepest tragedies of our existence. Intrinsic human goodness and decency prevail effectively all the time, and the moral compass of nearly every person, despite some occasional jiggling prompted by ordinary human foibles, points in the right direction. The oppressive weight of disaster and tragedy in our lives does not arise from a high percentage of evil among the summed total of all acts, but from the extraordinary power of exceedingly rare incidents of depravity to inflict catastrophic damage, especially in our technological age when airplanes can become powerful bombs. (An even more evil man, armed only with a longbow, could not have wreaked such havoc at the Battle of Agincourt in 1415.)

In an important, little appreciated, and utterly tragic principle regulating the structure of nearly all complex systems, building up must be accomplished step by tiny step, whereas destruction need occupy but an instant. In previous essays on the nature of change, I have called this phenomenon the Great Asymmetry (with uppercase letters to emphasize the sad generality). Ten thousand acts of kindness done by thousands of people, and slowly building trust and harmony over many years, can be undone by one destructive act of a skilled and committed psychopath. Thus, even if the *effects* of kindness and evil balance out in the course of history, the Great Asymmetry guarantees that the *numbers* of kind and evil people could hardly differ more, for thousands of good souls overwhelm each perpetrator of darkness.

I stress this greatly underappreciated point because our error in equating a *balance of effects* with *equality in numbers* could lead us to despair about human possibilities, especially at this moment of mourning and questioning; whereas, in reality, the decent multitudes, performing their ten thousand acts of kindness, vastly outnumber the very few depraved people in our midst. Thus, we have every reason to maintain our faith in human kindness, and our hopes for the triumph of human potential, if only we can learn to harness this wellspring of unstinting goodness in nearly all of us.

For this reason, a documentation of the innumerable small acts of kindness, the good deeds that almost always pass beneath our notice for lack of "news value," becomes an imperative duty, a responsibility that might almost be called holy, when we must reaffirm the prevalence of human decency against our preeminent biases for hyping the cataclysmic and ignoring the quotidian. Ordinary kindness trumps paroxysmal evil by at least a million events to one, and we will not grasp this inspiring ratio unless we record the Everest of decency built grain by grain into a mighty fortress taller than any breakable building of mere concrete and steel.

Our media have stressed—as well they should—the spectacular acts of goodness and courage done by professionals pledged to face such dangers, and by ordinary people who can summon superhuman strength in moments of crisis: the brave firefighters who rushed in to get others out; the passengers of United Flight 93 who drew the grimly correct inference when they learned the fate of the Twin Towers, and died fighting rather than afraid, perhaps saving thousands of lives by accepting their own deaths in an unpopulated field. But each of these spectacular acts rests upon an immense substrate of tiny kindnesses that cannot be motivated by thoughts of fame or fortune (for no one expects their documentation), and can only represent the almost automatic shining of simple human goodness. But this time, we must document the substrate, if only to reaffirm the inspiring predominance of kindness at a crucial moment in this vale of tears.

Halifax sat on the invisible periphery of a New York epicenter, with 45 planes, mostly chock-full of poor strangers from strange lands, arrayed in two lines on the tarmac, and holding 9,000 passengers to house, feed, and, especially, to comfort. May it then be recorded, may it be inscribed forever in the Book of Life: Bless the good people of Halifax who did not sleep, who took strangers into their homes, who opened their hearts and shelters, who rushed in enough food and clothing to supply an army, who offered tours of their beautiful city, and, above all, who listened with a simple empathy that brought this tough and fully grown man to tears, over and over again. I heard not a single harsh word, saw not the slightest gesture of frustration, and felt nothing but pure and honest welcome.

I know that the people of Halifax have, by long tradition and practice, shown heroism and self-sacrifice at moments of disaster—occasional situations that all people of seafaring ancestry must face. I know that you received and buried the drowned victims of the *Titanic* in 1912, lost one in ten of your own people in the Halifax Explosion of 1917, and gathered in the remains of the recent Swissair disaster.

But, in a sense that may seem paradoxical at first, you outdid yourselves this time because you responded immediately, unanimously, unstintingly, and with all conceivable goodness, when no real danger, but merely fear and substantial inconvenience, dogged your refugees for a few days. Our lives did not depend upon you, but you gave us everything nonetheless. We, 9,000 strong, are forever in your debt, and all humanity glows in the light of your unselfish goodness.

And so my wife and I drove back home, past the Magnetic Hill of Moncton (now a theme park in this different age), past the reversing rapids of Saint John, visible from the highway, through the border crossing at Calais (yes, I know, as in Alice, not as in ballet), and down to a cloud of dust and smoke enveloping a mountain of rubble, once a building and now a tomb for 3,000 people. But you have given me hope that the ties of our common humanity will bind even these wounds. And so, Canada, although you are not my home or native land, we will always share this bond of your unstinting hospitality to people who descended upon you as frightened strangers from the skies, and received nothing but solace and solidarity in your embrace of goodness. So, Canada, because we beat as one heart, from Evangeline in Louisiana to the intrepid Mr. Sukanen of Moose Jaw, I will stand on guard for thee.

29

Apple Brown Betty*

THE PATTERNS OF HUMAN HISTORY MIX DECENCY AND depravity in equal measure. We often assume, therefore, that such a fine balance of results must emerge from societies made of decent and depraved people in equal numbers. But we need to expose and celebrate the fallacy of this conclusion so that, in this moment of crisis, we may reaffirm an essential truth too easily forgotten, and regain some crucial comfort too readily foregone. Good and kind people outnumber all others by thousands to one. The tragedy of human history lies in the enormous potential for destruction in rare acts of evil, not in the high frequency of evil people. Complex systems can only be built step by step, whereas destruction requires but an instant. Thus, in what I like to call the Great Asymmetry, every spectacular incident of evil will be balanced by ten thousand acts of

*This *New York Times* op-ed piece ran on September 26, 2001. They used a different title; I now restore my original and intended version.

kindness, too often unnoted and invisible as the "ordinary" efforts of a vast majority.

Thus, we face an imperative duty, almost a holy responsibility, to record and honor the victorious weight of these innumerable little kindnesses, when an unprecedented act of evil so threatens to distort our perception of ordinary human behavior. I have stood at Ground Zero, stunned by the twisted ruins of the largest human structure ever destroyed in a catastrophic moment. (I will discount the claims of a few biblical literalists for the Tower of Babel.) And I have contemplated a single day of carnage that our nation has not suffered since battles that still evoke passions and tears, nearly 150 years later: Antietam, Gettysburg, Cold Harbor. The scene is insufferably sad, but not at all depressing. Rather, Ground Zero can only be described, in the lost meaning of a grand old word, as "sublime," in the sense of awe inspired by solemnity.

But, in human terms, Ground Zero is the focal point for a vast web of bustling goodness, channeling uncountable deeds of kindness from an entire planet—the acts that must be recorded to reaffirm the overwhelming weight of human decency. The rubble of Ground Zero stands mute, while a beehive of human activity churns within, and radiates outward, as everyone makes a selfless contribution, big or tiny according to means and skills, but each of equal worth. My wife and stepdaughter established a depot to collect and ferry needed items in short supply, including respirators and shoe inserts, to the workers at Ground Zero. Word spreads like a fire of goodness, and people stream in, bringing gifts from a pocketful of batteries to a ten-thousand-dollar purchase of hard hats, made on the spot at a local supply house and delivered right to us.

I will cite but one tiny story, among so many, to begin the count that will overwhelm the power of any terrorist's act. And by such tales, multiplied many millionfold, let these few depraved people finally understand why their vision of inspired fear cannot prevail over ordinary decency. As we left a local restaurant to make a delivery to Ground Zero late one evening, the cook gave us a shopping bag and said: "Here's a dozen apple brown bettys, our best dessert, still warm. Please give them to the rescue workers." How lovely, I thought, but how meaningless, except as an act of solidarity, connecting the cook to the cleanup. Still, we promised that we would make the distribution, and we put the bag of twelve apple brown bettys atop several thousand respirators and shoe pads.

Twelve apple brown bettys into the breach. Twelve apple brown bettys for thousands of workers. And then I learned something important that I should never have forgotten—and the joke turned on me. Those twelve apple brown bettys went like literal hotcakes. These trivial symbols in my initial judgment turned into little drops of gold within a rainstorm of similar offerings for the stomach and soul, from children's postcards to cheers by the roadside. We gave the last one to a firefighter, an older man in a young crowd, sitting alone in utter exhaustion as he inserted one of our shoe pads. And he said, with a twinkle and a smile restored to his face: "Thank you. This is the most lovely thing I've seen in four days—and still warm!"

30

The Woolworth Building*

The astronomical motto of New York State—*excelsior* (literally "higher," or, more figuratively, "ever upward")—embodies both the dream and the danger of human achievement in its ambiguous message. In the promise of the dream, we strive to exceed our previous best as we reach upward, literally to the stars, and ethically to knowledge and the pursuit of happiness. In the warnings of danger, any narrowly focused and linear goal can drift, especially when our moral compass fails, into the zealotry of "true belief," and thence to an outright fanaticism that brooks no opposition.

As a naturalist by profession, and a humanist at heart, I have long believed that wisdom dictates an optimal strategy for proper

---

*From *Natural History* magazine. I wrote this piece after the other three of this section. The chronology is important for my general intention and feelings, but, for obvious reasons in the context of this book, the subsequent piece must come last.

steering toward the dream and away from the danger: as you reach upward, always festoon the structure of your instrument (whether conceptual or technological) with the rich quirks and contradictions, the foibles and tiny gleamings, of human and natural diversity—for abstract zealotry can never defeat a great dream anchored in the concrete of human warmth and laughter.

For all my conscious life, I have held one object close to my heart as both the abstract symbol and the actual incarnation of this great duality: upward thrust tempered by frailty, diversity, and contradiction. Let me then confess my enduring love affair with a skyscraper: the Woolworth Building, the world's tallest at 792 feet from its opening in 1913 until its overtopping by the Chrysler Building (another favorite) in 1929. This gorgeous pinnacle on Lower Broadway—set between the Tweed Courthouse to the east (a low artifact of human rapacity) and, until the tragedy of September 11, 2001, the Twin Towers to the west (a high artifact of excelsior in all senses)—represents the acme in seamless and utterly harmonious blending of these two components that must unite to achieve the dream, but that seem so inherently unmixable.

The Woolworth Building surely reaches high enough to embody the goals of excelsior. But its lavish embellishments only enhance the effect, giving warmth, breadth, and human scale to the height of transcendence. The outer cladding of glowing terra-cotta (not stone, as commonly stated) reflects the warmth of baked clay, not the colder gleam of metal. The overtly gothic styling of the lush exterior ornamentation marries an ecclesiastical ideal of past centuries with the verticality of modern life (thus engendering the building's wonderfully contradictory moniker as "cathedral of commerce"). The glorious interior, with a million tiny jewels in a mosaic ceiling, its grand staircase, murals of labor and commerce, and elegantly decorated elevators, inspires jumbled and contradictory feelings of religious awe, technological marvel, and aesthetic beauty, sometimes sublime and sometimes bumptious. Meanwhile, and throughout, high grandeur merges with low comedy, as the glistening ceiling rests upon gargoyles of Mr. Woolworth counting the nickels and dimes that built his empire, and the architect Cass Gilbert cradling in his arms the building that his image now helps to support.

When I was young, the Woolworth Building rose above all its neighbors, casting a warm terra-cotta gleam over lower Manhattan. But I have not seen this optimally tempered glory since the early 1970s because the Twin Towers, rising in utter metallic verticality just to the southwest, either enveloped my love in shadow, or consigned its warmer glow to invisibility within a metallic glare.

There can be no possible bright side to the tragedy of September 11 and the biggest tomb of American lives on any single day since the Battle of Gettysburg nearly 150 years ago. But the fact of human endurance and human goodness stands taller than 100 Twin Towers stacked one atop the other. These facts need symbols for support, so that the dream of excelsior will not be extinguished in the perverse utilization of its downside by a few evil men.

I returned to my beloved natal city, following an involuntary week in Halifax (as one of 9,000 passengers in 45 diverted airplanes on September 11), on a glorious day of cloudless sky. That afternoon, my family and I went to Ground Zero to deliver supplies to rescue workers. There, I experienced the visceral shock (despite full intellectual foreknowledge and conscious anticipation) of any loyal New Yorker: my skyline has fractured; they are not there! But then I looked eastward from the shores of the Hudson River and saw the world's most beautiful urban vista, restored for the worst possible reason, but resplendent nonetheless: the Woolworth Building, with its gracious setbacks, its gothic filigrees, and its terra-cotta shine, standing bright, tall, and alone again, against the pure blue sky. We cannot be beaten if the spirit holds, and if we celebrate the continuity of a diverse, richly textured, ethically anchored past with the excelsior of a properly tempered reaching toward the stars.

When Marcel Duchamp moved from Paris to New York as a young and cynical artist, he also dropped his intellectual guard and felt the allure of the world's tallest building, then so new. And he decided to designate this largest structure as an artwork by proclamation: "find inscription for Woolworth Bldg. as readymade" he wrote to himself in January 1916.

The Reverend S. Parkes Cadman, dedicating the Woolworth Building as a "cathedral of commerce" at its official opening on April 23, 1913 (when President Wilson flipped a switch in Washington and illuminated the structure with 80,000 lightbulbs), paraphrased the last line of Wordsworth's famous ode on the "Intimations of Immortality" in stating that this great edifice evoked "feelings too deep even for tears." But I found the words that Duchamp sought as I looked up at this human beauty restored against a sky blue background on that bright afternoon of September 18. They belong to the poem's first stanza, and they describe the architectural love of my life, standing so tall against all evil, and for all the grandeur and all the foibles of human reality and transcendence—"Apparelled in celestial light,/ the glory and the freshness of a dream."

31

September 11, '01*

"To everything there is a season, and a time to every purpose under the heaven. A time to be born and a time to die: a time to plant, and a time to pluck up that which is planted." (Ecclesiastes 3:1–2).

I have a large collection of antiquarian books in science, some with beautiful bindings and plates, others dating to the earliest days of printing in the late fifteenth century. But my most precious possession, the pearl beyond all price in my collection, cost five cents when Joseph Arthur Rosenberg, a thirteen-year-old immigrant just off the boat from Hungary, bought the small volume on October 25, 1901. This book, *Studies in English Grammar,* written by J. M. Greenwood and published in 1892, carries a little stamp identifying

*This op-ed appeared in the *Boston Globe* on September 30, 2001.

the place of purchase: Carroll's book store. Old, rare and curious books. Fulton and Pearl Sts. Brooklyn.

The arrival of Joseph Arthur Rosenberg, my maternal grandfather Papa Joe, began the history of my family in America. He came with his mother, Leni, and two sisters (my aunts Regina and Gus) in steerage aboard the SS *Kensington,* sailing from Antwerp on August 31 with 60 passengers in first class and 1,000 in steerage. The passenger manifest states that Leni arrived with $6.50 to start her new life in America. Papa Joe added one other bit of information to the date of purchase and his name, inscribed on the title page. He wrote, with maximal brevity in the most eloquent of all possible words: "I have landed. Sept. 11th 1901."

I wanted to visit Ellis Island on September 11, 2001, to stand with my mother, his only surviving child, at his site of entry on my family's centennial. My flight from Milan, scheduled to arrive in New York City at midday, landed in Halifax instead—as the great vista of old and new, the Statue of Liberty and adjacent Ellis Island, with the Twin Towers hovering above, became a tomb for 3,000 people, sacrificed to human evil on the one hundredth anniversary of one little lineage's birth in America. A time to be born and, exactly a century later, a time to die.

Papa Joe lived an ordinary life as a garment worker in New York City. He enjoyed periods of security and endured bouts of poverty; he and my grandmother raised four children, all imbued with the ordinary values that ennoble our species and nation: fairness, kindness, the need to persevere and rise by one's own efforts. In the standard pattern, his generation struggled to solvency; my parents graduated from high school, fought a war, and moved into the middle classes; the third cohort achieved a university education, and some of us have enjoyed professional success.

Papa Joe's story illuminates a beacon that will outshine, in the brightness of hope and goodness, the mad act of spectacular destruction that poisoned his centennial. But his story will prevail by its utter conventionality, not by any claim for unusual courage, pain, or suffering. His story is the tale of nearly every American family, beginning with nothing as strangers in a strange land, and eventually prospering, often with delayed gratitude several generations later, by accumulated hard work, achieved in decency and fairness.

Especially in a technological age, when airplanes can become powerful bombs, rare acts of depravity seem to overwhelm our landscape, both geographical and psychological. But the ordinary human decency of a billion tiny acts of kindness, done by millions of good people, sets a far more powerful counterweight, too often invisible for lack of comparable "news value." The trickle of one family that began on September 11, 1901, multiplied by so many million similar and "ordinary" stories, will overwhelm the evil of a few on September 11, 2001.

I have stood at Ground Zero and contemplated the sublimity of the twisted wreckage of the largest human structure ever brought down in a catastrophic moment. And I recall the words that we all resented when we had to memorize Lincoln's Gettysburg Address in fifth grade, but that seem so eloquent in their renewed relevance today. Our nation has not witnessed such a day of death since Gettysburg, and a few other battles of the Civil War, nearly 150 years ago: "We here highly resolve that these dead shall not have died in vain."

The third chapter of Ecclesiastes begins, as quoted to open this piece, with contrasts of birth followed by death. But the next pair of statements then reverses the order to sound a theme of tough optimism. Verse three follows destruction with reconstruction: "A time to kill and a time to heal: a time to break down and a time to build up." And verse four then extends the sequence from grim determination to eventual joy: "A time to weep, and a time to laugh: a time to mourn and a time to dance."

My native city of New York, and the whole world, suffered grievously on September 11, 2001. But Papa Joe's message of September 11, 1901, properly generalized across billions of people, will triumph through the agency of ordinary human decency. We have landed. Lady Liberty still lifts her lamp beside the golden door. And that door leads to the greatest, and largely successful, experiment in democracy ever attempted in human history, upheld by basic goodness across the broadest diversity of ethnicities, economies, geographies, languages, customs, and employments that the world has ever known as a single nation. We fought our bloodiest war to keep our motto, *e pluribus unum* (one from many), as a vibrant reality. We will win now because ordinary humanity holds a triumphant edge in millions of good people over each evil psychopath. But we will only prevail if we can mobilize this latent goodness into permanent vigilance and action. Verse seven epitomizes our necessary course of action at my Papa Joe's centennial: "A time to rend, and a time to sew: a time to keep silence, and a time to speak."

Illustration Credits

Grateful acknowledgment is made to the following for permission to reproduce the images herein:

page 58	Courtesy of the Daughters of the Republic of Texas Library.
page 65	Republished with permission of Globe Newspaper Company, Inc.
page 92	Courtesy of the Granger Collection, New York.
page 94	Courtesy of Art Resource, New York.
page 101	Courtesy of the Granger Collection, New York.
page 114	Courtesy of the Natural History Museum, London.
page 117	Courtesy of AKG, London.
page 122	Dave Bergman Collection
page 198	Courtesy of the Granger Collection, New York.
page 244	Courtesy of Ron Miller.
page 262	Private Collection
page 264	Courtesy of Scala/Art Resource, New York.
page 266	Courtesy of Alinari/Art Resource, New York.
page 273	Courtesy of Cameraphoto/Art Resource, New York.
page 274	Courtesy of Cameraphoto/Art Resource, New York.

page 275	Courtesy of Cameraphoto/Art Resource, New York.
page 288	Courtesy of Tracie Tso.
page 294	Courtesy of American Museum of Natural History.
page 316	Courtesy of the Ernst Mayr Library of the Museum of Comparative Zoology, Cambridge, Massachusetts.
page 317 (top and bottom)	Courtesy of the Ernst Mayr Library of the Museum of Comparative Zoology, Cambridge, Massachusetts.
page 322	From "Fighting Dinosaurs," American Museum of Natural History. Photograph by Denis Finnin.
page 328	After Terry D. Jones, *Nature,* August 17, 2000.
page 330	From "Fighting Dinosaurs," American Museum of Natural History. Photograph by Denis Finnin.
page 374	Courtesy of American Museum of Natural History.
page 380	Courtesy of Joe LeMonnier.
page 381	Courtesy of Joe LeMonnier.

All other images appearing throughout are from the author's collection.

Index